REFERE

WITNEY LIBRAR
WELCH WAY
WITNEY OX8 7HF
TEL. WITNEY 70365

3 180243 145

FRONTISPIECE

JOHN KNIBB of Oxford. Lantern clock 8¼ inches high, ca. 1675. (Photo by courtesy of W. Summers).

Dr C. F. C. Beeson (1889-1975).

▲ Dr Beeson viewing the display of watches in the Beeson Room, Museum of the History of Science, in 1972. Behind him are Oxfordshire clocks from his own collection, and to the right is the Wadham College turret clock.

◀ Twelve-inch clock face by George Margetts, c.1770.

CLOCKMAKING IN OXFORDSHIRE 1400-1850

BY

C. F. C. BEESON

Third Edition
with a new introduction and index
by A. V. Simcock

OXFORD
MUSEUM OF THE HISTORY OF SCIENCE
1989

Published by the Museum of the History of Science
Old Ashmolean Building, Broad Street
Oxford OX1 3AZ

Distributed by Rogers Turner Books Ltd
22 Nelson Road, Greenwich
London SE10 9JB

© *Museum of the History of Science 1962, 1967, 1989*
Introduction © *A. V. Simcock 1989*

ISBN 0 903364 06 9

First published in 1962 as Monograph No.2 of the Antiquarian Horological Society and as Records Vol.4 of the Banbury Historical Society

Reprinted with the addition of Part Three in 1967 by the Museum of the History of Science

Third edition 1989

Printed in Great Britain
by Cheney & Sons Ltd, Banbury, Oxon.

CONTENTS

Preface to the Third Edition	v
Preface to the First Edition	vi
INTRODUCTION	1
Part One: Mainly Historical	
GUILDS AND CRAFTS	13
HISTORICAL REVIEW	17
TURRET CLOCKS	25
SUNDIALS AND SAND-GLASSES	76
Part Two: Mainly Biographical	
BIOGRAPHICAL DICTIONARY	84
Topographical List	155
Bibliography	157
List of Illustrations	159
Part Three: Addenda	
TURRET CLOCKS	162
BIOGRAPHICAL DICTIONARY	176
Supplementary List of Illustrations	192
Supplementary Topographical List	193
INDEX	195

FRONT COVER Hand from a lantern clock by George Harris of Fritwell (1614-1694), c.1670.

BACK COVER The north front of the Ashmolean Museum in 1760, from the second edition of Edward Lhuyd's *Lithophylacii Britannici Ichnographia*.

Published with financial assistance from:

Banbury Historical Society
British Horological Institute, Oxford branch
Cheney & Sons Ltd, Banbury
Christ Church, Oxford
Merton College, Oxford
St John's College, Oxford
John Smith & Sons, Midland Clock Works, Derby, Ltd

and
Thames & Chilterns Tourist Board

Financially assisted by the
English Tourist Board

English Tourist Board

in co-operation with the
Thames & Chilterns Tourist Board

PREFACE TO THE THIRD EDITION

Dr Beeson's *Clockmaking in Oxfordshire 1400-1850* was first published in 1962, jointly by the Antiquarian Horological Society as Monograph no.2 and by the Banbury Historical Society as Records vol.4. The second edition, a reprint of the first with the addition of a supplement (Part Three), was published in 1967 by the Museum of the History of Science, to which the author had recently given his collection of Oxfordshire clocks. It has now been out of print for some years, though there has been a steady demand for it both from horological historians and collectors and from local people.

The present, third edition has been prepared to meet this demand, and to mark the centenary year of the author's birth. To a reprint of the second edition are added an Introduction – including an account of the author's life – and an extensive new Index. Otherwise, apart from some very minor adjustments, the text and pagination of the main part of the book (pages 11-193) are unchanged from preceding editions.

Although no attempt has been made to revise or expand Beeson's work, a few issues are broached in the Introduction, while the Index incorporates clarifications, corrections, and some new information, largely in respect of the identities and dates of clockmakers and related craftsmen (583 of whom are listed). I hope that the Index, by making the book's wealth of detail fully accessible, will give it a new lease of life for antiquarian horologists and for local historians alike.

As Beeson's book is the essential companion to the Beeson Collection of Oxfordshire clocks (or *vice versa*), a list of this collection, as well as a very rapid summary of the Museum's general horological holdings, are included in the Introduction. A proper catalogue of the Museum's clocks is being prepared by Mr A. J. Turner. A catalogue of the watches, published in 1973, is still available.

Mrs M. A. B. Beeson has been most gracious in allowing and encouraging the reprint of her late husband's work, and in assisting me with biographical information for the Introduction. I am also grateful for the advice and comments of Mr C. K. Aked, Dr D. J. Boullin, Mr F. R. Maddison, and Mr A. J. Turner. Acknowledgement is due to those institutions and individuals who have given encouragement to this project; and not least to those who have given financial support, who are listed on a separate page.

Beeson Room A. V. S.
Old Ashmolean Building
September 1989

PREFACE TO THE FIRST EDITION

This book is written, not because it is finished, but because delay is unwise. Those who are long past middle age should print their material, if it can be of use to others, and not wait to make it more perfect

H. E. Salter

This is an incomplete survey of the history of clockmaking in Oxfordshire.

Incomplete – although much ground has been covered – because it is now certain that many more sources of information remain to be investigated and analysed. In parish church records and in the archives of towns and of the University much horological history is still hidden. That such sources exist does not seem to have been appreciated previously – at any rate, use has not been made of them for a study of Oxfordshire clockmakers. Now we have documentary evidence that there were clocks in Oxfordshire churches at the beginning of the fifteenth century and we know more about the men who made and mended the turret clocks and domestic clocks of later periods.

Of surviving clocks and watches in private ownership it has been possible to record particulars owing to the response evoked by circular letters and appeals in the press. Contacts so made have confirmed that the owner of an old clock may be as much interested in the maker as in his craftsmanship. This curiosity should be largely satisfied by the mass of new biographical data here presented.

An ephemeral but rewarding hunting-ground has not been neglected; auction sales of the contents of local houses have, unexpectedly, yielded more items of new information than have museums or the shops of antique dealers.

The facts collected in this book should form the basis of a county history of clockmaking to which much more can be added. It is hoped that the first reactions of a reader will be the detection of omissions and the recognition of his opportunity to put new facts on record for the benefit of others.

ACKNOWLEDGEMENTS

Grateful acknowledgement is made to the undermentioned persons who have kindly allowed the use of photographs, blocks, and drawings. Photographs: Mrs C. A. Allitt, Mr Oliver Bentley, Mr L. K. Bomford, Mr J. Brice, Mr A. W. Cox, Mr Percy G. Dawson, Dr P. R. Latcham, Mr Ronald A. Lee, Dr R. Vaudrey Mercer, Mr David Miller, Mr L. S. Northcote, Mr W. Summers, Mr J. M. Surman, Dr N. Watson, and the Ashmolean Museum. Blocks: *Antiquarian Horology*, Mr Percy G. Dawson, and Watlington Parish Council. Drawings: Mr L. S. Northcote (particularly for devising the map) and Mr J. M. Surman.

For reading proofs and MS the author is much indebted to Dr E. R. C. Brinkworth, Mr J. S. W. Gibson, and Mr L. S. Northcote.

In addition to those clock-owners who are mentioned in the Biographical Dictionary thanks are due to the following correspondents: Mr G. H. Bell, Mrs B. A. Benwell, Mrs B. M. Bishop, Mr J. H. Bonham, Mr W. J. Brooke, the late F. K. Challen, Mr Cecil Clutton, Miss G. H. Dannatt, Mr J. A. K. Fergie, Mr M. L. Dix Hamilton, Mr P. G. Hester, Mr E. R. Hicks, Mr E. J.

Lainchbury, Mr Eric L. Lee, Mr Bernard Mason, Mr R. H. Miles, Messrs Payne & Son (Oxford), Mr F. D. Price, the late R. Rowntree, Mrs D. K. Shirley, Messrs John Smith & Sons (Derby), Mr G. Surman, Messrs J. Taylor & Co. (Loughborough), Mr C. W. Tozer, Mr E. Trundell, Mrs F. E. Wells, the Ashmolean Museum (Mr Taylor), the Museum of the History of Science (Dr Josten), the Town Clerk of Oxford, and the Oxfordshire Rural Community Council.

For permission to examine and to photograph the clocks in churches and colleges mentioned in the section on Turret Clocks thanks are due to the vicars and rectors concerned, the heads of colleges, and their bursars or archivists.

Appreciation is expressed for the facilities granted to consult the account books of the University and colleges, the churchwardens' and overseers' account books, the minute books of vestries, parish registers, the archives of the Oxford City Council, and the County Record Office, and also for assistance given by the Bodleian Library and staff.

Documents consulted are listed in the Bibliography at the end of this book. To have substantiated each and every fact or extract by reference to a folio, page or line in the original source would have needed an abundance of footnotes, abbreviations, and repetitions of little interest to most readers.

Adderbury
10 February, 1962

C. F. C. B.

INTRODUCTION

By A. V. Simcock
Librarian of the Museum of the History of Science

There is something about a pioneering book – be it R. T. Gunther's *The Astrolabes of the World* or H. E. Salter's *Oxford City Properties* – that makes it endure even when, standing on its shoulders, we have discovered more about the subject, and can point to errors and omissions in the original. Not only does our respect for the pioneer increase with time, but his work inevitably retains information and insights which guarantee its place on our shelves. Dr C. F. C. Beeson's *Clockmaking in Oxfordshire 1400-1850* is, in a modest way, this kind of book. It is a classic not only in its own geographical area but in the historiography of English provincial clockmaking generally. One of the first county-based studies, it appeared in 1962, about the same time as H. Miles Brown's study of Cornwall and J. K. Bellchambers's of Devon. There has been a fertile tradition of them since, all following the example of these pioneers.

What they did was to open up an unexplored dimension of horological history – that of workaday provincial clockmaking illuminated through fieldwork and local history research. A world of village blacksmiths, town guilds, family and apprenticeship traditions, local stylistic features, and minor on-the-job innovations emerged from records that had not been studied before, and from the examination and comparison of the clocks themselves. No one had realised the extent to which church and college towers, and even cupolas over stable blocks, housed clockwork mechanisms often several centuries old, with their origins and the long history of their maintenance traceable in documents. Nor had many attempts been made to relate clockmakers' names culled from archives, advertisements, or directories to their actual products, which were not generally in museums but in private houses, rich and poor – often still ticking away at the foot of the stairs or on the windowsill.

THE AUTHOR

Cyril Frederick Cherrington Beeson was born in Oxford on February 10, 1889, the son of Walter Thomas and Rose Eliza Beeson. His father was Surveyor to St John's College and a lifelong college employee. 'Scroggs' Beeson attended the City of Oxford High School for Boys, where his best friend was 'Ned' Lawrence (later known as Lawrence of Arabia). In the autumn of 1907 he entered the University as a Non-Collegiate Student, and the following year obtained an Exhibition (a form of scholarship) which allowed him to transfer to St John's College. He read geology, graduating in 1910. He went on to study for the diploma in forestry, took his M. A. in 1917, and in 1923 obtained the Oxford D.Sc. During the War he was a Captain in the Royal Army Medical Corps, and served in the Mesopotamian campaign. He was first married in 1922, and had one daughter.

From 1911 until 1941 he worked as a research officer, conservator of forests, and forest entomologist in the Imperial (later the Indian) Forest Service. After secondment to study tropical and forest entomology in London and in Germany, his first duties were in the Punjab. In August 1913 he succeeded to the post of Forest Zoologist of India (renamed Forest Entomologist in 1922), which he held until retirement. It was based at Dehra Dun near the head of the

Ganges, where he helped to establish a new Forest Research Institute, opened in 1926. Over a period of thirty years he published more than sixty articles and reports on the taxonomy, ecology, and control of tropical forest insects. He edited an Indian forestry journal, and was a Fellow of various scientific societies. In 1941 his book *The Ecology and Control of the Forest Insects of India and the Neighbouring Countries* rounded off his career. It is still in use as the standard work on the subject.

He retired in the same year, 1941, and was made a Companion of the Order of the Indian Empire (C.I.E.) for his services to Indian science. He returned to Oxford, where from 1945 to 1947 he held a further post as Director of the Imperial Forestry Bureau. It was about this time that he bought his first antique clock, a grandfather clock which his wife suggested was an appropriate item of furniture for an old house they were moving into at Adderbury. Finding its tick too loud, however, they bought another one – and so it continued. Unfortunately his wife, who had been unwell for some years, died in 1946. But he stayed at Adderbury, and threw himself into his new pastime with the philosophy that one's retirement should be occupied by an activity completely different from one's previous profession.

He built up a respectable collection of domestic clocks, mostly from Oxfordshire, finding them both locally and in the London auction rooms. Not content merely to collect, he embarked upon serious research into the history of clockmaking, studying archives, looking at specimens in museums and in private hands, and boldly climbing towers to discover and examine turret clocks. In 1953 he was a founder member of the Antiquarian Horological Society. Its second meeting, at the Science Museum, London, in February 1954, heard him lecture on 'Early Oxford Clockmakers' – the first intimation of his work in progress. It anticipated the historical sketch published in his book, summarised his investigations into the Knibb family, and introduced the audience to the Wadham College turret clock, suggesting the possibility that its anchor escapement and pendulum were original features dating from 1669 – making it the earliest surviving anchor escapement, and probably the first. Later he was to conclude that Joseph Knibb was its maker, and many people now accept his interpretation.

A summary of this lecture was one of many articles which he contributed to the society's journal *Antiquarian Horology*, and to other periodicals, dealing with various aspects of horological history and antiques. Turret clocks were his favourite theme. Those described include little-known continental types seen on several of the society's foreign expeditions from 1957 onwards. In 1959-60 four issues of *Antiquarian Horology* were edited by Beeson. He also sat on the society's publications committee, and helped launch its series of monographs: his own book on Oxfordshire clockmaking was the second, issued in 1962. He dated its Preface on his 73rd birthday.

It was published jointly with the Banbury Historical Society, which he had joined soon after its formation in 1958. In November of that year he lectured on 'North Oxfordshire Clockmakers' to its fourth meeting. He was Chairman of this society in 1959-60, committee member until 1967, and founding Editor of its journal *Cake & Cockhorse* from September 1959 until 1962. He enjoyed local history quite apart from its relevance to clocks, and made forays into such subjects as seventeenth-century innkeepers, an Elizabethan inventory, and the strange history of the panelling from the Reindeer Inn's Globe Room.

In 1966, in the wake of the interest engendered by *Clockmaking in Oxfordshire*, Dr Beeson presented his collection of local clocks to the Museum of the History of Science, Oxford, and prepared a new edition of the book, published the following year. He was already writing up the subject of turret clocks; and *English Church Clocks 1280-1850: History and Classification* appeared in 1971, opening up this relatively unexplored world and providing a systematic basis for further study. Among its consequences, D. F. Nettell, T. O. Robinson, and other enthusiasts formed the Antiquarian Horological Society's turret clock group in 1973, of which Beeson became Chairman.

By this time he was deeply committed to the study of a document recounting the manufacture and installation of a clock and bell for the castle at Perpignan (in southern France) in 1356. This presented almost insurmountable difficulties, not least the language of the original, which was medieval Catalan! Remote though it seems, the project had many things in common with the history of clocks in Oxfordshire, entering as it did into a world of local craftsmen and the minutiae of their work; the document is, moreover, the oldest account of the making of a clock to contain this type of detail. It was first described by Beeson in a lecture in October 1969, the published version of which has been called by David Landes, a leading economic historian, 'one of the richest, most informative pieces I know of in the history of medieval technology, or any technology'.[1]

Dr Beeson was 80 when he gave this lecture – but they were still early days for the course he had determined upon, of turning the manuscript, with translation and commentary, into a substantial book. *Perpignan 1356: The Making of a Tower Clock and Bell for the King's Castle* was not published until 1982, some years after Beeson's death. He died on November 3, 1975, at the age of 86. His second wife, Mrs Margaret Beeson, whom he married in 1971, survives him.

He was a man honoured in two quite distinct fields. Charles Aked has called him 'the outstanding antiquarian horologist of the mid-twentieth century'; and Francis Maddison wrote in 1975 that 'the quality of recently published horological history would be much diminished without his contribution'. That he was a Vice-President of the Antiquarian Horological Society alongside Cecil Clutton and Humphrey Quill will indicate his standing to the horologist; while the historian will not dismiss the judgement of David Landes, who concluded: 'Beeson deserves well of History'.[2]

But his earlier scientific colleagues regretted the consequent loss to their subject – for to many who are unaware of his horological work, Beeson's name is known as an authority on tropical forest entomology. P. K. Sen-Sarma has summarised his career as a scientist, and concludes that he 'laid the foundation of Forest Entomology in India on a very solid ground'.[3] Such is the esteem in which he is held in this field that the road to his former residence in India is now called Beeson Road, and various relics are lovingly preserved there, including the car that he drove in the 1930s!

He has also been subject to a degree of fame by association, through his friendship with T. E. Lawrence. At the age of 15 the two were precocious local antiquarians and archaeologists. They cycled round 'the three counties' (Berks., Bucks., and Oxon.) visiting almost every village church to see its antiquities and monuments, and to make brass rubbings. In Oxford itself they made a hobby of monitoring building sites and 'by incessant watchfulness secured everything of antiquarian value which has been found', as the Ashmolean Museum, the grateful recipient of the finds, reported in 1906.[4] In the summer vacations of 1906 and 1907 Beeson accompanied Lawrence in exploring the castles of France, as part of a survey which became Lawrence's undergraduate thesis. Many of the excellent drawings of English and French sites used in the thesis, and reproduced in the book *Crusader Castles* (published in 1936 after Lawrence's death), are by Beeson.

This was the apprenticeship for Dr Beeson's later achievements, providing him not only with the interest in local history and antiquities which characterised his retirement but also with some of the skills which he brought to his study of clocks. Others came from the discipline of entomology, with its infrastructure of descriptive taxonomy. True to both influences, he was committed to the possibility of classifying stylistic and technical features, and always urged this approach. The classification of turret clocks reached for in the Oxfordshire book and achieved in *English Church Clocks* was a particularly original contribution. His aim was to show that their 'great variety of form and function . . . can be analysed and systematically studied'[5] – a very scientific attitude.

THE BOOK

Clockmaking in Oxfordshire 1400-1850 contains a wealth of information which, as its author realised, could be the foundation for a continuous pooling of knowledge, and eventually a more comprehensive and rigorously classified study. The present volume seeks only to make the original work available. Nowadays to rewrite it one would have to assemble all those collectors, curators, and historians who have seen or acquired additional Oxfordshire clocks, or have looked more deeply into genealogical and other records.

Dr Beeson himself continued to collect information and clocks, the results up to 1966 being contained in Part Three, added to the second edition. His collection in the Museum of the History of Science contains further items which he acquired after 1966. Two of them are by makers who do not feature in the book at all: the insignificant watchmaker G. H. Osmond (c. 1840), and George Margetts (1748-1804), who started his career at Woodstock before becoming very distinguished in London. Beeson's omission of Margetts illustrates just how uncharted was the territory he set out to explore.

He also missed a clock recently mentioned by Brian Loomes, a longcase of about 1730 signed 'Zec Montfort, Glostr. Green, Oxon',[6] though this is obviously the 'Mr Mumford' who attended to Carfax clock (page 58). Zachariah Mountford was apprenticed in London in 1677, and is later found in St Albans. Too little is known of him to say whether he may have worked in Oxford as an old man, or have had a son who moved there. It is not my purpose to catalogue additions and corrections to Beeson's work, but a few such matters are worth mentioning, if only to indicate cautions and potentialities to be borne in mind when using the book.

One of the problems of name-specific history, for instance, is the conflation of namesakes, especially in successive generations. Beeson established that the clocks signed by William Ball of Bicester were by two men, presumably father and son. My own opinion is that there were three: the older one mentioned on page 176 was almost certainly a clockmaker, if two of his sons were, and if the dating of some of the clocks as about 1705 is accurate. A third Edward Hemins seems to have existed too, presumably the son (page 112) who was a child when the famous Hemins died in 1744; the bracket clock described on page 183 is signed in a more sophisticated style and would attract a date around 1770. Whether an earlier Edward Hemins 'senior' was really a clockmaker remains uncertain: further research has confirmed a man of this name in Bicester in 1699, but not his profession.

The opposite, as Beeson suspected, may be the case with John Nethercott, who though listed three times by Beeson (and indexed twice by me) could well be the same person moving from place to place. John Stone of Thame and John Stone of Aylesbury (page 142) are also one person, but the information given under the second embraces a son of the same name, who was also a clockmaker. Such questions of identity provide ample scope for further research by local horological historians; and some cases await it before anything sensible can be said about them at all. The May 'family' of Witney (and London) remains a stubbornly undifferentiated rabble, with the added confusion that the London John May whom Beeson assumes to have hailed from Witney is regarded in other quarters as a Dutchman!

A further caution concerns some pitfalls of terminology. I am not always convinced that references to 'dials' in manuscript records are to clock faces rather than to sundials, which would be the more common meaning of the word well into the eighteenth century. Even 'making' can be a tricky word (see page 55), and 'new making' clearly means the same as renovating (as on page 65). It must also be remembered how many other things a 'clockmaker', an expert on mechanical wheelwork, might be doing if the documents do not specify – he could be called upon to make or repair a roasting jack (see page 174), a windlass (page 56), an automaton (page 106), an orrery (page 133), a milling machine, a musical box, or any other clockwork or wheelwork mechanism. Finally, a word of Beeson's own choosing, 'colony', for the Quakers of north Oxfordshire, should not of course be taken literally.

Dates assigned to undated clocks remain open to question. One variable, when dating by style, is the uncertain degree in which provincial products represent conservative fashions, of later date than their London counterparts. The maker's identity, as we have seen, is not always the decisive guide one might expect. In the case of personal dates, I have tried in the Index to clarify and extend Beeson's information wherever possible. Dates of apprenticeship are especially useful as they allow an approximate date of birth to be deduced, apprentices nearly always being taken at the age of 14.

Beeson used apprenticeship records with other evidence to show, often implicitly, the transmission of skills and the refinement of the craft, in the process looking into the world of municipal guilds and 'freedom'. Further studies have been done since, especially of the London City Companies, and scholars have constructed apprenticeship 'genealogies'. True heredity blends with apprenticeship in passing on trades and skills; important too is the role of marriage. One of Oxford's earliest clock specialists, the French immigrant John de Saint Paul, a transitional figure between the blacksmith and the clockmaker, married Elizabeth Smyth, who at this date (1576) can be presumed to represent a native family of blacksmiths. A couple of generations later their skills had evolved into those of the Quelch family, who were fine watchmakers.

In the eighteenth century the blacksmith-clockmaker Thomas Reynolds was the son of a conventional blacksmith, and trained several pure clockmakers, including the talented John Hawting. If it was Hawting's sister (page 113) who married the clockmaker John Herbert, shortly before Hawting's apprenticeship, this may have been the link which drew him into the profession. Readers of Beeson's biographical entries will be able to spot many significant (and insignificant) linkages of this kind.

As a case study of provincial clockmaking Oxfordshire is both typical and a special case. The Quaker clockmakers, though unique in themselves, are typical in that most counties show traditions of clockmaking (and of other crafts) in localised family networks whose primary bond was one of religious affiliation or something similar. On the whole Beeson's theme is important for the very reason of its typicalness: it presents us with clocks and clockmaking that 'represent the average standard of the trade . . . the traditionalism of the local makers, and also their minor originality within it'.[7]

Yet in Oxford itself two special influences come into play: the proximity of London, and the presence of the University. London's international standing as a centre of craft and commerce has inevitably affected nearby towns. London-trained men like Michael Bird have been key figures in the history of Oxford clockmaking; but equally, clockmakers who had practised for a while in Oxford (or, like Margetts, in Woodstock) could establish themselves as leaders of the craft in London, as did Joseph Knibb. Responses to London developments in fashion or technology were fairly quick in Oxford, and might occasionally happen the other way round. Links were strongest in the seventeenth century, Oxford being the alternative capital in case of civil war or plague.

The University has affected clockmaking in several ways: by the demand for communal time measurement to regulate its scholarly and religious activities; by the concentration of so many churches and colleges; and by the academic interest in horology and mechanics, especially amongst mathematicians and astronomers. The very invention of the clock had resulted from the first of these factors (and its function was not at first to display time on a face but to signal it by striking a bell). The concentration of demand must have created a healthy fraternity of blacksmiths, and helped to channel their skills not just into clockmaking but into the kind of talents demonstrated by William Young, the seventeenth-century blacksmith and locksmith who was an expert on chiming mechanisms.

Such men found themselves respected by the academic and scientific community, and were drawn into innovative experiments and collaborations. Sundials which were masterpieces of mathematical ingenuity (see pages 76-80), elaborate locks (such as those on the doors of the Museum of the History of Science), waterclocks (see page 39), and, on a grander scale, challenges of architectural engineering (like the roof of the Sheldonian Theatre), were among the things which fascinated scientists of the seventeenth century and were constructed or studied in collaboration with the craftsmen. Christopher Wren became the best-known example of such a scientist. His mentor, as a student at Wadham College in the 1650s, was John Wilkins, an authority on 'natural magic' and an enthusiast for mechanical contrivances, among them experimental flying machines and talking statues (none of which survive). It is less surprising when seen in this context that the construction of a new clock for Wadham, sponsored by Wren, should have been the occasion for a sophisticated innovation.

THE COLLECTION

During his studies Dr Beeson became closely associated with the University's Museum of the History of Science, which, amongst a general collection of horological instruments of all kinds, had preserved some key specimens of local clockwork – notably the Wadham College clock, the church clock from Combe, the Hawting astronomical regulator from the Radcliffe Observatory, and an exquisite miniature lantern clock by John Knibb. In 1966 Beeson presented his own collection. A small room was converted in order to centralise the Museum's clock and watch displays; and the Beeson Room was opened by the Vice-Chancellor of the University on February 14, 1972.

Founded in 1924 around the Lewis Evans Collection, which chiefly consisted of astrolabes and portable sundials, the Museum had always contained clocks and clockwork, illustrating how 'the skills of the horologist marked the technical frontier in small-scale engineering'.[8] The source of much of it before Dr Beeson's gift was a bequest by T. G. Barnett in 1935, including most of the magnificent display of watches. Among the finer clocks which the Beeson Collection joined are a sixteenth-century French globe clock; an astrolabe clock by Bommel; one of the first longcase clocks, by Fromanteel; bracket clocks by Quare and by Joseph Knibb (London); and a Napoleonic astronomical clock by Janvier and Breguet. Turret clocks include a wood-framed example from Onibury, and an ancient wooden chiming barrel from Amesbury. The orient is represented by a small group of Japanese clocks; and the modern world by an early electro-magnetic clock by Detouche, and an electric longcase clock invented by Alexander Bain.

Of special-purpose clocks, and applications of clockwork beyond time telling, the Museum possesses chronometers, astronomical regulators and alarms, telescope driving mechanisms, heliostats, a watchman's clock, a factory punch clock, orreries, gramophones, musical boxes, a bird scarer, and even a clockwork fly-catcher! One of the orreries is the very first, made about 1710 by the clockmakers Thomas Tompion and George Graham. In contrast, non-clockwork timekeepers include the sundials, nocturnals, and other astro-horological instruments at the heart of the Museum, as well as sandglasses, a metronome by Johann Maelzel, a waterclock, Chinese fire and incense clocks, and equally curious devices which measure time using oil and atmospheric pressure.

The picture of the actual craft of clockmaking, suggested particularly by the Beeson Collection, would not be complete without some of the tools used, ranging from heavy wheel-cutting engines to the tiny hand tools of the watchmaker. Also displayed are watchmaker's lathes, a lantern pinion drilling machine, 'depthing' and 'uprighting' tools, and various gauges. A group of instruments used by the Oxford watch repairer R. B. Bennett earlier in the present century represents the local trade after the period covered by Beeson, when it had become largely a matter of retail and repair.

The Beeson Collection in the Museum does not coincide exactly with the items described in this book as belonging to Dr Beeson. Sixty-one pieces are thus described, plus an unspecified number of Richard Gilkes clocks. Twelve of the former are not in the Museum – including, unfortunately, the Samuel Aldworth bracket clock, the balloon-decorated longcase by Joseph Williams, and the unique longcase with miniature turret clock movement by the Finmere surgeon James Clarke. Two of the longcase clocks described in a complete state are now represented by their faces only; and four others have been divested of their cases.

At least four items which appear in the book unattributed or belonging to other people were subsequently acquired by the author, and are in the Museum. There are also two anonymous movements with 'Quaker' faces (as on pages 100-101). Eleven further clocks and watches not featured in the book at all are now part of the Beeson Collection, most of them added by Beeson between 1966 and the opening of the Beeson Room in 1972. The most significant, as mentioned before, is the clock face which provides evidence of George Margetts's early practice in Woodstock.

Also described in the book are five clocks (possibly six) which were already in the Museum, and four seen elsewhere by Beeson which the Museum has since acquired – the Reynolds longcase of R. C. Righton, and the turret clocks from New College, Steeple Aston, and Idbury. The latter was restored by Beeson, and transferred at the same time as his own collection. One curiosity which has come to the Museum since – a degree plate for measuring the swing of a large pendulum, from the old clock in Tom Tower – is a relic of the trouble which Richard Rowell was having with this clock in 1841-2 (recounted on page 50). Finally, although not mentioned by Beeson as it fell beyond his cut-off date, it is pleasant to note that the flat-bed turret clock from his old school, installed by Rowell & Son in 1881, is now preserved by the Museum.

To make the preceding paragraphs a little clearer, I conclude this Introduction with a synoptic list of the Beeson Collection (both as recorded in the book and as actually present in the Museum), and of other Oxfordshire clocks in the Museum of the History of Science.

NAME, PLACE	TYPE	DATE	PAGE	MUSEUM
	m = movement	f = face		

Beeson Collection of Oxfordshire clocks

NAME, PLACE	TYPE	DATE	PAGE	MUSEUM
Samuel Aldworth, London	bracket	c.1700	85	—
Anon. (Quaker)	longcase m	c.1755	—	√
Anon. (Quaker)	longcase m	c.1765	—	√
William Ball, Bicester	longcase	c.1710	87	√
William Ball, Bicester	longcase	c.1770	87, 177	√
William Ball, Bicester	wall	c.1735	176	√
John Blundell, Horley	longcase	1700	89	√m
William Buckland, Thame	Act of P	c.1780	90	√
Henry Carter, Oxford	watch	c.1850	91	√m
James Clarke, Finmere	longcase	c.1850	92	—
John Clements, Oxford	watch	1808	177	√
Abraham Davis, Oxford	watch	1858	177	—
Samuel Denton, Oxford	bracket	c.1770	177-8	√
James Drury, Banbury	longcase m	c.1780	95	√
William Drury, Banbury	bracket	c.1790	—	√
William Drury, Banbury	bracket	c.1790	178	√
William Drury, Banbury	longcase	c.1790	178	—
William Drury, Banbury	watch m	1798	178	√
John Fardon, Deddington	longcase m	c.1770	97	√
Thomas Fardon, Adderbury	longcase m	c.1787	98	—
Thomas Fardon, Deddington	Act of P	c.1800	97	√
Thomas Fardon, Deddington	watch	c.1805	98	√m
Thomas Fardon, Woodstock	longcase	c.1840	180	√
John Ford, Oxford	lantern	1695+	—	√
John Gilkes, Shipston	longcase m	c.1750	101	—
Richard Gilkes, Adderbury	longcase	c.1750		√
Richard Gilkes, Adderbury	longcase	c.1765		√
Richard Gilkes, Adderbury	longcase m	1736		√
Richard Gilkes, Adderbury	longcase m	c.1740	102-3	√
Richard Gilkes, Adderbury	longcase m	c.1750		√
Richard Gilkes, Adderbury	longcase f	c.1745		√
Richard Gilkes, Adderbury	longcase f	c.1760		√
Henry Godfrey, London	longcase	c.1700	105	—
William Green, Milton-u-W	longcase m	c.1750	107	√
John Harris, Oxford	watch	c.1680	109	√
John Hawting, Oxford	longcase	c.1770	182	√
Edward Hemins, Bicester	bracket	c.1770	183	√
Edward Hemins, Bicester	lantern	c.1720	112	√
Edward Hemins, Bicester	lantern	c.1720	112	√
Edward Hemins, Bicester	lantern	c.1725	183	—
Edward Hemins, Bicester	longcase	c.1735	113	√f
Edward Hemins, Bicester	longcase m	c.1725	—	√
Edward Hemins, Bicester	turret	c.1735	31, 113	√
Thomas Jordan, Stadhampton	longcase	c.1780	—	√
John Knibb, Hanslope	longcase	c.1712	122	√
John Knibb, Oxford	bracket	c.1675	120	√

8

NAME, PLACE	TYPE	DATE	PAGE	MUSEUM
John Knibb, Oxford	bracket	c.1685	121 (various)	√
John Knibb, Oxford	longcase	c.1685	?121(Clark)	√
John Knibb, Oxford	wall	1685	120	√
John Knibb, Oxford	watch	c.1690	184	√
John Knibb, Oxford	watch	c.1690	184	√
John Knibb, Oxford	watch	c.1690	184	√
John Lamprey, Banbury	longcase	c.1750	125	—
William Lawrence, Cuddesdon	longcase	c.1750	126 (unattrib.)	√
William Lawrence, Thame	longcase	c.1745	186	—
Edward Lock, Oxford	watch	1774	—	√
George Margetts, Old Woodstock	longcase f	c.1770	—	√
John May, Witney	longcase m	c.1750	129	√
Edward Moore, Oxford	longcase	c.1730	130	√
John Nethercott, Dry Sandford	longcase	c.1750	131	—
John Nethercott, Long Compton	longcase	c.1750	131	√m
John Nethercott, Standlake	longcase f	c.1750	—	√
James Oakley, Oxford	watch m	c.1740	131	√
John Oakley, Oxford	lantern	1704	132	√
A. & D. Ortelli, Oxford	watch m	c.1825	132	√
G. H. Osmond, Oxford	watch m	c.1840	—	√
William Peacock, Banbury	Act of P	c.1800	133	√
William Peacock, Banbury	longcase	c.1790	188	√f
Thomas Pinfold, Banbury	longcase	c.1765	134 (unattrib.)	√m
Reynolds & Earle, Oxford	longcase	c.1795	189	√
George Rowell, Oxford	wall	c.1815	?138 (Maclean)	√
George Rowell, Oxford	watch m	c.1785	138	√
R. S. Rowell, Oxford	watch m	c.1900	—	√
Samuel Simms, Chipping Norton	bracket	c.1840	—	√
William Simms, Chipping Norton	longcase f	c.1820	140	√
John Sowter, Oxford	bracket	c.1820	141	√
John Sowter, Oxford	wall	c.1820	141	√
John Sowter, Oxford	watch	c.1830	189	√m
William Tasker, Banbury	watch	1829	143	√
George Tonge, Oxford	bracket	c.1770	145	√
J. G. Walford, Banbury	longcase	c.1840	—	√
George Walker, Oxford	lantern	c.1700	147 (Watson)	√
Francis Webb, Watlington	longcase m	c.1710	149	√
Francis Webb, Watlington	longcase m	c.1720	190	√
George Wentworth, Oxford	longcase	c.1735	150	√
John Westcott, Oxford	longcase	c.1780	190	√m
Joseph Williams, Adderbury	longcase	c.1800	151	—

Other clocks given to the Museum by Beeson

| Anon. (German) | tabernacle | c.1600+ | — |
| Abram Develay, Lausanne | turret | 1789 | — |

Other Oxfordshire clocks in the Museum

Anon. (Combe)	turret	17th C	35-6 (MHS)
Anon. (Idbury)	turret	17th C	166-170 (*in situ*)
Anon. (Oxford, Christ Church)	degree plate	1841/2	—
Anon. (Oxford, New College)	turret	c.1700	52-3 (*in situ*)
Anon. (Oxford, Wadham College)	turret	1669	64-66 (MHS)
Anon. (Steeple Aston)	turret	18th C	28 (*in situ*)
John Hawting, Oxford	regulator	c.1780	111 (MHS)

NAME, PLACE	TYPE	DATE	PAGE
John Knibb, Oxford	lantern	c.1690	—
Edward Moore, Oxford	bracket	c.1770	—
Edward Moore, Oxford	lantern	c.1720	130 (MHS)
Thomas Reynolds, Oxford	longcase	c.1775	?137; 188 (Righton)
Rowell & Son, Oxford	turret	1881	—
L. Wangler & Co., Oxford	wall	1881	?148 (unattrib.)
Francis Webb, Watlington	wall	c.1715	148 (MHS)

REFERENCES

1. David S. Landes, *Revolution in Time* (Cambridge, Mass. & London, 1983), p.193.
2. Aked and Maddison in Beeson, *Perpignan 1356* (London, 1982), pp.viii and xvi; Landes, *loc.cit.*
3. P. K. Sen-Sarma, 'Cyril Frederick Cherrington Beeson (1889-1975)' in *Biographical Memoirs* of the Indian National Science Academy, 1985, pp.34-40.
4. Ashmolean Museum, *Annual Report* for 1906.
5. Beeson, *English Church Clocks 1280-1850* (London, 1971), p.7.
6. Brian Loomes, *The Early Clockmakers of Great Britain* (London, 1981), p.402.
7. A. J. Turner, 'The Beeson Room in the Museum of the History of Science, Oxford' in *Museums Journal*, LXXII, 1972, pp.23-4.
8. Turner, *loc.cit.*

PART ONE
MAINLY HISTORICAL

GUILDS AND CRAFTS

MEN who played some part in the making of clocks in Oxfordshire belonged to a variety of crafts or trades. There were individuals who could be described as: (1) blacksmith and turret clockmaker, (2) bellfounder and turret clockmaker, (3) bellfounder, turret and domestic clockmaker, (4) whitesmith, turret and domestic clockmaker, (5) clock and watchmaker, (6) goldsmith and silversmith and clock repairer, (7) retailer of clocks and watches and a variety of other goods, (8) congregational minister, and (9) professor of astronomy. Nothing is known of the men who made wooden cases for clocks.

To appreciate the industrial and social environment in which they worked a short account of the organisation of guilds and crafts in the county may be useful. For this purpose four types are chosen — the townships of Henley on Thames in the south and of Banbury in the north, the city and university of Oxford in the middle of the county and lastly the Society of Friends.

1. HENLEY ON THAMES

The early records of the Borough of Henley provide an example of the mediaeval organisation of craftsmen and traders (Burn, 1861; Briers, 1960). At the beginning of our period, 1400, the Merchant Guild of Henley had become synonymous with the Borough itself. It elected the burgesses and from among them a warden, bailiffs, bridgemen and constables to administer the business of the town and the property owned by the Borough. Other officers dealt with the maintenance of the church fabric, its bells, clock and the chantry priests and lights. " There is no evidence of actual craft gilds at Henley but clearly some crafts were organised. They appear to have made regulations for themselves which the Merchant Gild (and therefore the community of burgesses) had the power to ratify or alter . . . At his election the new burgess paid a fine, usually 6s. 8d., subscribed to an oath and provided himself with pledges. If his children later sought election as burgesses they paid less — 1s. 4d. in the case of the eldest son and 1s. 6d. to 2s. 8d. for a second son." Foreigners, who certainly traded and held property in the town of Henley, were not compelled to acquire burgess-hood, which nevertheless conferred commercial privileges as well as other benefits, but they had to pay an annual fine for trading and for employing apprentices.

2. BANBURY

Mediaeval Banbury had no Merchant Guild. In the middle of the 16th century it was "*a greate towne replynysched with people and a great markett towne.*" When Queen Mary granted a Charter of Incorporation in 1554 there were at least 12 Trade Guilds or Companies of which the Smiths formed one. Each trade guild elected 2 wardens, who with their members had power to make orders and constitutions for the control of their trade, subject to approval by the bailiff and aldermen. The Council of the Borough of Banbury comprised 12 aldermen and 12 capital burgesses, the chief official being called a bailiff and later a mayor. Freedom of the Borough was obtained by qualifications of birth or apprenticeship or by redemption. No person could be made a freeman by redemption on a fine of less than 20s. payable to the Chamber and 12d. to the Town and 1d. for the relief of the poor and prisoners. Apprentices had to be enrolled and bound for not less than 7 years; each on taking up his freedom paid the Chamber 12d. and the Town Clerk 4d. Unfortunately the Banbury register of indentures no longer exists. Foreigners or inhabitants who were not freemen and settled for trading in the town had to compound with the trade company concerned and pay the required fee and 12d. to the Chamber. They were, however, allowed to sell merchandise in gross to a freeman or on market days, otherwise they were fined 10s. (Beesley, 1842; Potts, 1958). Some companies continued until late in the 18th century paying annual dues to the Corporation. Freedoms, besides those of members of the Common Council, were taken up as late as 1803.

3. OXFORD

The City: The freemen of the town of Oxford, known as burgesses in the Middle Ages, and later also as hanasters, formed a Merchant Guild. Only freemen were allowed to have shops in Oxford and those tradesmen who had not been admitted to the Guild had to settle outside the City boundaries. Artisans, such as masons, carpenters and slatters who did not produce articles for sale, were admitted to the Guild. Admission was obtained by purchase, by inheritance, and by serving an apprenticeship. The fees payable to the City varied considerably over the centuries. Freedom by purchase was not granted for less than 40s. and in some 17th century cases as much as £20 - 30 was demanded. The eldest son of a freeman could claim admission by his father's copy, while younger sons had to pay 9s. 6d. Admission by apprenticeship needed documentary proof of enrolment in the City's Register, and was refused or excused with a fine when the master had neglected to effect registration. In addition all 3 classes had to pay officers' fees amounting to 6s. 8d., and, in some years, a leather bucket or its value 3s. 6d. The parent or sponsor of an apprentice paid a premium when the boy was bound to his master, and in some cases the premium was subscribed in part from a fund. An apprentice served at least 7 years receiving board and lodging and clothing; if he remained as a workman from the 8th year onwards he received a wage.

The City Council comprised the mayor, aldermen and assistants, making an Assembly of thirteen, also bailiffs, chamberlains and the Common Council of twenty-four. There were various minor offices, e.g., of keykeepers, millmasters, constables, fairmasters and searchers of cloth, leather, fish, meat, etc. A clockmaker might hold any of these offices in the course of his career. The original records of the proceedings of the Council have been largely preserved in Oxford and form a valuable source of information, which has been consulted for the purposes of this book. Extracts from the same have been printed by the *Oxford Historical Society* in four volumes (Salter 1928; Hobson and Salter, 1933; Hobson, 1939 and 1954) and in an earlier work by Turner, 1880. Surveys, leases, rentals and taxes have also been published in two volumes (Salter, 1920 and 1926).

Besides the Merchant Guild there were various craft guilds subordinate to it. The more ancient craft guilds had privileges from the king and were perennial, but of other crafts there were several incorporations and new corporations at different times, particularly in the 16th century. Some trades never had a craft guild and others only from time to time. "When there was an energetic man in a trade he would suggest the formation of a gild. After a time it would die or be united with another gild; for when two gilds were weak they would unite to secure an adequate number for their dinners and meetings" (Salter, 1928). In February 1667 there were only 3 freemen watchmakers in Oxford, so in a petition to the Council in that month we find the Smiths and Watchmakers combining, but for another petition in the following October they call themselves the Clockmakers and Watchmakers of the City. This is the first reference to any kind of association of clockmakers in Oxford; assumptions that such existed earlier are based on errors in deciphering manuscripts. C. Moore (1896) stated that in the year 1604 a lease of the Town Wall near Bocardo Gate was granted by the City authorities to the Company of Clockmakers; actually this lease was a renewal granted to the Cordwainers, a guild dating from the time of Henry I. Again, in the Hanasters List for 1531-32, William Graunden is shown as a " clockmaker ", but he was actually a cloakmaker.

Early in the 18th century the Locksmiths, Gunsmiths and Farriers were combined in one Company and the Smiths formed another Company complete with a master and wardens. The clockmakers were never numerous enough to form a Company.

The University: Privileged Tradesmen: Besides the trade guilds there gradually developed in Oxford another special class of tradesmen. Under a Charter of Edward III granted to the Chancellor and Scholars of the University certain persons being beadles, stationers, manciples and cooks enjoyed the privileges and protection of the University and were outside the jurisdiction of the City Council. This privilege had been extended to individuals of all trades and all ranks employed by the University and was one of the several grievances in the almost perpetual controversy between University and City. The Charter granted to the University in 1523 at Cardinal Wolsey's request greatly increased its power over the City and although the Wolsey Charter was repealed in 1543, an Act of Parliament of 1582 practically restored its provisions and disputes continued.

One aspect particularly irritating to the City Council was the competition within the town between privileged tradesmen and those in the guilds. In 1609 the Council recorded " *that the Universitie doth of late much oppresse the Towne by taking so many Tradesmen into their priviledge by colour which doe worcke abroad in the Towne as well as in the colledge.*" In 1643 the City petitioned Charles I complaining that the University now extends its privileges " *to all sorts of people viz. esquires, counsellors at law, attorneys, bakers, brewers, apothecaries, inneholders, carriers, tailors, barbers, ironmongers, carpenters, slatters, joyners, masons and other mechannick trades, alehousekeepers, women and reteyners, whereas few of them be their meniall servants as they ought to be.*" And in 1650 the city still argued that privileges should be restricted to " *barbors, parchmentmakers, limners and booke bynders.*"

Another constant preoccupation of the Council was the suppression of strangers and foreigners who tried to set up shop and keep open trade in the city " *unless they have special authority from his Majesty to do so.*" Influential patrons of foreign craftsmen also sometimes tried to obtain the freedom for their protegés. In 1663 Lady Falkland, wife of Henry Viscount Falkland, the Lord Lieutenant of the county, requested freedom for the watchmaker Nicholas Pantin (apprenticed in London in 1651), but was refused on the strength of objections from the few watchmakers in the City. In 1667 the Smiths and Watchmakers had to combine in an attempt to prevent Joseph Knibb trading as a privileged servant of Trinity College, but because of other important political considerations, they were unsuccessful in stopping him from eventually becoming a freeman by purchase. As late as 1748 a remonstrance was presented to the Vice-chancellor concerning the matriculation of persons in the Company of Smiths.

Nevertheless University tradesmen continued to be created during the 18th century but the distinction was ceasing to have much force. In 1777 a clockmaker, Robert Haines, was registered as *horologium fabricator privilegiatus* yet occupied and traded in a house leased to him by the City. Thomas Hunt, also a clockmaker, was registered similarly in the same year. In the 19th century even the freedom of the City became less essential. A clockmaker, Samuel Denton, who was made free by Act of Council in 1802, resigned his freedom in 1827 in a joint petition with several others.

4. THE SOCIETY OF FRIENDS

In the north of the County the colonies of Quakers which originated during the lifetime of George Fox (1624-1671) gradually established permanent centres for their meetings. In Banbury in 1664 Edward Vivers purchased a close for the erection of a Meeting House and an enclosed burial ground. In Adderbury West Bray D'Oyley, of ancient lineage, built a Meeting House with burial ground in 1675. D'Oyley, Thomas Gilkes and Thomas Fardon with others bought land in 1681 for the Sibford Gower Meeting House. Other early centres for Friends' meetings were Charlbury, Chipping Norton, Milton under Wychwood, North Newington and Shutford. Many Quakers became clockmakers working in villages but excluded from the large towns.

A young Quaker had to be apprenticed to a Quaker to whom, in such a closed inter-marrying community, his family was usually related. These apprenticeships were rarely registered in the local county records and documentary proof of them is seldom available. The Quaker Records of births, deaths and marriages provide some information about trades and reliable deductions may be made. Bishop Secker's Visitation Returns for 1738 (*Oxf. Rec. Soc.*, 1957) also give an idea of the distribution of Quaker families at that time.

HISTORICAL REVIEW

ONE would expect the earliest evidence for the existence of time-keeping devices in Oxfordshire to be found in the archives of the University and of the City of Oxford. The few that have been preserved are in the form of isolated entries in the accounts of expenditure by colleges and churches.

The Thirteenth Century: For the 13th century only one is available — the oft-quoted record from the accounts of Merton College which begin in 1287 and state for the year 1288 — *ad opus orologii . . . iiijs. iiijd.* In the whole of England there are only two contemporary records of a horologium, viz., old St. Paul's Cathedral, 1286, and Canterbury Cathedral, 1292 — and there are none earlier. The clock-tower of Westminster Palace, which legend assigned to the year 1288, was actually built in 1365-7 (R. Allen Brown, 1960).

The term horologium at this period has usually been taken to cover a variety of time-keeping devices such as a water-clock, a sundial or an equatorium or planetary indicator rather than a weight-driven mechanical clock. But the most recent view does not dismiss the possibility that such records, at least in other European cities, may have referred to constructions that can rightly be termed weight-driven clocks with a mechanical escapement. In view of the high degree of perfection of mechanical clocks reached by the middle of the 14th century, Lloyd (1958) now places the date of their invention between 1277 and 1300. Previously Zinner (1954) had suggested the period 1271 to 1348.

Whether the Merton horologium of 1288 was a water-clock, or a dial, or an instrument constructed on the newly discovered principles, the college soon acquired some time-keeping device of which the function was to operate an alarm that would warn the sacrist when to ring a bell to summon the scholars to their studies or religious observances. It was kept in the Hall, as the phrase *orologium de aula* used 40 years later would seem to confirm.

The Fourteenth Century: For Oxford in the 14th century only 2 records have been preserved, and these again by Merton College. In 1327 the archives have an entry, *ad renovandum orologii de aula, viiij d.,* which, if it does not refer to the 13th century clock, means it was one worth perpetuating and probably not an out-moded water-clock. Sixty years later in 1387 a reference to iron and ropes, apparently for a clock brought from London, suggests the acquisition of something incorporating the latest horological developments about which the College would certainly be well-informed. Whether or not the clock was still sited at ground-level, the bell-chamber was not added to the tower until 1451.

The progressive outlook of Merton in astronomical and horological knowledge was summed up by D. J. Price in his lecture on *Clockwork before the Clock* given in 1955 to the British Horological Institute and the Antiquarian Horological Society. By about 1300 planetary calculating instruments were introduced into Europe as the equatorium, a companion instrument to the astrolabe. Richard of Wallingford worked on one just before he began to build his great clock for St. Alban's, and after his death in 1336 the tradition was handed on to his colleagues at Merton. One of these, Simon Bredon, may have been the constructor of the only surviving specimen of a mediaeval instrument combining an astrolabe and an equatorium. Another associate of the Merton College astronomers was the poet, Geoffrey Chaucer. In 1391 he wrote the first scientific book in English, a treatise on the astrolabe and in the following year a sequel on the equatorium. Zinner (1954) finds evidence in Chaucer's work on the astrolabe that the 12-hour system, i.e., from midday to midnight, was known in Oxford several years before 1391. For astronomical purposes the modern equal hours had been in use throughout Islam and during the entire 13th and 14th centuries in Europe. But as D. J. Price points out (*in litt*), Chaucer's use of the Oxford tables as a constant is no indication that equal hours were used for mechanical clocks or even that they existed.

The Fifteenth Century: With the onset of the 15th century the picture is clearer owing to increasing documentary evidence. In fact C. Cox (1913), an authority on English churchwardens' accounts, goes so far as to write, " there was hardly a clockless church to be found in either town or country in the fifteenth century. The majority of the clocks, even in large town churches in the fifteenth and sixteenth centuries had no outside faces, and were in clock-houses within the church." For example, the early 15th century clock in the church of St. Laurence, Reading, had a jack with a bell and was in a wooden clock-house against an inside wall. It was still in position in 1510 when a parishioner paid rent for his seat " under the clock house." Similarly at St. Edmund's, Sarum, there was a pew under the clock in 1532.

Although there is no proof that by 1400 in Oxfordshire there were hour or quarter-striking clocks, the fame of clocks with carillons and automata had already spread and had become sufficiently familiar for comprehension by mediaeval sermon audiences. The moralist, Alexander Carpenter, uses a clock with automata as a similitude in his *Destructorium Viciorum*. He wrote this treatise between 1426 and 1429 and is said to have been an associate of Balliol College, Oxford. His horological example combines the features of a non-repeating alarm and a monumental clock with carillon and horizontally rotating jack-wheels.

" As in the case of a clock, the clock-keeper places one pin (*cavillam*) in a certain wheel, and when it reaches a certain point in the clock, immediately the mechanism is released; and then all the bells strike and the figures in the semblance of clerks and priests pass by in procession chanting. But how long do these things last ? Assuredly, until the weight reaches the ground and no longer; for, after the weight has grounded, everything immediately stops." (Translation by G. R. Owst, 1952). Carpenter continues his simple analogy to show that the din and tumult of tolling bells and chanting priests cease immediately a dead man's body has been buried and passes into oblivion. However, that ignores the continuing use of church bells at anniversaries and obits, and the role of chantry priests to perpetuate the memory of the deceased person.

The life-history of the clock is linked with that of the bells; an increase in the number of a ring, the rebuilding of the frame and the recasting of old bells often involved new work on the clock and sometimes accelerated the acquisition of a new movement and a quarter-striking or chiming train. The clock-keeper was usually keeper of the bells also, collector of fees for their use and parish clerk in general.

The early Assembly Books of the Borough of Henley on Thames have survived from the time of Richard II, the oldest entry being dated September 1395 (Briers, 1960). Here we have the first definite evidence of a mechanical clock on 3 November, 1410, viz., *pro emendatione cuiusdam instrumenti vocati Clokke . . vjs.;* evidently " some sort of instrument called a clock " was still a mystery to the scribe. By 1494 the clock-keeper was on a regular annual stipend to look after it and its chimes, the clock is then called *horecudium*, i.e. striking the hours. Five new bells had been bought in 1470, the fees from which and from the chimes had become an important source of revenue (See Thomas Harington).

The Churchwardens' Accounts of the parish of St. Mary, Thame begin in 1442 with an entry, *for makying of ye clocke . . vid.*, which should be interpreted as a repair charge. The sexton was then paid 10s. for three years' arrears, *ffor kep'g of ye klokke*. And in 1443 the churchwardens received, *as for kepyng of ye kloke . . iiijs., ixd.* indicating regular fees for a clock already functioning some years previously. In 1474 a payment, *for mendying of ye clok ye wyche ye smyth of tryng mad for owr parte*, suggests a replacement of an earlier clock (See Thomas Cotteswold, Robert Smith).

At Oxford the Churchwardens' Accounts for St. Michael's Church for the year 1470-71 have one entry mentioning a horologium.

The Proctors' Accounts for the University Church of St. Mary the Virgin, Oxford, reveal that annual payments were made to the clerk, *pro custodia horilogii*, from 1469 to 1497, while major repairs in 1471 and 1477 imply that the clock had already given considerable service.

The Sixteenth Century: It is not until the beginning of the 16th century that one finds the names of the makers or contractors who supplied the clocks installed in the bell-towers of Oxford. The earliest example is that which was made for the newly built tower of Magdalen College in 1505. A master mason of Abingdon and Burford, William Este, who worked for Oxford colleges in the first quarter of the century, contracted to supply a one-hand clock for £10; in this assignment he was associated with Martin Williamson, a brewer of Oxford and Louis Foose, a painter of Abingdon, presumably as guarantors for the satisfactory completion of the work by an unnamed smith.

Merton College ordered a new clock in December 1509. The record states, *novum horolegium de novo fieri ut clerici possunt surgere in aurora hora debita*, that is, so that the scholars might rise at the appointed morning hour. Nothing is known of Merton's clock during the 15th century but presumably it was by 1509 quite out-of-date, and an hour-striking turret clock was overdue.

The old clock in St. Mary the Virgin was repaired in 1510 and the clock-house was apparently pulled down about this time when the church was extensively rebuilt. In 1523 a new quarter-striking clock was installed which lasted into the 17th century.

St. Martin's church at Carfax had a clock for which a clock-keeper was regularly employed by 1540. Its chimes and quarter jacks are first mentioned in the City Council Acts of 1592 but were evidently in existence much earlier (see Plate 2 fig. 2).

When Oseney Abbey was demolished in 1545 its old clock was taken down and re-erected in the Priory church of St. Frideswide which became part of Christ Church Cathedral. A clock-house was built for it in the steeple by John Wesbourne, carpenter, and a clock-keeper was appointed at 6s. a year. Frequent repairs were needed right into the 17th century.

St. John's College had a clock before 1569 for the maintenance and repairs of which Johan de St. Paul was engaged.

Documentary evidence of 16th century turret clocks in places outside Oxford is scanty although there are some anonymous examples still existing and more or less altered, which may have started life in that century. They have some features similar to those of the quarter-striking and chiming clock in East Hendred church, Berks., which Henry Seymour made in 1525.

It is no longer possible to discover how old was the clock in Cropredy church in 1512 when the Vicar left a sum of £6 13s. 4d. in trust to the churchwardens to find someone to keep the clock going hourly and to toll daily the Aves bell at 6 a.m. and the curfew at night.

Thame had a clock in its Moot Hall in 1543 and Thame church obtained a new clock about 1562.

The Borough Records for St. Mary's church, Henley, extend to 1543 and during that period the 15th century clock and chimes continued to function.

Stonesfield church is said to have had a clock made in 1543 which had orginally been in a near-by manor house.

South Newington church had a clock before 1560.

The early Churchwarden's Accounts for Spelsbury deal only with revenue and include an item about a piece of clockwork in 1574.

Yarnton church had a 16th century clock which was replaced by a new one in 1641.

Until the last decades of the 16th century the men who made and repaired large clocks were smiths. However, the term "clockmaker" is used in Thame church accounts as early as 1530. The first record in the Oxford City archives of a man then described as a clockmaker occurs in 1588 when Robert Harvey was made free on 19 September paying an outsider's fees of 40s. and 4s. 6d. Four years later the City Council were prepared to admit another clockmaker, Thomas Bull, for only 4s. 6d. provided he bound himself to the churchwardens of St. Martin's parish to look after the clock, jacks, chime and dial of the City church at Carfax. But this arrangement was never ratified as the churchwardens preferred to employ smiths for their clock and in fact continued to do so until 1704 (see Hugh Corbett).

The Seventeenth Century. 1600-1660.

The first half of this century saw the establishment in Oxford of true clock and watchmakers as distinct from smiths and locksmiths. Triumph de St. Paul, born in Oxford the son of an immigrant Frenchman engaged in repairing college clocks, was made a freeman on 18 February, 1601. In August 1608 he was able to take an apprentice, Richard Quelch, who was admitted free in 1616. These two men were the pioneers of domestic clock and watchmaking in Oxford and were for many years the only local craftsmen supplying the demand for balance-wheel lantern clocks and watches (see Plate 19, fig. 34. Quelch's son Richard, apprenticed to his father, was free in 1652 and took his first apprentice then. His son John continued the business.

Another family business was started by Michael Bird, who was apprenticed in London in October 1648, and gained his freedom in Oxford in September 1654.

Work on turret clocks still remained the right of smiths. At St. Martin's church Thomas Ranklyn and John Raye kept the clock and chimes in order and the parish clerks did the winding for the first 30 years. St. Mary Magdalene church acquired a clock in 1604. The University church of St. Mary the Virgin was equipped in 1640 with a new quarter-striking clock made by the smith, John Raye; the churchwardens continued to employ smiths on their clock for another hundred years. From documentary evidence it seems that the college turret clocks were maintained by the same Oxford smiths who were regularly employed for the iron work of college buildings and bells; these old clocks continued to function for long periods under their care.

In the villages and small towns of the county the few surviving account books supply little information about the purchase of their church clocks but repairs and maintenance are recorded and were usually done by local men. One rare entry of a purchase is for Yarnton's new clock in 1641 which cost the parish only £5 18s. 0d. as the maker took the old ironwork in part payment. Fritwell church had a clock before 1645 and parts of it have survived. Burford Tolsey had a clock before 1653 but that has disappeared. From their structure it may be concluded that some of the still existing but undocumented church clocks were made in the early 17th century. For example, the Combe church clock now in the History of Science Museum with its restored foliot balance, the much repaired clock in Claydon church, that in Wigginton church with its unusual upsidedown striking train, those at Langford and Church Hanborough still with crown wheel and verge escapements, and some just over the county border at Kings Sutton (before 1630) and Farthinghoe (Northants).

The Seventeenth Century. 1660-1700.

This period was one of unusual developments which, as elsewhere, resulted from the application of the vertical pendulum to the verge escapement by Christian Huygens, and, a few years later, from the invention of the anchor escapement.

In Oxford until about 1660 not more than 2 or 3 craft clock makers were working at the same time in the City and occasionally none had an apprentice. Thereafter the relatively new profession attracted many young men from neighbouring villages and counties who wished to be apprenticed to Oxford masters. This expansion reached its peak about 1685 when there were 5 freemen clockmakers and 12 apprentices — the full quota was absorbed and even exceeded.

During the next 8 or 10 years the apprentices completed their training and further recruitment fell off. By the end of the century the apprentices had disappeared, some to London, some to oblivion and only a few settled in Oxford as replacements of the older men. The boom in the clockmaking

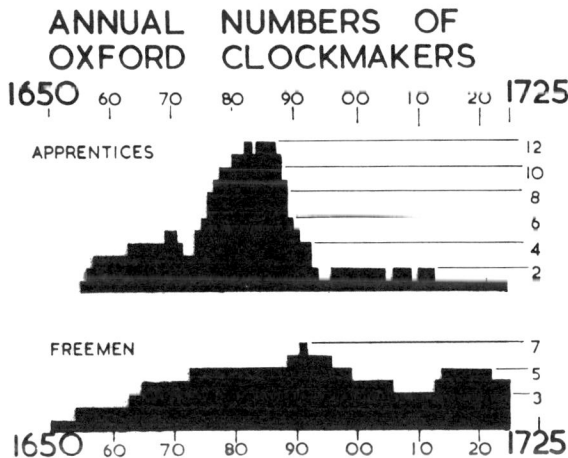

craft as a new career had finished and in the first quarter of the 18th century Oxford supported only 4 or 5 freemen clockmakers with at most 2 apprentices between them. This phase is shown in graph form in the adjacent text-figure.

As to the parentage of these apprentices it may be noted that out of 30 the father was a yeoman in 9 cases, watchmaker in 5, cleric in 4, gardener in 2, mercer in 2 and the rest were chandler, fisherman, gunsmith, joiner, mason, printer, surgeon and tailor. The freemen to whom they were apprenticed and who became free during the period 1660-1700 were: 1663, John Quelch, third of that family; 1667, Joseph Knibb and in 1672, John Knibb his brother; 1672, Anthony Hodges and in 1678, John Harris, both apprentices of Michael Bird; 1688, Hugh Broadwater, apprentice to Harris; 1689, Samuel Aldworth, John Knibb's journeyman; 1691, John Ford and in 1694, John Goweth, both apprenticed to John Knibb. The older freemen, Michael Bird and Richard Quelch, enrolled 8 other apprentices during the same period. Two of these masters, Michael Bird and Joseph Knibb, were doing enough business by 1668 to make it worth while to issue their own token coins. For his emblem Bird chose a cock and Knibb used a one-hand clock dial.

The earliest dated clock made in Oxford at the beginning of this period is a lantern clock inscribed, *Johannes Knibb Oxon fecit* 1669 — undoubtedly an apprentice piece. In the county the earliest dated clock is also a lantern made by George Harris of Fritwell in 1668. Turret clocks dateable on documentary evidence include that made for Wadham College by Joseph Knibb with an anchor escapement in 1670; at the same time he fitted an anchor escapement to the clock in St. Mary the Virgin church. Both of these escapements antedate that made for King's College, Cambridge, in 1671 (Beeson, 1961). George Harris also made a clock with a crown wheel and verge escapement for Hanwell church in 1671.

Elsewhere in the county there was only sporadic development. A dated lantern clock was made by William Kenning of Banbury in 1674 and there is no further indication of clockmaking in that town until Benjamin Lamprey appears at the end of the century. At Henley on Thames John Barton was trading about 1685 and Edward May made a longcase clock which Britten dates ca. 1680. No 17th century clocks or watches are known from the market towns of Bicester, Thame, Woodstock and Witney. By contrast clockmaking started towards the end of the century in a small village, Sibford Gower, which was the birthplace of a Quaker, Thomas Gilkes. He trained other Quakers and founded a monopoly of clockmaking in North Oxfordshire villages which remained in Quaker hands throughout the 18th century (Beeson, 1958).

The Eighteenth Century. 1700-1750.

Oxford was well served by watch and clockmakers during the first half of the 18th century. Among the few who survived from the 17th century the leading maker was John Knibb until he died in 1722. After a short time in business in Oxford, Samuel Aldworth and Hugh Broadwater left to settle in London, and later John Ford moved to Aylesbury. Newcomers continually arrived to replace them. The ten men whose names follow may safely be described as craftsmen and makers; the date associated with each name is that of the earliest record of his craft activity.

Greenaway Curtice, who was trained in London and in Oxford, became free in 1699 but died in 1702. John Oakley put the date 1704 on his lantern clock. John Free started business in 1709, George Wentworth in 1713 and Edward Moore in 1714. Humphrey Brickland became free in 1723. Robert Denton matriculated as a University tradesman in 1730. James Oakley, son of the earlier John, started in 1735. John Herbert put up a sign "The Dial" in 1743 and the long-lived Thomas Reynolds had his first workshop in Holywell in 1745. Edward Moore, the second, followed his father in 1751. For further details of these men see the Biographical Dictionary, Part II.

In the rest of the county businesses started in the market towns but also in small and remote villages. In the village of Horley John Blundell put his name and the date 1700 on the dial of a 30-hour longcase clock. At Watlington in the south Francis Webb made 30-hour clocks as early as 1710. John Nethercott was in Long Compton about 1707 but also worked in other places. In Bicester William Ball made longcase clocks early in the century but was overshadowed by Edward Hemins (died 1744), who also had an extensive business as bell-founder and maker of turret clocks. Banbury's first clockmaker, Benjamin Lamprey, died in 1744 and was followed by his son John (died 1759). Of the Quaker clockmakers John Fardon began in Deddington about 1725, Richard Gilkes in Adderbury in 1735 and his brother Thomas, the minister in Charlbury, about 1725. In Great Haseley Thomas Holloway was working as early as 1734. Thame does not appear to have had a clockmaker before 1744 when William Lawrence set up business but he also worked in Cuddesdon.

There is some proof that village clockmakers hammered out and engraved their dials, and cast the brass corner-pieces from patterns fashionable with the London makers. The iron work and brass work of 4-posted clock frames was certainly constructed in village workshops. Although indisputable evidence that they made wheel blanks and cut the teeth is not yet available it is reasonable to assume that they were self sufficient in this respect also. (See Quaker dials and movements, bird designs, under Gilkes and May, Biographical Dictionary). Chains were supplied by chainmakers and iron weights were cast in local foundries. The plain oak or pine longcases were made by local carpenters as is revealed by individual decorative details and by rather crude proportions.

The Smiths: The value of smiths for work on turret clocks rapidly declined. Colleges preferred to employ clockmakers for repairs and annual maintenance. The City of Oxford followed suit for St. Martin's church in 1704. The churchwardens of St. Mary Magdalene, however, continued to rely on smiths until 1722 and those of St. Mary the Virgin until 1741. In the villages a clockmaker was called in whenever available or the church clock was carted away to a clockmaker's workshop, sometimes at a considerable distance.

The Eighteenth Century. 1750-1800.

Clock and watch business in the later 18th century followed much the same trends in Oxfordshire as in cities and towns elsewhere in England. One cannot guess how much original fabrication of clocks and watches was done by men apprenticed as watchmakers, or to what extent they assembled movements and fitted cases. Importation of parts and of finished movements and dials was no doubt extensive. From scribings on movements it is clear that Oxfordshire men did a great deal of competent repairing and from documentary evidence it is known that they maintained private and public clocks satisfactorily. The value of cataloguing the names on clocks and watches in this period is not to prove that they made pieces indistinguishable from the products of London, but that there existed a thriving horological trade in provincial localities. Some indication of the amount of bought-in material stocked in their shops, and of the sources of their supplies can be learned from the bill-heads of tradesmen in 1760 to 1770 preserved by John and Richard Stone of Thame and Aylesbury (q.v.). Their suppliers in London were Charles Blanchard, James Brogden, John Garland, Charles Howse, Benjamin Lamb, Robert Ward, Thomas Wild and James Willshire; the clock-founder Griffith Ellis, the clock-bell maker Robert Romley; and the engravers N. C. Gutknecht and John Thompson.

In Oxford the goldsmiths and jewellers, e.g. Edward Lock and George Tonge (q.v.) sold goods obtained from London. And George Rowell (q.v.) quite openly put his name and serial number on a watch movement the dial of which is inscribed, " Gregson à Paris." Recent evidence published in the *Horological Journal* shows that there was a regular supply of clock movements by Thwaites and Reed and by Handley and Moore to clockmakers whose names are far better known and more esteemed than those of the latterday Oxfordshire men. The painted dials of grandfather clocks were supplied by Walker and Hughes and also by Osborne.

Towards the end of the 18th century turret clocks were beginning to be obtained from firms specialising in their manufacture and were installed under the direction of local clockmakers, e.g., that at Rousham House by John Davis of Windsor, 1760, and those in University College, 1792, and in Garsington church, 1796, by the predecessors of Thwaites and Reed, as well as the Woodstock church clock by John Bryant of Hertford, 1792. In the first half of the 19th century Oxfordshire towers acquired clocks by Evans of Handsworth, John Moore and Sons, William Taylor, Thwaites and Reed and B. L. Vulliamy.

The Nineteenth Century. 1800-1850.

Tradesmen classified as clock and watchmakers in this period usually were in business also as jewellers, silversmiths, opticians and even as ironmongers or general hardware merchants. Over a hundred names are listed for the county as a whole of which 26 were in Oxford and 22 in Banbury. Names of foreign origin, German and Italian, begin to become more numerous. From advertisements in a mid-century county directory we quote the standard prices for watches:

	£	s.		£	s.
" Silver horizontal watches, 4 jewels	2	2	to	3	15
Patent lever watches	4	4	to	7	7
Gold horizontal watches, 4-10 jewels	3	15	to	10	10
Gold lever watches, 10 jewels, compensating balance	7	10	to	16	16
Vertical silver watches, jewelled				3	10
Jerome's American clocks from					15

Secondhand watches in silver cases warranted and kept in repair for 12 months, from £1 each."

In the Dictionary, Part II, no names are included for businesses not in existence before 1850.

TURRET CLOCKS

A GREAT deal of information on the making and maintenance of turret clocks in Oxfordshire has been discovered in the financial accounts and archives of churches, colleges and town councils. In many cases annual accounts cover periods measureable in hundreds of years, some from the beginning of the 15th century. These sources have also yielded valuable biographical data about the craftsmen concerned, the smiths, clockmakers and clock-keepers. In fact, documentary evidence not only provides a more reliable history of turret clocks than is obtainable solely from the study of surviving examples, but also throws fresh light on the chronology of English clockmaking in general. For this reason the author considers it useful to include in this section detailed analyses of the financial accounts of certain parishes and colleges; from these it may be seen that events occurring in one place confirm or supplement events occurring in another place and help to explain them.

When more documentary records have been traced and analysed they may help to identify the anonymous makers of village church clocks and add them to the company of worthy craftsmen such as Harris of Fritwell, Hemins of Bicester and Hawting, Reynolds and Young of Oxford.

An interesting comparison of variations in design may be made from the clocks at Aynho, South Newington and St. Mary the Virgin, Oxford, and the work at St. John's College, all for the year 1740-41.

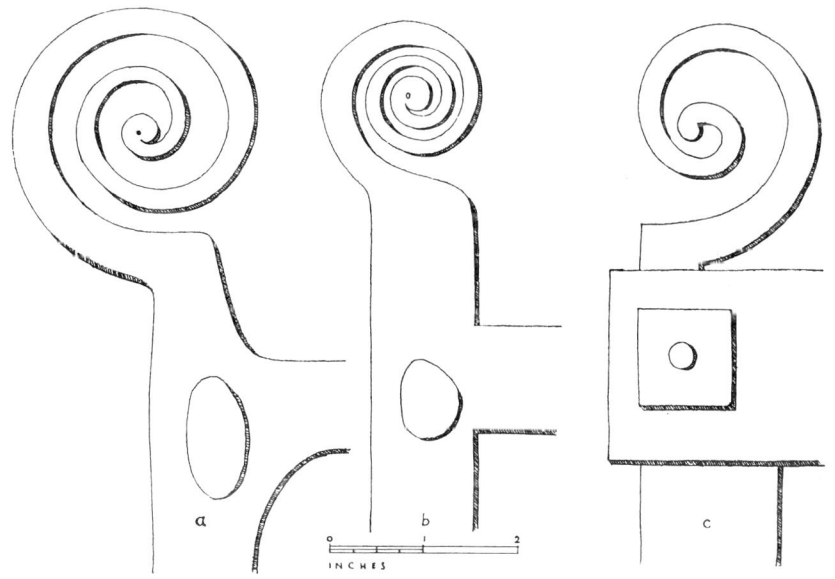

SCROLL FINIALS: (a) Thomas Reynolds, Swalcliffe Church, ca. 1755 — (b) Joseph Knibb, Wadham College, ca. 1670 — (c) Anonymous, Hornton Church, ca. 1740

ADDERBURY HOUSE

A disused quarter-striking clock of 3 side-by-side trains. The iron bar frame, 30 × 19 × 15 inches, is fastened together by square nuts; there are 3 pairs of pivot-bars and no cross-ties. All the wheels, barrel plates and locking plates are of brass; all barrels of wood for rope-drive. The finials on the corner-standards are simple scrolls, resembling fig. *a* of the adjacent text-figure, but not snub-ended.

The going train in the middle has the great and 2nd wheels with 4 straight arms; the escape wheel has its arms widened in a curve to the centre. Solid pinions of 10 and 6 leaves. The semi-dead beat anchor has no neck and spans 7 teeth; its arbor at the crutch end is pivoted in a simple detachable cock fixed by a hexagonal nut and washers. The upper end of the extended pivot-bar is turned over to take the pendulum spring which is held by a pin; there are no chops or up-and-down adjustment. The pendulum rod, about 6ft. long, is looped to clear the winding square (the bob is missing). The opposite pivot-bar is forked at the top into 2 opposed scrolls. On the end of the great wheel arbor is a plate with 8 pegs and a lantern pinion which turned the hour wheel (missing).

The hour-striking train great wheel has 8 hammer-pegs and a pinion turning the locking-plate wheel which is attached to the disc locking-plate proper. The 2nd wheel has a long stop-peg and a pinion of 10 leaves. On the fly arbor is a pinion of 6, a stop-peg and a fly with 2 vanes set at right angles.

The quarter-striking train has on the great wheel arbor the quarter locking-plate with lifting-pegs for 2 hammers. These hammer levers are mounted on one transverse arbor, one loosely and longer, producing a quick ting-tang. The 2nd wheel has a pinion of 8 leaves and a lifting-peg. On the fly arbor are 2 stop-pegs and a 2-vaned fly. The full striking control is by the peg and lever method.

On the gable of the stables building of Adderbury House is a circular dial with painted hour numerals and one counter-poised hand. It is contemporary with the masonry of the wall, and the clock was evidently installed when the building was erected. One of the water spouts bears the date 1722.

ADDERBURY — St. Mary.

The Oxford Archdeaconry Papers record an order in 1684 to provide, among other things, a clock.

The Churchwardens' Accounts begin in 1713 and in that year is an entry, *for bell ropes & clock ropes as p. a bill . . . 19s. 3d.* Thereafter purchases of clock ropes were frequent and there was an almost annual charge of 6d. for wire for the chimes. Sixpence was the price of half a pound of wire and, as it was purchased in these small amounts for 35 years, the charge, perhaps, should not be interpreted as actual consumption. In 1719 is an item, *Pd. ye Ropers bill for bell ropes & Chime gabil rope . . . £1 4s. 0d.* The gabel rope is mentioned later in 1758, *for a new Gabel Rope for ye Chimes Gabel . . . 7s. 6d.*, but its particular function is not clear. Mr. J. Fergie considers that a gabel bell only is meant, at the junction of the nave and chancel where externally one may see the site, roughly above the stair to the rood loft. The chimes were evidently an independent unit, possibly for quarter-striking or for playing a tune, as in 1723 it was sent away for repairs, viz., *Tho. Williams for mending the chimes . . . £1 12s. 0d. — for carring the chimes to king Sutton and fetching them . . . 2s. 6d. — for Cleaning the Clock . . . 1s. 10d.* Thomas Williams appears also in the Kings Sutton church accounts as clock-repairer.

26

From 1738 to 1744 Williams was paid further sums annually but the charge of six pennyworth of wire for the chimes continued to be passed. In 1747 is an entry, *Paid ye man for Comming our about ye Chimes & yr Neglect &c by order at Vestry . . . 2s. 6d. — for mending ye Clock and Chimes . . . 5s.* In 1748 the charge for chime wire was abolished and the Adderbury clockmaker, Richard Gilkes (q.v.), was appointed clock-keeper, a post he held for 38 years until 1786. His bills amounted to only a few shillings each year, and apparently included the cost of winding. Expenditure on the upkeep of the chimes was accounted as separate payments annually to William Edwards, until 1776 when they suddenly ceased and do not reappear in the Accounts.

Richard Gilkes died in February 1787; his Friend, Thomas Fardon of Deddington (q.v.) came to Adderbury to wind up Gilkes' business. While there in 1787 Fardon repaired the clock for £1 9s. 0d.

Next year, 1788, Joseph Williams, also a Quaker, became the local clockmaker and was given annual charge of the church clock, which service he performed for 40 years at a fee that rose to £1 a year with extra for repairs. He was also responsible for the chimes which were refitted when the bells were recast, as recorded by the following entry for 1789, *To Mr. Briant of Hertford for taking down the old Peal of 6 bells and recasting them into 8 with extra New Metal hanging and all Materials to complete the Peal Work to Clock & Hammers to Do . . . £100.*

Under Joseph Williams' care the clock functioned satisfactorily and economically. The only unusual event was recorded in 1792, *To Mr. Lovell for painting and Gilding ye Church hand . . . £2 12s. 6d. — For Carrige and Charges With the Church Hand . . . £1 0s. 9d.* This is the first mention of a dial in the Accounts, but from pictures of the church before 1797 it is known that the dial had two hands and was set diamond-wise over the upper part of the east window of the tower. William Williams took over from his father in 1828 at an annual stipend of £1, and also looked after the fire engine.

In 1839, after the extensive restoration undergone by the church in previous years, Thomas Strange, clockmaker of Banbury, was employed to restore the clock and bring it up to date as the following entries show: *Mr. Strange Repairing Church Clock . . . £7 10s 0d. — Ditto for two Setts of new works for the minute hands and fixing . . . £12 10s. 0d. — Ditto for two New Copper Dials (6 feet) . . . £10 0s. 0d. — Richd. Bannard for work at Clock outside . . . £4 5s. 3d. — Beer for Men while at Clock . . . 10s. 9d.* The clock dials of 1839 were fixed above the apices of the tower windows.

In the same year the job of keeping the clock was taken away from Williams and given to Strange at £1 1s. 0d. a year. It is most unlikely that Strange came out from Banbury daily to wind the clock and, presumably, he subcontracted the work to Williams who was on the spot and had charge of the fire engine each year until his death in 1862.

The clock was in the charge of Strange until 1866, although since 1851 his annual fee had been reduced to 15s. 9d. Later in the 19th century the Banbury clockmakers, Burditt and Walford, were successively paid for cleaning and repairs. This clock has disappeared, except for a shaft of the motion work and the remains of a hand.

The present $\frac{1}{4}$-striking and tune-playing clock is by John Smith & Sons, Derby.

AMBROSDEN — St. Mary

A disused clock in a frame, 36 × 18 × 25 inches, fastened throughout by nuts; no finials. The wheels of the 2 side-by-side trains are of iron, 4-armed, the escape wheel of brass. The anchor is original and its pendulum (removed) probably beat one second. A brass setting dial operates 2 opposed brass bevel wheels, one above the other, and an hour wheel with a clutch. The striking train has an external iron locking-plate, peg and lever control with the lifting-piece raised by the hour wheel, and an adjustable 2-vaned fan. A plate on the frame records, " THOMAS HARRIS, HENRY COOPER, OLIVER PANGBOURN, CHURCHWARDENS IN YE YEAR OF OUR LORD 1711. VINCENT SMITH AT POUNDON FECIT."

The name of Thomas Harris occurs on two of the bells cast in 1703 and 1716. The names of Cooper and Pangborn are also on the 1703 bell.

Vincent Smith was paid £9 and an outside dial was added in 1713 for £5 12s. 0d. The use of iron for the secondary wheels in both trains is unusual so late as 1711. Poundon is a small hamlet near Marsh Gibbon just across the county border in Buckinghamshire. Presumably this clock from a blacksmith's workshop was cheaper than one obtainable from the clockmaker in Bicester.

An hour-striking clock with a double 3-legged gravity escapement by Potts of Leeds is now in use since 1929.

ASTON, NORTH — St. Mary

Richard Rawlinson who visited the church about 1720 recorded that Lady Howard, lessee of the estate, had given " a Clock and hand on the outside of the Tower." This has disappeared and the present clock, which was installed when the church was restored in 1866, was made in Theobald's Road, London. The cylindrical bob of the pendulum is signed, Tucker, London 1867. There is one skeleton dial.

ASTON, STEEPLE — St. Peter

A disused clock is preserved in its cupboard in the tower. The frame, 32 × 15 × 22 inches, is of flat bars and 2 cross-members fixed by nuts; the corner-bars have degenerate buttress mouldings and finials inclined outwards ending in subquadrate knobs. There are two pairs of pivot-bars fixed with nuts for the side-by-side trains. All the wheels are of brass, the barrel-plates of iron. The great wheels have 4 straight arms, the 2nd and escape wheels 4 arms narrow flask-shaped. The anchor has no neck and spans 7 teeth. On the front frame member a pendulum cock is fixed with cube-headed screws and the pendulum rod has a loop in it to clear the winding square; its length is 47 inches from the chop to the centre of the disc bob. The only regulation is by means of a wing nut below the bob. The striking great wheel has 10 hammer-pins and a small iron ratchet-plate with the spring and pawl on the barrel-plate, as in the going train. The fly arbor has a curved peg and a fly with 2 adjustable vanes. The lifting-piece, which is raised by one of a pair of pins on the great wheel, has one arm for contact with the fly peg and one arm for raising the locking-piece. The iron locking-plate is internally toothed with offset spokes. Transmission for a 2-hand dial has been added outside the going train apparently in the 19th century.

The old dial was removed and a new convex copper dial with 2 hands was fitted in 1819 at a cost of £5 15s. 0d.

From the features of the crossing-out of the wheels it is probable that the clock was made by Edward Hemins of Bicester about 1720-1725. The only references to the clock in the Churchwardens' Accounts are for cleaning at 2s. 6d. to 5s. a year from about 1739.

The present hour-striking mechanism by Potts is entirely electrically driven and wound.

AYNHO — St. Michael (Northants)

This clock is included as it is just across the county border and is by an Oxfordshire maker. The setting dial is signed, *Edwd Hemins* BISITER FECIT, and the rim of the great wheel of the striking train is inscribed, ED HEMINS BISTER FECIT 1740. (See Biographical Dictionary under Ball and Hemins for variations in the spelling of Bicester).

The frame, 36 × 26 × 18 inches, has the side members and pivot-bars fixed by nuts; the vertical brass finials are somewhat spinning-top shaped with an axis extending above and below. All the wheels of the two side-by-side trains are of brass, as also the locking-plate, the wheel driving the contrate wheel on the vertical leading-off arbor and the barrel ratchet plates. The crossing-out shows the typical, slender, flask-shaped, welded arms of Hemins. Anchor and dead-beat escape wheel and crutch are modern replacements; the pendulum with a wooden rod and heavy disc bob beats $1\frac{1}{4}$ seconds, but the clock runs for only 4 days although the fall of the weights is through 2 stories of the tower. Let off and stop for the striking is by peg and lever.

The clock serves 2 circular dials and a carillon movement with pins for 7 tunes each using the ring of 8 bells. It was installed by Gillett and Johnson in 1913 and runs for about 12 hours with one winding if set to play at 3-hour intervals.

BAMPTON — St. Mary

Overseers' Records for Bampton, 27 January 1733, " At a Vestry this day held and application being made to the said Vestry by John Reynolds of Hagbourne in the county of Berks (sic) Blacksmith for the payment of the Sum of Thirty Four pounds due to him for making a new Church Clock with chimes in the parish church of Bampton, he having performed his said work according to his agreement of this Vestry therefore it is ordered by this Vestry that the Churchwardens of this parish for the time being do forthwith pay unto the said John Reynolds the said sum of Thirty Four Pounds according to the agreement of this Vestry for that purpose except Fourty Shillings which is to be left as a Caution till the Clock is further proved."

The present clock is by John Smith and Sons, Derby.

BANBURY — St. Mary

The predecessor of the present church was demolished ca. 1790; its chiming clock was made by Joseph Hemmins in 1741 and was preserved until 1897 and then was sold to the parish of South Newington (q.v.) and erected in that church as a Jubilee commemoration. The tower of the new church of St. Mary was not completed until 1822 and its present clock is by Gillett and Johnson, 1897 and has Westminster quarters. A carillon also by these makers plays every three hours and has 7 tunes (originally 21). There is a ring of 10 bells and 2 additional semi-tone bells for the chimes.

BANBURY — The Town Hall

The second Town Hall was a timber and plaster building supported on pillars which stood in the Market Place from about 1663. The only reference to its clock is in the Borough Vestry Book entry for 9 October 1771 to the effect that *The Overseers of the Poor shall pay at all times hereafter for the repareing of the Town Hall clock and also pay John Lamprey his present*

Bill for Repares thereof Amounting to One pound and two shillings. This hall was demolished in 1800, and a brick building replaced it which had a bell-turret and a quarter-clock (illustrated in Potts, 1958, opp. p. 108). The 4th Town Hall was erected in 1853-4 and its clock was supplied later by John Smith and Sons, Derby.

BANBURY

The location of this clock has not been traced; it is specified in an invoice of Thwaites of London dated July 1797 to Messrs. Perigal and Son.

" To a new 8 Day Turret Clock to strike the hours on a Bell of about 90lb and to shew 2 Inside dials of 14 In. Diameter each the striking Great wheel 9 in Diameter the watch Great wheel 8 in Diameter and the lesser wheels of proper proportion and all of brass the pinions and pallats and pins for drawing the hammer work all hardened and all in a strong Iron Frame with weights Ropes Pullies Dial Plates Hammer Work &c and everything compleat except Bell or Package or fixing.

To a new Bell wt 1c 1q 11lb 8oz.

To 2 Packing Cases for the above 1 to contain the clock with false Cover for the Hammer work &c & 1 for the Dial works.

To a Man to Banbury to fix the Clock 12 Days with expenses of Living on the Road & Carriage there & Back.

To 5lb of the best Hemp Line."

Within twelve months the clock had to be resited in a higher storey and a further invoice was made out through Messrs. Perigal & Son.

" To moving the Turret Clock from the first floor to the loft made 4 New 8 in Pullies made a New Universal Joint & a set of Rowlers lengthened the Centre Arbor & made larger fly fans & 25lb. of New Lines with Expenses of Living Carriage &c for Man & Expenses for Self Going down."

BARFORD ST. MICHAEL — St. Michael

The Oxford Archdeaconry Papers record that the Barford clock was out of order in 1753, the clock, chimes and bell out of order in 1769, the clock and chimes needing repair in 1770-1 and the bell wheels, and chimes out of order in 1801. The chiming clock referred to on these dates is not the one now in the church.

According to the *Oxfordshire Archaeol. Soc.*, Rept., 1907, p. 13 the Barford church clock was transferred from South Newington church in 1897 and originally came from Bloxham church. However, local tradition is firm that Wigginton church has had the Bloxham clock since about 1880. There are no records at South Newington after 1684, which might have covered the period when the clock was made. The iron frame 30 × 22 × 23 ins., has corner-bars ending in finials projecting at right angles to the side of the frame and ending in a pear-shaped body with a square top. The main vertical bars are fastened with square nuts carved in leaves and other bars are tenon-wedged. The two trains are side-by-side; the going train great wheel has 2 lifting pins as there is no separate hour wheel, an escape wheel with a recoil anchor, and a pendulum about 42 inches long. The striking train has an internally toothed locking-plate, an early type fan with slipping pawls, and peg and lever control. All arbors and iron bars are roughly octagonal and all pinions are integral with their arbors, those for the going train of 10 and 8 leaves and those of the striking train of 8 and 8. Each barrel has a spring hook and pawl.

BICESTER — St. Edburg

A clock was placed in the tower about the middle of the 18th century (Blomfield, 1884). Its history is very scanty but there were also chimes in 1766 when they were mended for £14. One would expect it to have been made by Edward Hemins, the bellfounder and clockmaker of Bicester (q.v.).

A turret clock by Edward Hemins was found in a scrap metal dealer's yard in Bicester a few years ago. It is illustrated in Plate 4 and described below.

The iron bar frame, 25 × 13 × 16 inches, with 3 cross-members and 3 pairs of vertical pivot-bars, is held together by square nuts on threaded pins. The corner-standards have pseudo-buttress mouldings and outward inclined finials which end in depressed sub-pyramidal knobs. All the wheels, barrel plates and ratchet plates are of brass; the wheels show the typical Hemins crossing-out of 4 narrow flask-shaped arms. Both great wheels have a 5 inch spring and pawl fixed to the rim. In the going train a 10-leaf pinion drives the 2nd wheel and a pinion of 7 the escape wheel. The recoil anchor has no neck and spans 9 teeth. The pendulum cock is bolted to the top bar, has 2 chops and a wing nut with threaded rod for regulation. From chop to centre of the heavy lenticular bob the pendulum measures 47 inches. The striking great wheel has 10 pins and a 10-leaf pinion for the internally toothed locking-plate which is of iron with offset spokes. The 2 pinions are of 8 and 6 leaves. On the fly arbor is a stop peg and a 2-armed fly with a ratchet wheel and 2 small adjustable vanes.

The indicator dial, numbered 1-60 in fives is signed, " Edward Hemins FECIT." On its arbor is a hand-grip for setting the iron clutch on the hour wheel. It also has a pinion of 10 turning a contrate wheel on a vertical arbor journalled in 2 cross-members. This leads to the motion work. Striking control is by the peg and lever method, the action of which is common to many turret clocks and may be described here.

A lifting lever and a locking lever are pivoted on transverse arbors in one pair of corner-standards. The lifting piece is raised by a pin on the clutch of the hour wheel. A peg set at right angles on it lifts the locking lever until its catch is free of the notch in the locking-plate. The train then partly revolves or warns until a peg on the fly arbor is caught by another peg on the lifting piece. When the hour wheel has turned far enough the lifting piece drops freeing the peg on the fly arbor so that the locking-plate can advance. The catch on the locking piece rides on the rim of the moving locking-plate until it falls into the next hour notch and the train is abruptly halted. The freewheel ratchet of the fly allows it to over-run and stop.

This small clock was made ca. 1735.

BICESTER — Town House and Shambles

The Town House was built in 1622, altered and enlarged in 1686 and provided with a bell-turret and a clock. It was pulled down in 1826.

BLENHEIM PALACE — Woodstock

The bell tower built by John Townsend, an Oxford mason, in Blenheim Palace houses a turret clock made by Langley Bradley of London in 1710. The contract signed in August 1710 with Sir John Vanbrugh and others on behalf of the Duke of Marlborough included the clock and its weights for £150, three bells for £107 16s. 3d., a copper dial for £27 surmounted by *a large Coronet of copper with a large head and wings representing Time, Chais't bold and strong, £12.* The whole account came to £303 16s. 3d. but was later abated by a hundred pounds. The dials with gilded coronets above them and Gibbons' garlands below survive but Time with his large head and wings has disappeared. The original chime of three bells was replaced by one quarter bell in 1753, the other quarter bell and the hour bell in 1842; they are fitted with dead stocks and clock-hammers (Green, 1951). Langley Bradley also specified four sundials on pedestals for the purpose of checking clock time by the sun; these were designed by John Rowley and the stone work was carved by Henry Banks (see Sundials).

The clock was later repaired and modified by B. L. Vulliamy.

BLOXHAM — St. Mary

The old clock was transferred to Wigginton Church about 1880. The present clock is by Gillett and Bland, Croydon, 1880.

BODICOTE — St. John the Baptist

A clock dated 1700 of two trains side-by-side with an anchor escapement. The iron frame, 35 × 17 × 31 inches, has rectangular corner-standards with buttress mouldings at the top and bottom and recessed swan-neck finials ending in rimmed brass balls. The principal wheels are of iron with welded spokes, and the new wheels of brass except for the offset wheel train to the hand setting dial. The going train has integral pinions of 8 and 8, a semi-dead beat escape wheel and an anchor with fabricated pallets. The pendulum is about 5 feet long with a broad wooden rod. Originally intended for half hour striking with two pins on the great wheel (but now disused) the striking train has an external locking-plate with offset spokes and peg and lever release; pinions of 8 and 14 are integral.

This is one of the oldest turret clocks in Oxfordshire bearing a date and the maker's name. Two plates on the frame are inscribed *Tho. Bradford & Rich. Wise Church Wardens. John Wise Londini 1700* and *Improved by T. Strange Banbury*. The original motion work for half hour strike and a single hand was mounted immediately below the main barrel axis on the frame. The new motion work, setting dial and repairs to the escapement are probably the work of Thomas Strange in 1843. An early diamond-shaped one-hand dial has been replaced by a circular two-hand dial.

John Wise, a member of the London Clockmakers' Company from 1683 to 1723, was related to the Wise families of Bodicote. Another relative, Richard Wise, put up a dial on the mutilated pedestal of the Weeping Cross in Bodicote, now disappeared, which was inscribed *Given by Mr. Richard Wise Clockmaker in London, Anno Domini 1730.*

The Churchwardens' Accounts are available only for the period 1768 to 1802. Repairs to the clock were done by Thomas Pinfold (q.v.) from 1768 to 1789 and by William Peacock (q.v.) thereafter. The clock-winder was paid 12 to 16 shillings a year. The ring of 5 bells was cast in 1843 and is mounted in a bell-frame constructed at that date. There is no record of earlier bells except the saunce of 1624 which was recast in 1900.

BOURTON, GREAT — All Saints

A clock by J. Smith and Sons, St. John's Square, Clerkenwell (1835-1842) in a detached campanile containing one bell hung for ringing on which Messrs. Smith also inscribed their name (Sharpe, 1949).

The County Directory for 1852 mentions " an ancient clock " attached to the Chapel School, which has since disappeared, and the building has become two dwelling houses.

BUCKNELL — St. Peter

The clock has a pendulum beating 2 seconds and had only one hand until 1894 when a minute hand was added (Blomfield, 1894).

BURFORD — St. John the Baptist

In 1635 4 bells of a previous ring of 6 were recast for the church by Henry Neale who established a bell-foundry in Burford. Presumably a clock was acquired some time in the following 30 years but possibly as late as 1668. The Churchwardens' Accounts record in 1669 that £2 11s. 11d. was paid to Richard Vincent *for settinge up a house over the Clocke*. Next year, 1670, is the item, *Pd Mr. Yonge of Oxford for the Chime and materialls belonging to it, £16*. William Young, the smith of Catte Street, made chimes for other churches in Oxfordshire and Berkshire. A sundial was erected over the church porch in the same year, 1670.

The subsequent history of the clock is scanty. According to the Churchwardens' Presentments to the Archdeacon in 1753, 9 October, *The Church clock out of order but the same was to have been regulated on the 8th inst. and was not on account of the Parson (with whom the agreement was made) not coming according to his appointment*. In 1756 nothing was presentable so presumably repairs were done, but in 1769 the clock and chimes and bells were again out of order. In 1770 and 1771 the clock and chimes were still in need of repair. By 1789 the church was in a very bad state, *the Church rents being in the Minister's hands who refuses to account to the Churchwardens for the reparation of the same*. This presentment led to the issue of orders at a Court held at St. Mary's, Oxford; three years later it was reported all was well.

BURFORD — The Tolsey

The Tolsey building dates from the 15th century. The earliest records of its clock occur in the folio of the Bailiff's Accounts for 1651 to 1658; repairs were done in 1653, 1659, 1664, 1666 and 1668, from which one may conclude that it was important as a town clock. Some later notes on expenditure apparently kept by the Clerk to the Corporation of Burford (and dated by Dr. B. E. A. Batt as 1709 to ca. 1730), show that the clock-keeper, Hastings, was paid 12s. rising to 16s. a year with extra for ropes and repairs. There is nothing to show if or when the early 17th century clock acquired a long pendulum. About the middle of the 19th century a new, hour-striking clock with a semi-horizontal bed was installed to serve 2 dials on a drum-shaped case mounted externally on a wooden bracket. This was replaced in 1960 by Smith and Sons, Derby.

CHALGROVE — St. Mary

The clock was under repair in 1801 (Ox. Archdeac. Papers).

CHARLBURY — St. Mary

The clock and chimes were under repair in 1800 (*ibid*).

CHARLTON ON OTMOOR — St. Mary

An old clock in going order, wound daily. The bar frame, 36 × 24 × 19 inches, has the corner-standards, one pair of cross-members and two pairs of pivot-bars all tenon-wedged. The finials are inclined outwards and end in a round bun. Of the 2 side-by-side trains the going great wheel is of iron with 4 straight welded arms, and 2 lifting pins. The second and escape wheels are of brass with 4 narrow flask-shaped arms characteristic of Edward

Hemins (q.v.). Anchor without a neck. The pendulum cock is fastened to the top frame bar with wedges. The crutch is cranked laterally to avoid the tenon in the pivot-bar which it straddles. A circular loop in the pendulum rod enables it to clear the winding square of the going barrel. In the striking train the wheels are of iron 4-armed, and the first pinion is a lantern type; the iron locking-plate is outside the frame. Striking control is by peg and lever and a 2-vaned fly. There is no top regulation for the pendulum and no setting dial.

It seems that the going train has been renewed by Edward Hemins in order to bring the pendulum from an inaccessible position at the back of the frame to the front. The second and escape wheels, the shape of the pendulum cock and the provision of a loop in the rod are very similar to the arrangement adopted in the Steeple Aston clock (q.v.).

The frame and striking train probably belong to the late 17th century. The 4th and tenor bells were cast in 1681.

CHIPPING NORTON — The Market House

An entry in the Minute Book of the Old Corporation dated 27 July 1793 ordered John Cheney, Borough Architect, to survey the frame on the top of the Market House to see if it was suitable to receive the Church Clock, and if it were not to estimate the cost of repairing the frame to receive the Church Clock and making 4 dial plates one on each side. Nothing further was recorded about this project.

The tower of the church of St. Mary, Chipping Norton, was rebuilt in 1823 at which time it had no clock dials.

The present Town Hall was built in 1842 and provided with a clock ca. 1849 by Samuel Simms. This clock disappeared after a fire in 1950 and was replaced by an electric clock, automatically wound, by John Smith and Sons, Derby.

CHURCH HANBOROUGH — See Hanborough, Church

CHURCHILL — All Saints

A ring of 6 bells was made for the church by Robert Taylor and Sons, Oxford in 1826, when the church was built. The 8-day clock was made by William Taylor (q.v.) of the same firm. The pendulum beats approximately 3 seconds and hangs from the clock room through two floors. The following extract from the notebook of William Taylor is supplied by Messrs. John Taylor & Co. of Loughborough:

"Nos. Diameters &c of the Wheels and Pinions Churchill Clock.

Watch

	No.	dia.		No.	dia.	dis. of ccn.
Gt. Whl.	90	10·9	1st Pinn.	18	2·33	6·425
2nd	80	8·85	2nd	16	1·909	5·2057
3rd	70	7·5	3rd	14	1·634	4·409
Swing Wl.	30	6·57	Scapes 9 teeth			4·375
Center Pinion	30			30	3·759	7·139
Snail Whl.	72	7·5	Pinion	12	1·386	4·2795

5 × 5 = 25 × 5 = 125 = No. of turns Swing Wheel makes to one of Gt. Wheel. Gt. Wheel goes round once in 6 hours, therefore 125 ÷ 6 = 20·833 = No. of turns Swing Wheel makes in one hour also the No. of Vibs. the Pendulum makes in one minute.

Length of Pendulum 27ft. 1·23in. Swings 3ft. 7in. Wt. of Ball 1 cwt. Weight of Watch part — double line Roll 6·75 in dia.

Striking Part

	No.	dia.		No.	dia.	dis. of cen.
1st Whl	96	13	1st Pinn.	20	2·876	7·725
2nd	96	10·3	2nd	12	1·435	5·7
3rd	80	8·5	3rd	10	1·208	4·68
Roll Wl	4.5	9·33	Pinion	14	3·1256	5·9

Pin Wheel makes 156 turns in 8 days. Then as 20: 156::96:32·5 the number of turns the Gt. Wheel makes in 8 days Roll 7·2 dia. Weight."

CLAYDON — St. James the Great

A small early clock of uncommon design. The wrought iron frame 24 × 22 inches, is a lantern clock type consisting of 3 vertical pivot-bars joined by 2 horizontal members. The vertical bars extend through an oak bolster on 4 legs and are bolted below by 3 large wing-nuts. Similar smaller wing-nuts fasten the top cross-member. One finial remains, its stalk bent outwards and ending in a sub-quadrate pear-shaped boss. The two trains are end-to-end both pivoted in the middle bar. Originally with a crown wheel, and foliot the going train retains its barrel and windlass of 3 iron spikes but now has later main, 2nd and escape wheels with an unusually flat anchor bar — the latter resembling the anchor used in 18th century 4-post wall clocks. The pendulum, about 42 inches long, is suspended from an inverted U-shaped bracket with wing-nut regulation, bolted in place of the original finial. A loop was made in the external pivot bar to pass the crutch arbor through it In the striking train the great wheel has 8 hammer studs and an external locking-plate; warning is by two pegs on the fly arbor and a lever, with the lifting piece pivoted in a pair of curved arms projecting from the vertical bar. A device incorporating an outrigger for the bell-hammer wire is mounted in its original place on the oak bolster. The fly now has a free wheel and two fixed vanes on double cranked arms.

There is no outside dial and 18th century references to painting the dial apply to a sundial.

The Churchwardens' Accounts are available for the period 1746-1859. Repairs were done by John Lamprey of Banbury (q.v.) in 1746. An entry for 4 August 1751 *Pd to Mr. Pinfold for altering ye clocke £1 10s. 0d.* cannot be interpreted with certainty as an escapement conversion. Thomas Pinfold (q.v.) did further repairs in 1756 and 1758 and thereafter maintenance was undertaken by the local smith and the winders. In 1801 the clock was sent to Banbury to William Peacock (q.v.) for repairs costing £1 11s. 6d. and he was again employed in 1809; a new clock cupboard was made in 1802. James Durran (q.v.) did repairs in 1834 and in 1852-4. Voluntary subscriptions in lieu of a church rate for a clock fund raised £9 12s. 0d. out of which £2 19s. 0d. was forthwith spent on repairs, a cord and weights. The clock-keeper was paid annually 10s. 6d. from 1746 to 1764, 15s. from 1765 to 1800, 17s. 6d. to 1812 and £1 from 1813 to 1835.

The oldest of the three bells in the church is dated 1609 and it may be that the clock was made about this time, unless it was transferred from Cropredy church at the end of the 17th century.

COMBE — St. Laurence

The tower of Combe church was damaged by fire in 1918 and the clock was discarded. It was rescued by H. Minns of Cassington and restored to what he considered to be its original design. It is now in the History of Science Museum.

The frame, 32 × 21 × 26 inches, has corner standards with buttress mouldings above and below, and ending in outwardly curved finials tipped with hemispherical bosses ringed basally. All parts are joined by wedged tenons. The two trains are end-to-end pivoted in a middle bar also wedged. In the going train the great wheel has a windlass for winding, the spokes are bifurcated and dovetailed into the rim. A circular spring detent operates on the spokes. The crown wheel, verge and foliot are modern restorations. In the striking train the great wheel spokes are not bifurcated and the barrel carries an extra toothed wheel on the outside which engages with a lantern pinion to the arbor of which a winding handle can be applied. It also has a windlass and circular spring as in the going train. There are two lantern pinions of 7 and 10. The 2nd wheel has 4 paired spokes dovetailed to the rim and also a hoop with a wide gap, despite the peg and lever release. The locking-plate is internally toothed with offset spokes. There is an indicator dial of problematical date with a clutch plate and pointer, traces of punched hour marks and a circular plate with 12 pins.

The earlier ring of 5 bells, damaged by the fire and recast in 1924, were dated 1621 to 1629, which may indicate the age of the clock. The present clock is by John Smith and Sons, Derby, 1948.

CROPREDY — St. Mary

This church had a clock at the beginning of the 16th century. In an indenture dated 26 August in the 4th year of Henry VIII, 1512, the Vicar of Cropredy, Roger Lupton, provided £6 13s. 4d. in trust with the churchwardens of Cropredy and Bourton binding them and their successors that they *shall fynde or cause to be founde at their p'per coste and charge immediately after date hereof and so contynew for e'r more oone p'soone to kepe dewly the clock of Cropredy aforesaid goyng hourely and to ring dayly booth wynter and somer at foure of the clok in the mornying the grettest or the myddle bell by the space of a quarter of an houre and to toll dayly the avees bell at sex of the clok in the mornying at xij of the clok at noone and at foure of the clok at afternoone. And to toll also in Wynter at viij of the clok in the nyght thre tolls and immediately after tolling to ring Curfew by the space of a quarter of an houre and in somer to toll and ring Curfew in lyck manner betwene viij of the clok and vi at the nyght.*

For failure to observe these conditions they were liable to a fine of 6s. 8d. a month, and, if defaulting for a whole year, they would be fined £10 and would have to buy land within seven years of value to provide sufficient income for the upkeep of the clock and ringers. By wise administration this custom has been maintained by the trustees in essentials — except for the prohibition during war years — until today.

It is regrettable that documentary history of the 16th to 19th centuries is not available for Cropredy church. It is, however, probable that a new clock was obtained at the end of the 17th or early in the 18th century and this was in use until the middle of the 19th century, when it was transferred to Horley (q.v. for description). It is possible that Lupton's clock was transferred to Claydon which was a chapelry of Cropredy.

The present clock is by John Moore and Sons, Clerkenwell, 1831; it has 3 trains, striking and ting tang chiming the quarters, with a semi-dead beat escapement and a 2-seconds pendulum swinging partly below the bell-loft floor. The dial is on the west wall of the tower.

CUDDESDON — All Saints

A new clock was made for this church in 1776. A print of 1823 shows that it had a diamond-shaped dial on the west wall of the tower. The present clock is by John Smith and Sons, Derby.

DEDDINGTON — St. Peter and St. Paul

The present clock was made by William Taylor of W. and J. Taylor, Oxford, and installed in 1833. It is in a massive cast iron frame with cylindrical pillars and ball finials; two trains, rack striking and maintaining power. The rack is vertically disposed, toothed on both sides with the gathering pallet below; striking control is by means of a peg and lever; the hammer pegs are semi-circular in section; the fan is 3-bladed. The escapement is a recoil anchor with a wooden rod pendulum about 18ft. overall. On the indicator dial is inscribed *Thomas Fardon, Clockmaker, Deddington.*

Messrs. John Taylor and Co., Loughborough, supplied the following extract from the notebook of William Taylor:

"Numbers & diameter of the Wheels & pinions of Deddington Clock.

Watch Part

	No.	dia.		No.	dia.	dis. of cen.	thickness of tooth
Gt. Wheel	96	12·96	1st Pinn.	30	4·157	8·3	2·07
2nd	88	10·842	2nd	12	1·628	6	1·88
3rd	80	9·21	3rd	11	1·41	5·09	1·757

Swing Wheel 30 Escapement the same as Wst. Bromwich & Sedgley makes 1600 Vibs. in an hour or 26·666 in a minute. Pendulum 16 feet ·555 inches long. Watch Barrel 6¼ inches dia. Striking Barl. 6½ dia. 58 feet 4 inch fall for Wts Weight & Pulley 2·4=56 feet. Watch Weight 1·00 Striking 20·0 single line Pendulum Ball, Hammer 38.

	No.	dia.		No.	dia.	dis. of cen.
Snail Whl.	96	10·773	Pinion	16	2·03	6·2
Wheel on 1st Pinion Arbor	72	8·588	Index Wl.	36	4·35	6·2

Striking Part

	No.	dia.		No.	dia.	dis. of cen.	thickness of tooth
Gt. Wheel	96	15·36	Pinion	20	3·37	9·062	2·454
2nd	88	11·9	2nd	11	1·657	6·525	2·07
3rd	80	9·57	3rd	10	1·35	5·235	— "

A tablet in the church records that the clock was the bequest of Mr. Hudson who died in October 1832 "so that the old church clock should give place to a better." There is no history of the earlier clock as the Accounts are missing, but Deddington church tower collapsed in 1634 and Charles I requisitioned the bells in 1643; it is not until the 18th century that a ring of four bells is recorded.

DORCHESTER — The Abbey Church of St. Peter and St. Paul

The Oxford Archdeaconry Papers record an order to repair the clock chimes in 1625. The subsequent history of this clock has not been traced. Moore and Son, Clerkenwell, supplied a clock with Westminster chimes in 1868.

DUCKLINGTON — St. Bartholomew

A mid-18th century clock is recorded.

EWELME — St. Mary

A clock was made for the church by Joseph Stockford of Thame some time before 1770. The present clock is by John Smith and Sons, Derby.

FINMERE — St. Michael

The clock was installed in the tower at Michaelmas 1697 its cost, £8 10s. 0d., being defrayed by the subscriptions of 22 persons. According to the Rector's Book the balance of 2s. 6d. was given to " Mr. Ford's man." This may mean that the clock was made by John Ford, freeman of Oxford in 1691. An inventory of goods belonging to Finmere church in 1733 includes 3 bells and 1 clock. Thomas Chanels, parish clerk from 1710 to 1761 wound the clock for 5s. a year. It was altered in 1858. Dr. James Clarke, M.D., of Finmere House bore the cost and invented the escapement and had the alterations and additions done under his own care (Blomfield, 1887). It was re-erected in 1859 when £10 was paid to William Bayliss for work on the clock.

At a sale of the contents of Water Stratford House in 1953 a collection of clocks, organs and other musical instruments was sold. Many of the long case clocks were made in the mid-19th century by Dr. Clarke and William Bayliss of Finmere and were fitted with small turret clock movements and a variety of escapements including the gravity types, and with chimes.

GARSINGTON — St. Mary

The clock, which strikes on the tenor bell cast in 1788, bears an inscription " Made by John Thwaites Clerkenwell 1796, For the Parish of Garsington under the direction of Richd. Pearson, Oxford, by Donation from the late Mr. Willm. Harper of this Parish." It cost £172 4s. 0d. There is a ring of 6 bells the 4th and 5th of which are since recast. The invoice of John Thwaites to the Parish of Garsington specifies it as:

" A New Thirty Hour Church Clock to Shew Two Outside Dials Hours only & to strike the Hours on the Tenor Bell the Striking great wheel 14 inches the Watch great Wheel 12 inches & with a set of Three Hour Work to Discharge a Set of Chimes & work to hold the Flirt untill the Clock had done Striking . . . Two Copper Dial Plates of 4ft. 6ins. Diameter each with Hammers Moulding round the Edge & Painted Black with Gilt Figures & Moulding & with Proper Bolts & Brasses to fasten them up . . a Compleat New Set of Chimes to Play one Tune upon 6 Bel!s with two Hammers to each Bell with a Strong Oak Barrill & with two Iron great Wheels & with Proper Locking & Discharging Work the Pins in the Chime Barrill all Harden'd & all in Strong Iron Frame."

The cost of fixing was paid by Pearson. The dials still have only an hour hand.

GREAT HASELEY — See Haseley, Great

GREAT MILTON — See Milton, Great

HANBOROUGH, CHURCH — SS. Peter and Paul

This church has one of the very few clocks with trains end-to-end and a crown wheel and verge escapement in original condition.

The plain flat bars of the main frame are hammer welded together without finials, but the middle box-frame and the 4 vertical pivot-bars are tenon wedged. In the going train the great wheel turns a lantern pinion and the contrate wheel, and has on the inner end of its arbor a ratchet pinion checked by a spring click. The crown wheel is mounted in a footstep bearing offset from one middle bar and in a curved bracket fixed on the top cross-bar. This bracket also pivots one end of the verge, the other being held in a swan neck riveted outside the frame. In the striking train the great wheel has 8 hammer pins and turns a solid pinion on the 2nd wheel arbor, and that wheel engages a solid pinion on the fly arbor. The locking-plate is placed within the pair of middle pivot bars. An unusual feature is that the rods of the lifting and stop levers have been wrought with a barley-sugar twist. Both barrels have the remains of a 4-armed windlass in addition to a square for a crank handle.

The photograph Plate 4, fig. 6, shows the stone weights, wooden pulleys, the short pendulum and hammer-lifting levers.

No documentation is available to date the clock but the oldest bells are dated 1602 and 1603.

HANWELL — St. Peter, and Hanwell Castle

Sir Anthony Cope of Hanwell Castle appears to have been interested in clocks. He had in the castle grounds a water-clock which was described by Robert Plot in 1676 (*Nat. Hist. Oxf.,* p. 235, par. 49), " Nor can I pass by unmentioned, a *Clock* that I met with at *Hanwell,* at the House of the Right Worshipful Sir *Anthony Cope,* that moves by *water* and shews the *hours,* by the rise of a new guilded *Sun* for every hour, moving in a small *Hemisphere* of wood, each carrying in their *Centers* the number of some *hour* depicted black; as suppose of *one* a clock, which ascending halfway to the *Zenith* of the *arch,* shews it a quarter past one, at the *Horizon* three quarters past *one;* and at last absconding under it, there presently arises another gilded *Sun* above the *Horizon* at the other side of the *Arch,* carrying in its *center* the figure *two;* and so the rest. Which ingenious device, though taken out of *Bettinus,* who calls it, *aquarii Automatis Ingeniossimi horariam operationem*: yet being since improved by that *ingenious Person,* and applyed to other uses, particularly of a *Pseudo-perpetual motion* made by the descent of several guilt *bullets* upon an *indented* declivity, successively delivered by a wheel much of the same fabrick with the *Tympanum* of the *Water Clock,* so that they seem still the same."

Plot is quoting Mario Bettino, *Aerarii Philosophiae Mathematicae,* 1648. Tom. 2, Exodus 5 (see Baillie, 1951, p. 47).

The Cope library contained a copy of G. da Capriglia's *Misura del Tempo,* 1665, the earliest treatise on clocks, which describes and illustrates turret clocks. In 1671 Sir Anthony commissioned George Harris of Fritwell (q.v.) to make a clock for Hanwell church. This exists in a substantially original state with a crown wheel and verge escapement (Plate 1, fig. 1).

The corner-standards of the frame have buttress mouldings above and below and end in outwardly curved finials topped by a depressed pyramidal knob; they are fastened by square nuts to the side-members. The end pivot-bars and a pair of outer middle vertical bars are tenon-wedged but the two inner middle pivot-bars are welded to the upper and lower cross-members. In the going train the spokes of the great wheel fork at the middle into U shaped extensions but those of the striking great wheel are straight. Each at the opposite end of the barrel has a toothed wheel engaging with a lantern pinion pivoted in a separate vertical bearing-bar and a rectangular bracket attached to it; although the bracket is fixed by nuts it is part of the original design. All other pinions are solid. On one top cross-bar is a cock combining a swan neck pivoting one end of the verge and an angled branch taking the top pivot of the crown wheel. The external end of the verge, to which the bob pendulum is fixed, rests in a cock fixed to the back cross-bar and the vertical bar. The hour locking-plate is housed between the two trains. Outside the narrow side of the frame is a rectangular bracket tenon-wedged to the vertical pivot-bar of the going train holding a 4-armed wheel with lifting pegs and a setting dial. In the photograph Plate 1, fig. 1, the weight and link attached to the hammer-lever are for the purpose of putting the strike out of action when the bells are rung.

HASELEY, GREAT — St. Peter

The oldest bells in the church are dated 1641 and the first mention of a clock is in the Churchwardens' Accounts for 1666, earlier records being lost. Presumably a clock was installed between these dates. In this parish major repairs were beyond the skill of local artisans and the churchwardens called in a clockmaker or the clock was carted to him in Oxford, Chalgrove, Stadhampton or Thame.

Small repairs done in the years 1666 to 1673 cost a few shillings on each occasion. In 1675 one Parker was paid 10s. and 5s. for mending and 2s. was spent *for beer for ye clocke mender & helping him.* But his work was not of permanent value as next year are the following entries: *For carring the Clocke to Oxford . . . 2s. — Spent with the clockmaker . . . 6d. — again Spent with the Clocke Maker . . . 6d. — Spent uppon ye Clocke makar when he brote hom the Clocke . . . 1s. 4d. — Paid to Woten for mendin the Case and boards . . . 2s.* The clockmaker's charge for mending was £3 10s. 0d. plus 5s. for a clock line. There was also a dial which had to be replaced, according to an item for 1677-8: *pd for mending a Bell & fastning the Dyall . . . 3s. 6d.* At this date the only men in Oxford likely to have advised fitting an anchor escapement were John Knibb and William Young. But conversion may have been postponed until 1690.

Winding, setting and cleaning the clock were in the hands of the parish clerk. This post was held by members of the family Hinton for several generations who are mentioned by name, William, young William, Thomas and Samuel.

In 1690 the clock was again sent away for repairs, perhaps to Oxford as before: *Charges for the Clockmaker coming over . . . 4s. — For bringing home ye Clock in a Cart . . . 4s. — Charges when ye clocke came home & keeping ye Clockmaker two days and two nights &c . . . 8s. — for making ye dial frame . . . 12s. — for painting and guilding ye Dial . . . £1 16s. 0d. — for mending ye Clock . . . £5 1s. 0d. — Charges for the Clockmaker when he came for his money . . . 3s.* This and later references to the dial indicate an unusual pride in keeping the church dial in good condition; such records are uncommon in parish accounts. The two new ropes for the clock costing 5s. were charged up the next year, 1691. More repairs were needed in 1692, *For mending the Clock . . . 13s. and for his Charges (ye Clockmaker) . . . 7s. 2d.*

In 1703 the clock was once more sent away, this time to a village expert, *For Carridge of the Clock & bringing it back from Chalgrove . . . 2s. — The Clock Mans Bill . . . £1 15s. 0d.* Chalgrove is not far from Stadhampton where later the Jordans were resident clockmakers; whoever did the work it was well done and no serious breakdown occurred for many years. About 1734 Great Haseley acquired its own resident clockmaker, Thomas Holloway, and he kept his church clock in order at very little expense for about 25 years.

In the Accounts for Easter 1759 to Easter 1760 is the entry: *For a New Clock and Dial . . . £8 15s. 0d.* Holloway evidently served his village well with a very low priced clock; at the same time he supplied a weathercock for 10s. In 1764 he retired (or died) and was succeeded in business by Thomas Stockford (q.v.) who possibly later moved to Thame. At any rate when the clock failed in 1770 it was sent to Thomas Jordan at Stadhampton: *Jordan's Bill for mending the clock . . . £1 10s. 0d. — for carrying ye clock to Stadhampton . . . 5s.* During the next 20 years any repairs needed were done by Jordan (e.g., £1 4s. 0d. in 1779).

Another clockmaker appears on the scene in 1794, William Buckland of Thame (q.v.), who completely overhauled Holloway's clock for £7 12s. 0d.;

the Accounts do not disclose if this was done on the spot or if the works made another journey. At the end of the 18th century a watch and clockmaker, John Stevens, set up in Great Milton, a village only a mile or so from Haseley, and by 1800 he was engaged at 5s. a year to clean the clock and to do minor repairs for the next 20 years. However in 1814 more expert work was needed and was given to Buckland of Thame at a charge of £5 10s. 0d. In 1819 three guineas was paid to *John Cooper writing & guilding the dial to the Church Clock.*

In Thame and Bicester a Tomlinson family had businesses as watch and clockmakers and gunsmiths. One was employed to repair the Haseley clock in 1826 at a cost of £10 5s. 0d. and thereafter was frequently called in. Repair prices at this period now amount to pounds rather than shillings, thus Mr. Job Tomlinson's bills were in 1836 £1 7s. 11d., in 1843 £3 3s. 0d., in 1848 £1 10s. 0d. and in 1852 £6 4s. 6d. and a final 18s. 6d. in 1856.

R. R. Rowell of Oxford installed a new clock in 1865, the maker probably being Evans of Handsworth.

HEADINGTON — St. Andrew

An 18th century clock by Stone of Thame previously in this church and disused from the end of the 19th century was eventually renovated and installed in 1951 in St. Nicholas church, Old Marston (q.v.).

HENLEY ON THAMES — St. Mary

The administration of Henley Church in the 14th to 16th centuries was in the hands of the Borough. Its Assembly Books for the period 1395 to 1543 have been transcribed and printed as vol. XLI, 1960, of the Oxfordshire Record Society. From them something can be learned of the church clock and its maintenance. An entry for 3 November 1410 reveals the existence of a clock and gives some clues as to its age. *Vnde de predicta summa soluit predictus pro emendatione cuiusdam instrumenti vocati Clokke . . . vjs.* The phrase " some sort of instrument called a clock " suggests it was still a novelty. It must have been looked after by a local smith, John Atte Lee (q.v.), who was elected burgess in March 1412 and died in 1448; and thereafter by Thomas Hyde, another smith.

On 9 March 1470 five new bells were bought for £89 2s. 3d. and fees for the use of them at various ceremonies were then fixed but revised later in 1495. In January 1493 John Mitchel of Wokingham was paid £9 3s. 4d. for making the great bell. Some time between 1493 and 1499 chimes were added to the clock. When Thomas Harington (q.v.) was appointed clock-keeper in September 1494 the clock was described as *horecudium*, i.e. striking the hours. *Et eodem die electus est et admissus pro aquebaiulo Tho. Harington . . . et quod dictus Tho. custodiat horecudium et campanas et habebit pro labore suo xxs. per annum,* as well as fees for bell-ringing. He was not allowed anything extra for bell ropes or minor repairs to bells, and presumably also had to keep the clock in working order as there are no entries for repairs in the Accounts.

In December 1499 John Balam was elected an *aquebaiulus ad occupandum dictum officium simul cum Tho. Harington socio suo et accipiendum stipendium et emolumentum pertinentia dicto officio pro parte sua prout fuit ex antiquo vt patet in precedentis.* This appointment was further clarified in May 1501 to show that the two men shared the profits equally. *Tho. Clerke alias Harington electus est aquebaiulus vt fuit preantea tam pro sacramento et sacramentalibus quam pro campanis et horecudio capiendo pro vadiis suis vt preantea cepit . . . et quod permittet*

Joh. Balam ad occupandum cum eo capiendo dimidium valoris et proficuorum prouenientum ex officio predicto preter de le clocke et chyme. The final reversion to clerical dog-French makes it clear that the clock already had chimes.

This arrangement continued for about ten years until December 1509 when Thomas Lewes was appointed second aquebaiulus in place of Balam. In August 1510 Harington was granted additional emoluments from the sale of 10lbs. of wax used for the light of the Blessed Virgin Mary. He continued to draw the stipend of ten shillings *pro campanis et le clok et chyme* until April 1515 after which there is no reference to him by name; during all his period of office no charges for repairs are recorded.

The Borough Records contain no information about the clock or its keeper for the next 17 years although Thomas Lewes appears as a burgess holding various posts. Thereafter the cost of repairs and materials was allowed as an extra to the clock-keeper's wage. In September 1532 the chimes were mended at a cost of 8d. by John Archer, a smith. In January 1535 John Archer and Thomas Derrell were paid 13d. and 12d. respectively for mending the chimes. In September 1541 4d. was paid *pro le wyre pro le chime* and again the following month 14d. *pro wyrys pro le clocke and chyme.* The Records finish in September 1543 without further mention of repairs.

No records now exist for the next 180 years during which undoubtedly a new clock must have been required.

A Vestry Book of Henley on Thames exists for the period 1725 to 1815 but no detailed accounts are written in.

Another volume covers 1816 to 1865 and no reference is made to a clock in the consolidated tradesmen's bills. But from 1831 it is clear that John Grayson, clockmaker of Henley, had the care of the clock at an annual wage of £5; this was raised to 5 guineas about 10 years later. He continued to look after the clock until 1860. No charges for repairs are recorded except £2 1s. 0d. in 1859.

HEYFORD, LOWER — St. Mary

A church clock of iron (on an oak frame) said to be dated 1695 and worked by heavy stone weights, was taken out of the church at its restoration about 1867-8 (Blomfield). A stray record reveals that the parish clerk was clock-winder from 1763 to 1849 at 10s. a year. The church had 4 bells in the time of Edward VI but none of the existing ring of 6 is older than 1766.

HOOK NORTON — St. Peter

In 1731 the tower possessed 5 bells and a clock with a carillon which played Purcell's tune " Britons strike home " at the hours of two, six and ten. A tablet in the tower dated 1731 warns the ringers to make sure that the bells are at rest " when the chimes go." Twenty years later the clock was taken down for overhaul. In 1768 a new clock was installed which got a new dial in 1802; the charge was £4 4s. 0d. The present clock by John Smith and Sons, Derby, 1913, has a double 3-leg gravity escapement with 3 dials and Westminster chimes.

HORLEY — St. Ethelreda

The present clock was originally in the church of St. Mary, Cropredy, whence it was discarded in 1831 on arrival of a new one by John Moore and Sons. In 1849 the Rev. W. J. Pinwill, Curate of Horley, purchased it with the help of a parishioner Mr. Hitchcock. In 1850 two watchmakers in Banbury estimated repairs of £16 to £20 but Charles Webb of Hornton "a self-instructed watchmaker" agreed to do the work for £5 10s. 0d. plus £1 10s. 0d. for new work on a hand-setting extension and some other work, total £8. Drawing, gilding and painting of the dial was done by Mr. Pinwill. and for the whole project he seems to have received little support from the Vestry.

The clock of two side-by-side trains is in an iron frame 32 × 17 × 25 ins. joined by square nuts. The corner-standards end in rearward inclined finials with concavo-convex brass bosses. The going train great wheel and second wheel have morticed and welded iron arms, roughly octagonal arbors and solid pinions. The brass escape wheel is offset in an L-shaped bracket added to the vertical pivot bar; a slender anchor without a neck. Approximately one-second pendulum with a cylindrical lead bob. The striking train great wheel and second wheel are of similar construction to those in the going train. An internally toothed locking-plate with pinion is outside the frame. Striking release by lever and check pin on the going arbor.

The iron wheels are similar to those of the clock in Chalcombe church, Northants, and the frame has similar brass bosses. The L-shaped mounting of the escape wheel and anchor is similar to that in the clocks at Mollington and Hornton (q.v.) and is probably the work of the same repairer. The use of a Quaker ring and zig-zag dial for the hand-setting pointer is further evidence of a local craftsman's work. Mr. Pinwill's two-hand diamond-shaped dial is still in use.

HORNTON — St. John the Baptist

The clock of two side-by-side trains is in an iron frame 32 × 14 × 22 ins. with square section corner-posts having shaped bases. There are three pairs of snub-ended scroll finials and all bars are fastened with square nuts. The going train has a pin-wheel escapement, an anchor with a long neck and these are pivoted in an L-shaped addition to a vertical bar as in Horley and Mollington church clocks. The pendulum is about 41 ins. long with a large lenticular bob. See text-figure, page 25.

In the locking-plate striking train the 2nd wheel is a hoop wheel and the strike release is by lever and check pin on a wheel. Except for iron barrel plates all the wheels are of brass, and the pinions solid with 7 or 8 leaves.

There is no record of expenditure on this clock which may have the original trains, although of brass, with an alteration of the escapement at a later date. It is possibly associated with the ring of 5 bells made in 1741. The circular dial with fine pierced gilt hands is probably contemporary with the installation of the bells.

HORSPATH — St. Giles

The iron frame of the clock appears to be of late 17th century make, its cross-members are welded and the finials are outwardly curved, ending in a flat bun. The vertical pivot-bars are tenon-wedged. Both side-by-side trains have been modified. The great wheels and locking-plate are of iron and two lantern pinions remain but both second wheels and the escape wheel are later substitutes of brass. A seconds pendulum with a heavy lenticular bob is regulated only by a wing nut under the bob. Striking control is by levers and peg stops on the fly wheel arbor and second wheel. The original heavy stone weights are used. There is no dial.

There are no early churchwardens' accounts extant.

IDBURY — St. Nicholas

J. Kibble in 1928 observed the remains of an early clock in the tower which he considered to be worth preserving.

ISLIP — St. Nicholas

A possibly late 17th century clock in a frame of rather light iron bars, the corner-bars with a stepped corbel at the top to support a finial bent outwards and upwards and topped by a cubical boss. Two pairs of pivot-bars and the middle cross-bars are tenon-wedged. The two trains are side-by-side with all the wheels at the back of the frame, and of brass with 4 straight arms. The anchor is a flat arc without a neck and its arbor slides laterally for disengagement from the escape wheel. The pendulum beats 63 to the minute and its rod is expanded in a large circular loop to clear the leading-off rod to the motion work in the storey above. In the striking train only the 2nd wheel has 6 arms; control is by peg and lever; the locking-plate is outside the frame. The leading-off gear goes vertically upwards through a floor and out to a 2-hand circular dial on the east wall of the tower.

It is believed that the clock was made by Edward Hemins, senior, of Bicester, about whom there is very little information. It does not resemble turret clocks made by Edward Hemins, junior.

KIDDINGTON HALL

Over the stable yard gateway is a half-hour striking clock by B. J. Vulliamy.

KIDLINGTON — St. Mary

The Churchwardens' Accounts, which have been preserved from 1754, reveal that a clock existed in the first half of the 18th century. John Cosier, blacksmith, undertook repairs and cleaning. Clock lines were supplied in 1760, 1764 and 1779, and for the next 20 years there is no reference to the clock and no payments specifically for winding. In January 1806 Thomas Fardon of Deddington (q.v.) was paid £7 7s. 0d. for a clock and 11s. for putting it up. This was probably not a turret clock as local tradition refers to a clock on the east wall of the nave which was eventually removed as it distracted the attention of the congregation. Thomas Fardon had an annual fee to keep it in order, at first of 7s., but from 1815 onwards of 10s. 6d. The Fardons had charge of the clock for 45 years; i.e. up to April 1851. There was, however, an unusual expense in 1840 of £7 4s. 0d. of which there are no details. Nothing is recorded from 1851 to 1873, when James Powell cleaned it and continued to clean and oil it from time to time until 1881.

The clock, at present in going order in the tower, has no outside dial. Its iron bar frame measures 2ft. 11 inches × 2ft. and appears to be of early 19th century construction but again local tradition states it was installed in 1897. The going train has a dead-beat pinwheel escapement of 50 pins and a pendulum beating 32 to the minute. The striking train has a rack of unusual design and strikes on the tenor bell. Remains of leading-off work, unused as there is no dial, confirm that it is a second-hand movement.

KINGHAM — St. Andrew

The churchwardens' accounts record the repair and winding of a clock between 1765 and 1790. The present clock is by E. Dent and Co., Cockspur Street, London, 1875.

LANGFORD — St. Matthew

A much altered clock but still retaining a crown wheel and verge escapement.

The frame is of flat iron bars fastened at the corners by square nuts and braced with 2 cross-members in the middle; there are no finials. The 4 pivot-bars for the 2 end-to-end trains are tenon-wedged as are also 2 vertical bars on the front and back of the frame. All wheels, with the exception of the brass crown wheel, are of iron and all pinions are solid. The going train originally comprised the great wheel, contrate wheel, crown wheel and horizontal verge of normal construction. Later the barrel was removed from the great wheel arbor and a pinion was added to the arbor at the end opposite to the wheel. A new main wheel and barrel, pivoted in two bars bolted across two pairs of the vertical members of the frame, was added, the main wheel engaging with the new pinion. This alteration provides a longer period between windings but increases the labour of raising the heavier stone weight. The pallet arbor is fitted with a crutch and the pendulum is suspended by a spring; it beats approximately one second and has a small conical lead bob. This unusual design is apparently original.

In the striking train a similar reconstruction of the great wheel and barrel has been done. The locking-plate is not internally toothed for a pinion drive but has a complete wheel attached to it as in 30-hour lantern or 4-posted clock movements. There is also a hoop wheel in addition to peg and lever control of the striking. The fly is 2-vaned with a crude ratchet.

The clock is disused and has no dial; it was probably made in the later 17th century; there is no record of the church bells existing at that time as the present ring of 5 bells was cast in 1741. It is described and illustrated by T. R. Robinson, 1957.

LEWKNOR — St. Margaret

The church has a clock with a 3-seconds pendulum, about 30 feet long, reaching from the top of the tower to the ground floor.

LOWER HEYFORD — See Heyford, Lower

MAPLEDURHAM — St. Margaret

The clock was made by B. L. Vulliamy with a pendulum beating approximately 2 seconds. It bears the royal arms and initials of William IV who gave it to the church in 1832. His son, Lord Augustus Fitzclarence was Vicar from 1829 to 1854.

MARSTON, OLD — St. Nicholas

This clock was previously in the church of St. Andrew, Headington and after being disused for 40 or 50 years was sold to a scrap dealer and by him to an Oxford clock and watchmaker, who, in turn, sold it to the Vicar of Old Marston and he presented it to his church in 1951 after considerable alterations and renovations had been done.

The frame, 42 × 28 × 14 inches, is of flat iron bars with slight buttress moulding at the top and bottom; the two remaining finials are square with lateral grooves and topped by a flattened cone. The pivot-bars of the 2 side-by-side trains are tenon-wedged. The going train has an anchor escapement and an adjustable suspension to the pendulum which beats about 42 to the minute. In the striking train the locking-plate is toothed on the inner rim and mounted outside the frame. The hammer-pins on the great wheel are strengthened by an external rim. Transmission and motion work and dials are modern.

The setting dial originally bore the name, " Richard Stone Thame " (q.v.) which has been erased with substitution of " C. A. L. Lewis of Oxford." It strikes on the 2nd bell of a ring of 5 cast in 1823 by W. and J. Taylor of Oxford.

MERTON — St. Swithin

The clock has a lightly built bar frame, 27 × 21 × 12 inches, tenon-wedged, with small finials bent at right-angles and ending in a sub-cubical knob. The 2 side-by-side trains are pivoted in 2 pairs of wedged bars and there is a pair of middle cross-bars. The great wheels are of iron and the remaining original features suggest a late 17th century date. A reference to the clock in the Victoria County History gives it a William and Mary ascription. At some later date a middle pair of pivot-bars, fixed with nuts, was added for a setting dial and hour wheel and still later a segment of the dial was cut away to fit maintaining power with a shutter over the winding square. The present anchor escape wheel is 5-armed of brass, its anchor with a long neck, is pivoted in lengthened or replaced bars. The pendulum is suspended with a wing-nut for rise-and-fall and its long rod extends through the floor; the circular iron bob has 19th century features. Control of striking is by an external locking-plate, a lifting-piece raised by the hour wheel and peg and lever check work.

Churchwardens' accounts exist for the period 1779-1812. In 1779 William Ball, clockmaker of Bicester (q.v.), was in charge of the clock and was paid 5s. a year to keep it in order until about 1803 when William Jennings (? of Fritwell) took over the annual maintenance. He was succeeded by Richard Smith in 1810 also at 5s. a year. During this period of over 30 years no repairs were charged except in 1812 by Smith, £3 2s. 6d. The V.C.H. mentions a repair costing £5 15s. 0d. in 1817.

In 1867 a minute hand was added to the clock and a new dial was supplied.

MIDDLETON STONEY — All Saints

The Oxford Archdeaconry Papers for 1757 record orders to provide a clock.

MILTON, GREAT — St. Mary

In the plain iron bar frame of this William III clock the lower horizontal bars are placed so as to leave unusually long legs. The finials are laterally inclined swan-necks ending in spherical " door-knobs." Of the 2 side-by-side trains the going barrel has been replaced by one of much larger diameter and now takes an increased length of wire rope for a 4-sheaf pulley-system. The brass 2nd wheels in both trains are lightly built with 8 straight arms expanded near the rim. A plate is bolted to the escape wheel to reinforce it. Two vertical bars strengthen the frame at the back to support the pendulum cock. The pendulum is an iron rod carrying a heavy lenticular bob beating about one second. Three pairs of pivot-bars are attached by nuts, and on the front of the frame is an extra bar provided with 2 slots permitting it to slide up and down; a lateral projection on this bar rests on the squared arbor of the great wheel and the device presumably works as a maintaining power. In the striking train the fly arbor is offset and the locking-plate is outside the frame. Drive to the dial is taken from a contrate wheel vertically upwards to the storey above.

At the top of the pivot-bar, taking the indicator dial and clutched middle wheel, is a small plaque inscribed, " *Nicholas Harris de Fritwell Com't Oxon Fecit* 1699." As can be seen in Plate 6, fig. 9, the clock is mounted on stout wooden corbels projecting from the wall and is not enclosed in the usual cupboard. It is the oldest Oxfordshire turret clock with a date and the maker's name.

The churchwardens' accounts have disappeared and its 18th century history is unknown.

MIXBURY — All Saints

The frame of iron bars hammer-welded together, has 2 side-by-side trains in 2 pairs of pivot-bars fixed by tenon and wedge; cross-bars are also tenon-wedged; the finials are curved outwards and end in sub-quadrate knobs. All wheels are 4-armed. The recoil escape wheel and its almost semi-circular anchor work with a pendulum beating $1\frac{1}{4}$ seconds. The great wheel of the striking train has 8 pins for the hammer-lever; the locking-plate is internally toothed and external. For the ratchet of the fly a strong spring and a large pawl are sited on the arms of the fly the blades of which are an integral forging and not adjustable. There is no indicator dial or provision for time-setting. Striking let-off and stop are by peg and lever.

The youngest of the 3 bells is dated 1627 and they hang in a 17th century frame. The clock appears to have been made towards the end of the 17th century.

MOLLINGTON — All Saints

The iron bar frame, 35 × 15 × 22 inches, is joined by square nuts and has no true finials, the corner-posts being extended in upright spikes. In the 2 side-by-side trains all the wheels are brass and the pinions solid of 8 and 10 leaves. The going train has the anchor and escape wheel pivoted in an L-shaped extension of the vertical bar as at Hornton and Horley (q.v.). The pendulum is about 50 inches long with a lenticular bob. Striking gear includes an outside locking-plate, internally toothed, fly with a click and control by peg and lever. A time indicator dial is placed inaccessibly at the back of the frame and the accessories are partly dismantled.

Although this clock and those in Hornton and Horley churches are of different make and origin they have all had the escapements altered by the same repairer at some period in the late 18th century.

NETHER WORTON — See Worton, Nether

NETTLEBED — St. Bartholomew

The Oxford Archdeaconry Papers for 1723 note that the legacy of a clock fee was not executed. In 1845-6 the church was rebuilt.

NEWINGTON — St. Giles

The Oxford Archdeaconry Papers for 1719 record that the 3rd bell was cracked and the clock was out of order.

NEWINGTON SOUTH — St. Peter ad Vincula

Most of the annual Accounts of the Churchwardens of South Newington for 1560 to 1684 have been preserved.

A clock existed in 1560 and the first item about it is, *It for mendyng the clok twysse . . . iiijd*. In 1562 there is a payment for mending and in

1578 for wire. The year 1579 has the entry, *Imprimis for thamendinge of the Swaipe of ye clocke . . . vd.* The sweep is evidently the term used for the foliot. In 1582 and 1583 substantial repairs were needed — *Itm to the Smyth for Amendinge of the clocke . . . vis. viijd. — Itm to John Edes for repairinge the clocke house . . . xxd. — Itm in Nailes for the same . . . id. ob. — Itm for henges and hoile for the Clocke house Dore . . . vid. — Itm to Richard buggie for reparinge the Clocke . . . xiid. — Itm for reparacons About the Clocke . . . xiid.* In 1584 the smith provided " A Catche " for the clock (3d.) and in 1587 wire costing a penny.

In 1588 there were again expensive repairs, *It to Rich Buggin for mending the clock . . . viis. iid. — It' for a lock for the clock Doore . . . vid. — It' for a staple for the clock doore . . . id.* The smith, Richard Buggin, was thereafter engaged at 20 pence a year to look after the clock and this arrangement continued without extras for repairs until 1606. A pulley, 3d., and a ladder, 12d., were supplied in 1592 and the carpenter occasionally repaired the clock-house door. The accounts for 1607 to 1618 are missing but by 1619 Buggin's wage had been increased to 24d. a year.

Another gap occurs up to 1658 when John Calcott was in charge of the clock and a small repair cost 1s. 6d. Except for clock ropes no expenditure was recorded until 1669 when the following information is given, *Item spent at Garnors upon the nebers with the Clock man . . . 1s. — Item payd to the Clock man . . . £4 10s. 0d. — Item spent at the seting up of the Clocke . . . 1s. 8d.* The clock man was almost certainly George Harris of Fritwell (q.v.). The date 1669 is not too early for a conversion from foliot to Huygens verge pendulum by a man of the ability of George Harris. In 1672 the local man, John Colcutt, charged 1s. 6d. *for scowringe the Clocke,* but in 1674 George Harris was apparently again consulted and sent his young son, Nicholas, to South Newington. *It' to nicklas harris for coming ouer about the clocke . . . 8d. — Item for mending the clock and scouring the clock . . . 4s. 10d.*

The remaining entries in the Accounts are sparse. John Colcutt did repairs costing only a few pence between 1677 and 1681 and ropes were purchased for 7s. After 1684 there are no records extant and it is not known what clock was in use in the 18th and 19th centuries unless it is that which is now in Barford St. Michael church (q.v.). It was still existing but out of order in the 1880s.

In 1897 the clock from the old church of St. Mary, Banbury, was obtained and installed. This was made in 1741 by Joseph Hemmins of Banbury (q.v.) and has 3 trains in a screw-nutted iron frame, 59 × 25 × 33 inches, without finials. The going train has a recoil anchor escapement with a 45-inch pendulum and winged nut regulation on the cock. Maintaining power is arranged through a lever which covers the winding square at one end and has a pawl at the other end which engages the wheel teeth when raised. Striking is by means of a locking-plate with external quarter-hour pegs, lever and peg let-off action and fly with adjustable vanes. The chiming train has a squirrel-cage barrel of brass staves, originally provided with pegs for 6 bells, only 2 sets of which are now used; all 6 hammer-levers are present, 4 being disconnected. All pinions are integral with their arbors, of 10 and 7 in the striking train and of 8 and 7 leaves in the other 2 trains. Leading-off to the motion work and dials is through a contrate wheel and Hooke's joint. Two dials were placed on the south and west sides of the tower in 1897.

The indicator dial, numbered 0-60, is inscribed, *Joseph Hemmins Banbury fecit 1741,* and on either side of it, " Josh. Philipps removed this clock from St. Mary's Banbury to this church in Victoria Diamond Jubilee

PLATE 1
Fig. 1 HANWELL, St. Peter. Clock with verge and crown wheel escapement by George Harris of Fritwell, 1671.

PLATE 2

Fig. 2. Sixteenth century Jacks formerly on St. Martin's church, Oxford. Fig. 3, Inset, copies of the original Jacks and quarter bells now on Carfax Tower, Oxford. (Photo, fig. 3 by courtesy of *The Oxford Times and Oxford Mail*).

PLATE 3

Fig. 4. OXFORD. St. Mary the Virgin. Clock by Thomas Pars of Warwick, 1741.

PLATE 4

Fig. 5, below, BICESTER, Clock by Edward Hemins of Bicester, ca. 1735.
Fig. 6, above, CHURCH HANBOROUGH, SS. Peter and Paul, 17th century clock. (Photo by courtesy of A. W. Cox and J. Brice).

Figs. 7 & 8, OXFORD, Wadham College, Clock by Joseph Knibb with anchor escapement, 1670.

PLATE 5

PLATE 6
Fig. 9, GREAT MILTON, St. Mary, Clock by Nicholas Harris of Fritwell, 1699.

PLATE 7
Fig. 10, above, OXFORD, Magdalen College, Early 18th century carrillon barrel.
Fig. 11, below OXFORD, Trinity College, Clock by John Hawting of Oxford, 1787. (Photo by courtesy of L. S. Northcote).

Fig. 12, THAME, St. Mary, Pendulum casing and anchor escapement of the church clock. (Photos by courtesy of L. S. Northcote).

Fig. 13, WATLINGTON, Town Hall, Late 17th century clock. (Block by courtesy of the Watlington Parish Council).

PLATE 8

June 22/97 ". The movement should be compared with that made by Edward Hemins in 1740 at Aynho church.

NEWTON PURCELL — St. Michael

The clock at the west end of the church is curious and of considerable age. It has heavy stones for weights (Blomfield 1889). The church has only 2 bells hung for chiming in an open western turret.

OXFORD — Balliol College

There is no record of a clock earlier than that which was installed in 1838.

The frame has cast iron corner-pillars with ball finials, and vertical bars fastened by nuts. An 8-day rack-striking movement uses double pulley-blocks to overcome the restricted fall for the weights. The anchor escapement has a slight recoil and the heavy pendulum beats 40 to the minute. On the brass hand-setting dial is " Thwaites & Reed Clerkenwell LONDON 1838 " and on the going train barrel-plate " THWAITES & REED CLERKENWELL LONDON ". A tablet fixed to the frame reads, " HOC HOROLOGIUM *grati animi* D D RICARDUS JENKYNS STP. hujus Collegii *Magister* MDCCCXXXVIII ".

The skeleton two-hand dial is on a small bell turret above.

OXFORD — Christ Church Cathedral

The first clock of Christ Church Cathedral came from Oseney Abbey. When the Abbey was being demolished in 1545 an unnamed smith took down the clock, which had last been mended in 1544, and set it up in the Priory church of St. Frideswide to which the See of Oxford was transferred by Henry VIII in 1546. The west end of the Priory church was closed to form part of the new Cathedral. This smith was not Thomas Wynkyll, who was then employed at Oseney for various ironwork and to make the great bell clapper; he may have been " the smyth of Abynton," who made casements for windows in the Hall in St. Frideswide's which were glazed by John Nicholson — in both cases the work was done " upon his bargen ". John Wesborne, carpenter, who took down the Oseney bells, made the bell frame and set up the bells in the steeple of St. Frideswide, also made the clock house in 1546 (Turner 1880).

Payments for keeping the clock appear in the Accounts as early as 1548 and were continued at 6s. a year in the 16th century (Hiscock 1946). Repairs were done in 1583, 1586, 1587 and 1595 (see Richard Cakebread, Thomas Rolewright and Johan de Saint Paul). Practically nothing is known about the clock during the 17th century. In 1730 the Cathedral clock was in a ruinous condition and repairs were done by George Wentworth (q.v.); again in 1737-1740 by Edward Moore (q.v.) and finally in 1743-1746 a considerable reconstruction was undertaken by Thomas Reynolds (q.v.) for £35 14s. 0d.

B. L. Vulliamy told Capt. W. H. Smyth in 1848 that the Cathedral clock was very similar in style to that at Hampton Court Palace. It was cleaned in 1838 by order of the Dean and preserved in the tower (*Archaeologia*, 1849, p.14) but it disappeared during Gilbert Scott's restoration in 1870, as has also Vulliamy's description of the movement. In 1848 Capt. Smyth, Octavius Morgan, Mr. Johnson of the Radcliffe Observatory and Richard Rowell visited the clock room and inspected the clock. Smyth's description in *Archaeologia,* op. cit. is — " It certainly has a curious train, although there is neither maker's name or date; yet from the second and scape wheels being made of brass, from the form of the engraved numerals, and from the driving regulation being by a pendulum, I cannot but consider it to have been made about the time of Huygen's isochronal discovery was made public. The quarter, striking and going great wheels are, as in all the first clocks, made of iron ". However, we should now conclude that the escapement conversion was done in the 18th century by one of the clockmakers mentioned above.

B. L. Vulliamy supplied a new clock in 1838 for £396.

OXFORD — Christ Church, Tom Tower

W. G. Hiscock (1946) has recorded the following information about the clocks in Tom Tower.

In 1685 a new clock was installed in the newly completed Tom Tower, the first item of information concerning it appearing in a bill of that year: " For 2 hundred waite of lead cast in waites for the clock for greate tom, £1 17s. 4d." In 1686 William Young, the smith, was paid £2 10s. 0d. for work on the clock. He may well have made it, although chiming work was his speciality. In 1740 the clockmaker, Edward Moore (q.v.), was paid £10 11s. 6d. for repairs which evidently comprised major replacements. Nevertheless Thomas Reynolds, clockmaker of Holywell Street, five years later was able to find considerable work to do on the Tom Tower clock as he did on several other clocks in Oxford about this period. His bill came to £20 11s. 6d. and presumably he was thereafter employed on its maintenance on an annual wage.

In 1841 the suspension spring was found to be broken. Richard Rowell of Oxford (q.v.) had increased the weight of the pendulum bob which, B. L. Vulliamy said later, was right in principle. He went on to say " In my opinion the increase of the weight of the Pendulum Bob from 90 to 222lbs. is much too great for so imperfect a clock. That circumstance and the action of the spring not being in accordance with the centre of motion of the verge, I apprehend to be the cause of the spring breaking." In 1842 Vulliamy wrote that Rowell suggested " a new weight for the going part made in different pieces in the same manner as the weights of the clock in the Cathedral. That will afford him the opportunity of keeping the arc of vibration of the pendulum constantly the same. Mr. Rowell is certainly right."

These adjustments lasted nearly 50 years until a new clock was made in 1888 by J. B. Joyce of Whitchurch, Salop, at a cost of £170. Joyce's clock was made to the specification of Lord Grimthorpe and comprised: gun-metal wheels, double 3-legged gravity escapement, bolt and shutter maintaining power, and a pendulum bob weighing 3 cwt. The hammer weighed 300lbs. with a lift of 9 inches.

Mr. Hiscock adds that it is interesting to note that Joyce said, when submitting that his estimate, that the clock would be " the same as Worcester Cathedral excepting the quarters " — Great Tom being struck only at the hours.

The life-history of Great Tom and his several recastings has been published several times and more recently by Hiscock 1946 and Sharpe 1951. The bell was rehung for ringing by Mears and Stainbank in 1953 (Hughes 1960).

OXFORD — Exeter College

On 10 June 1820 a new porch for the Hall surmounted by a clock was completed; the porch cost £103, the clock £125 12s. 3d. (Boase 1894).

The 1820 clock was replaced in 1912 by another which is now superseded by an electric model.

OXFORD — Jesus College

The clock in the First Quadrangle was given by Principal Foulkes in 1831 and cost £110. It was made by Moore and Son, Clerkenwell and installed by George Rowell, Oxford.

OXFORD — Magdalen College

The first clock of which there is any record was ordered in 1505 in a bond with William Este, master mason then of Abingdon, Louis Foose, painter of the same town and Martin Williamson, brewer of Oxford. Presumably the clockwork was sub-contracted to a smith and the three men were sureties. It was a striking clock with a one-hand dial and the contract price to include a year's guarantee was £10. If the dial was placed outside it was on the new tower then being built in which bells were installed also in 1505; it would therefore be one of the earliest outside dials. In 1516 one of the Fellows, John Symonds, was appointed clock-keeper at 15s. a year. The only note of repairs now traceable is for 1602 when Triomphe de St. Paul (q.v.) was paid 6s. 8d.; a new bell was then cast for the tower. This clock has disappeared.

On the first floor of the tower there is now a rusted three-train turret clock in an iron bar frame, 52 × 30 × 23 inches, fastened with square nuts. Each corner-standard has incised line decoration and buttress mouldings at top and bottom and is surmounted by a finial with an out-curved stalk topped by a hexagonal boss. The 3 side-by-side trains have 6-armed brass wheels and a recoil anchor escapement, the arbor of which can be moved for setting; the pendulum is missing. External to the striking train is an iron locking-plate with bifid curved arms. On the barrel arbor of the chiming train is the quarter locking-plate and also a narrow barrel with pegs for lifting 8 hammer-levers, which are mounted in the narrow side of the frame. Both flies are 4-vaned.

In the middle of the 17th century there were 6 bells which were increased to 8 in 1712. The construction of the frame has features of the sixteen seventies but in view of the number of chime hammers the clock in its present form probably dates from 1712.

Up in the third storey of the tower is a large, damaged, wooden carillon barrel with 16 hammer-levers, mounted in a stout oak frame about which there is no history. Possibly the complete equipment was used when the Magdalen chimes first rang out in 1713. Plate 7, fig. 10.

The present clock is by John Smith & Sons, Derby, 1908.

OXFORD — Merton College

The early history of time-keeping devices in this College in the 13th and 14th centuries has been given in the Historical Review, pp. 17, 18. For the 15th to 17th centuries no information is readily traceable. In 1657 the ring of 5 bells was recast into 8 but the result was not satisfactory and they were recast in 1680. There must have been a clock during this period, even if its use was interrupted occasionally, and it probably survived throughout the 18th century. At any rate the clock which was in use up to 1813 was considered to be worn out according to a decision minuted on 27 February of that year.

A new 8-day clock was supplied by John Thwaites & Co., London, and is still preserved in its original site in a loft over the Upper Bursary in the

Fellows' Quadrangle. It is in very good condition. The frame, 32 × 20 × 16 inches, with cast iron corner-pillars has 3 trains with quarter-striking gear mounted in 5 pairs of vertical pivot-bars. The one-second pendulum is hung from a braced cock on the frame and has no wing-nut regulator above. The going train barrel-plate is signed, " Jno Thwaites & Son, Clerkenwell London " and on the indicator dial is " John Thwaites & Co,, Clerkenwell London 1813 ".

It is specified in the firm's invoice of May 1813 as: " An 8 day Quarter Turrett Clock to strike the hours on a bell of 1 Cwt. & the Quarters upon 2 suitable bells & to shew One Outside Dial hours & Minutes the Quar'r Gt wheel 9ins. Strik'g Gt Wheel 9ins. & watch Gt wheel 8ins. diam with dead Scapement & wood Pendulum . . . a Copper dial plate 3ft. 9½ins. diam'r hammerd Convex with Mould & round the edge & painted black gilt figures & Mould'g & bolts &c . . . 2 bells for the Quarters wt 1cwt. 11qr. 17½lb."

It took a man 22 days to bring it to Oxford and erect it, with two visits by John Thwaites. The circular dial is on the wall of the quadrangle close to the clock. In October 1932 a Committee advised its replacement by an electrical system which now operates a master clock, the above-mentioned dial and Canterbury chimes on the bells which are at a considerable distance in the Chapel tower.

OXFORD — New College

The bell-tower standing alongside the City Wall was built about 1400 on the site of a former bastion. According to Rashdall and Rait, 1901, it served as a receptacle for a clock and the 5 bells given by the founder of the college. Unfortunately the early rolls of college accounts do not give details from which the history of a clock can be constructed. The 5 mediaeval bells were recast in 1665 into a ring of 8, the first of this number in Oxfordshire. There is a tradition that a quarter-striking clock was used with this ring but the rolls do not mention expenditure on it. William Young, the smith and expert on chimes, was regularly employed by the college at this time and certainly for work in the tower when the bells were rehung. Another bell was recast in 1672 and after this occurs the first entry about the clock — *1672-3, Custus ad Intra. 4° Term'no. Solut to Young for mending ye Chimes &c . . . £5 — Sexton for 5 ropes . . . 14s. — Solut to Young ye Smith for keeping ye Chimes for ye Quarter ending at Michs . . . 2s. 6d.* Then in 1676-7, *Domorum Solut Young for ye Clocke & worke done at Severall times ut p. Bill . . . £11 7s. 6d.* And in 1677-8, *Custus ad Intra Solut for A Rope for ye Chymes . . . 6s. 5d.* The later 17th century accounts give the totals of bills only. Young was paid in 1679-1682 also.

In the tower clock-room is a damaged rusted clock in an iron bar frame, 56 × 31 × 20 inches. The corner-standards are plain flat bars welded at the junctions with the cross-members; each of the 4 finials has a vertical stalk 4 inches long topped by a depressed pyramidal boss. The pivot-bars of the 3 side-by-side trains are fastened by nuts and a fourth pair takes a wide, iron, chime barrel with cams for 8 hammer-levers and also the quarter locking-plate. The wheels of the 3 trains are of brass, 4-armed, and the hour locking-plate is external. Lantern pinions are used for the great wheel and the chime barrel wheel, the other pinions are solid. In a short extension above the top member of the frame an inverted V-shaped anchor with semi-dead-beat pallets is pivoted and the pendulum has wing-nut suspension. Clearly this is not the clock mentioned in the accounts for 1672-3 which infer repairs to chimes previously existing, and it is not likely that a new clock is included in the bill for £11 7s. 6d. in 1677. It was probably obtained within the next 20 or 30 years. In 1712 the ring of 8 bells was increased to 10 hence it is likely that the clock was in place before 1712.

The vertical finial with a square boss is unique for the county; vertical tapered finials without bosses occur on the clocks in Mollington and Wigginton churches.

New College ordered the 8-bell chiming clock now in the same tower room from J. W. Benson, Ludgate Hill, London in 1885. No alteration was made to the tune of the chimes.

OXFORD — Oriel College

The clock dial is on the east side of the west range of the Front Quadrangle. The clock bell, in an open turret above the roof of the gatehouse tower, is hung dead and is dated 1820.

On the indicator dial of the clock is inscribed " Vulliamy. Clock Maker to the King. London. AD 1820 No. 748 ". The two sides of the flat, iron, bar frame are joined by 4 cast iron pillars and 6 vertical pivot-bars, all nutted. Vulliamy has used a pinwheel escapement of the Amant type strengthened by uniting the external ends of the pins with a circular rim; the pallet arms are adjustable. Simplified hour-striking has been achieved by means of one lifting-piece and a series of spaced hour pins on the great wheel instead of a locking-plate. The pendulum has a very short suspension spring, a heavy iron rod and a disc bob and it beats $1\frac{1}{4}$ seconds. The drive to the dial is short and direct.

It was presented to the College by Edward Copleston, Provost from 1814 to 1828. Dr. Copleston wanted to have a double blow struck as a warning 2 or 3 minutes before the hour but Vulliamy discouraged the idea.

There is no record of an earlier clock.

OXFORD — The Queen's College

The clock in the restored cupola on the north range of the Front Quadrangle designed by N. Hawksmoor in 1714, is by Gillett & Bland Steam Clock Factory, Croydon, 1880. Some publications erroneously date the clock 1714. There is no record of an earlier clock.

OXFORD — St. Clement

The old church of St. Clement was abandoned in 1824. The only reference to its clock in the surviving Churchwardens' Accounts is for 1820. " Winding the clock £2 2s. 0d. — For gilding the clock dial, £2 12s. 6d." Old prints show that the dial on the east tower wall was diamond-shaped.

OXFORD — St. Giles

Extracts made by A. Wood from the church records include, *1693, Received of the Parishioners of St. Giles towards a new Clock, Pulpit-cushion and Fringe, for the Tower Stairs and three Gates in the Church Yard . . . £15 1s. 0d — Of St. John's College and other Benefactors . . . £28 16s. 11d.* Rawlinson's engraving of 1754 shows a one-hand diamond-shaped dial on the south wall of the tower (Peshall, 1773).

OXFORD — St. John's College

The college was founded in 1557 and a clock was acquired very shortly after, either made or supplied by John de St. Paul, a French clockmaker settled in Oxford. The earliest reference to it in the surviving Computus Annuus (accounts) is in 1569 when two new cords were required. St. Paul looked after the clock for 7s. a quarter and was paid extra for material and repairs until 1596. He was then succeeded as clock-keeper by Hugh Corbet, a smith (q.v.) on the small wage of 5s. a year with extra for repairs, but Corbet did not continue the apparently unprofitable work after 1603. One Richard Weller was then found to look after the clock at 20d. a quarter, until Triumph de St. Paul (q.v.), son of the earlier John, had established his business as a freeman of Oxford and had taken an apprentice. From about 1608 to 1630 Triumph de St. Paul was the college clock-keeper at 3s. 4d. a quarter; nothing was charged for repairs during this period. In 1631 John Raye, a smith (q.v.) took over charge for a few years at the same wage. After 1636 the clock-keeper is not named but Raye was paid 8s. in 1640 for repairs. By 1645 the keeper's wage was doubled to 6s. 8d. a quarter and the sexton wound the clock among his other duties, which sometimes included writing the Year Book.

This arrangement continued until 1690; the cost of any repairs or renewals was not separated from the annual bills of the college smith; but the clock by this time must have become obsolete. In the third term's accounts for 1690-91 is the entry, *It to Mr. Knibb for ye Clock ut p't bill . . . £24 16s. 0d.*, and in the following year, *It Pd. for ye carved Supportes for ye Clock Case . . . 16s.* This suggests that the new clock was sited where the present clock stands in the organ loft at the end of the Chapel and needed a wooden screen to conceal it. Having installed the clock, which struck the hours, John Knibb was retained at £1 a year to keep it in order, and the sexton continued to wind it at 6s. 8d. a quarter. For the next 30 years no extra charges for repairs are recorded.

John Knibb died in 1722; the care and repair of the clock was then undertaken by his apprentice and later journeyman, Humphrey Brickland (q.v.), at £1 a year until 1750. The sexton continued to wind the clock at 6s. 8d. a quarter. In 1741 Thomas Paris of Warwick, son of Nicholas, a celebrated maker of turret clocks, wrought iron work, gates and firearms, was in Oxford installing a new clock in the church of St. Mary the Virgin. He was consulted by St. John's College and undertook an extensive overhaul of the clock. His detailed bill has fortunately been preserved and is transcribed as follows:

" A Bill dated 13 July 1741 For Clockwork done at St. John's College by Tho. Paris Warwick.

Work done to the Watch part making one New Iron Barrel for the cords to wind on one New Rash Wheel and fore Plate and one Spring and catch and 3 Screw Pins £2 0s. 0d.

For a New Detting for to keep the Clock going to Loose know time while winding up two New Standreds for the Pivets a New Detting to put the Strikeing Part of and 2 new pins to Lift up the Detting £ 12s. 6d.

for makeing a cock Head a Spring and 2 Screw pins and New Turn the top of the teeth of the Varge Wheel new cutting down all the teeth and makeing a new of Pallets £1 15s. 0d.

for makeing a New Pendulum Rod takeing the Brass Ball to peices and putting 6 pound of new lead to it £ 10s. 0d.

Work done to the Strikeing part for makeing one New Iron Barrel for the cord to wind on one new Rash Wheel and fore Plate and Spring and catch and 3 Screw Pins £2 0s 0d.

for makeing a New Iron Hoop to the Second Wheel a new Detting to Lock into it and fileing the Wheel of Account and makeing a New Steel

Pinion Instead of the Brass one that was Broke and three Iron Studs put in
£1 10s 0d.

for makeing a New fan a New Rash Pinion a new Spring and catch 2 new Brasses for the fan that it should make know Noise after Strikeing
£0 12s. 6d.

for my Self and 2 men puting the Clock up Satterday my Self on Monday altering the Hammer £0 10s. 0d.

July the 15 1741 Received payment Tho. Paris Total £10 10s. 0d."

This bill is of interest, not only for the names used for components, but as revealing the considerable amount of replacement, etc., that can be covered by a brief entry in the accounts, which actually was, *To Paris for mending the Clock . . . £10 10s. 0d.*

After looking after the clock annually Brickland died in 1750 and Robert Denton was called in; he was one of the few clockmakers who matriculated as privileged tradesmen of the University. His services were used only to the 3rd quarter of 1753 when Thomas Reynolds of Holywell Street secured the maintenance contract at the increased wage of £2 a year. Reynolds had charge of several other turret clocks in Oxford all of which he retained until his death in 1799. If any repairs to the St. John's clock were done during his period of 46 years the costs were not separately accounted but were included in general smiths' bills. Clock-winding continued to be the sexton's job at 6s. 8d. a quarter; in view of rising costs for labour it is remarkable that this wage remained exactly the same until the end of the 19th century, that is for over 250 years.

After Reynolds' death his partner Thomas Earle (q.v.) continued the contract at £2 12s. 6d. a year but only until 1805. George Rowell of Broad Street (q.v.) was appointed in 1806 at an annual fee of £3 3s. 0d. plus £2 2s. 0d. every 2 or 3 years for cleaning the clock. Richard Rowell succeeded him and maintenance was entrusted to the firm of Rowell throuhgout the 19th century. After about 1867 bills for repairs became heavy but it was not until 1919 that a new clock was ordered from Thwaites & Reed, London. This is a small hour-striking two-hand movement with a gravity escapement and a one-second pendulum. The setting dial is inscribed " 1919 Rowell & Son, Oxford ". Owing to the short fall for the weights, winding twice a week is needed. The bell, which was cast in 1910, is in an enclosed central turret and is hung dead.

OXFORD — St. Martin (or Carfax Tower)

St. Martin's or Carfax Church was the official church of the City of Oxford.

The Churchwardens' Accounts for the period 1540 to 1794 have survived. A clock was already in existence in 1540 and 8s. a year was the keeper's wage. The first mention of repairs was in 1547 when George Samson received 4s. 4d. In 1554 are the following items, *to George Samson for mendying the clokk . . . xiijd. — for a pully for the clokk . . . iiijd. to the workman thatt made the clokk . . . xs. id. — for thre wekes bord to the workman thatt made the said clokke and mendyd the said belle . . . vijs. — to Wyllm Gomes for kepeying the clokk for the half yere . . . iiijs.* Making in this case must mean repairing. Eightpence was spent on a repair in 1557 and further repairs were done in 1560 viz., *paid for me'dinge of ye cloke spring . . . iiijd. — mendinge ye cloke . . . id. — me'dinge ye cloke . . . viijd.* In 1564 whipcord and nails were charged; in 1569 a pulley at 3d.; in 1570 more whipcord at 2d.; in 1575 a rope only 2d.; in 1577 the spring of the clock was again mended; in 1589, *It for mending plummets for the cloke . . . iis.* and in 1590, *It pd for mending the wheele of the clocke . . . iis.* During the whole of the foregoing period the clerk was paid 8s. a year for keeping the clock.

In November 1592 the City Council sanctioned the freedom of Thomas Bull, clockmaker (q.v.), on condition that he repaired and maintained the clock, jacks, chimes and dial at Carfax and made a service agreement with the Churchwardens, but apparently they could not agree and Hugh Corbet, a smith (q.v.), was appointed clock repairer. This is the first mention in the city records of jacks, chimes and a dial. Later prints illustrating St. Martin's church show the dial and jacks on the wall at the east end of the south aisle. The circular one-hand dial was on a diamond backing between two masonry pilasters with basal corbels and moulded capitals, joined above by a slight moulded canopy supporting two bells. The carved and painted oak jacks stood on corbels on either side of the upper part of the diamond and struck the quarters with axes. According to 19th century accounts the clock and its accessories were remounted in this position in 1624 (Plate 2).

Hugh Corbet did some repairs to the chimes and the clock (6s.) in 1594 but in 1595 more work on the chimes costing 22s. 11d. was needed and 18 lbs. of lead was used for the ' plummets'. He was then appointed at 20s. a year to keep the clock and chimes in working order. The clerk, Jennings, was paid 20s. a year from 1595 for winding and setting the clock and chimes.

In 1601 23s. 4d. was *paid for newe amendinge the chyme being greatlie decayed,* and in 1611 the *watch wheele of the clock* was mended for 3s. But maintenance was not being carried out by the Churchwardens to the satisfaction of the City Council and on 20 December 1621 the following minute was written, *It is agreed as touching the proposition for contribution towards repayreing making and finishing of the Clock Jacks dyall and chymes at Carfax, in regard to the inabilitie of St. Martyn to doe the same and for that it is a common church whither the Mayor and his Brethren and the rest of the cittizens doe resort for sermons and at other times of solemnity, there shall be taxation made viz. the Thirteen 10s. each, those with bailiffs places 6s. 8d., Chamberlains 5s., Common Council 3s. 4d. and a reasonable taxation for this work shall be laid upon the Commons that be of ability* There is no note of the amount raised or of immediate expenditure on the clock.

The wage of the clock-keeper, Matthew Jennings, had been £2 a year from 1601 and repairs were done by Thomas Ranklyn, the smith (q.v.), until 1626. Then John Raye began to be employed and his bills appear more frequently, thus, *Pd to Raye for mending the clock and Chymes . . . iiijs. — oil for bells locks and chymes . . . xxiiijd. — ropes for the chyme & trebel . . . viis. iiid. — for Iron barres and mendying the chymes . . . xviis. — 1627 Pd John Raye for mending the clock and chymes . . .xxiiis. — 1628 Pd Wray the Smith for mending ye clocke & chymes . . . xis. — wyre for the clock & chymes & oyle for the whole year . . .iiis. iiijd. — 1629 Pd Thomas Slatford for mending of the wyndeles . . . Iron for the chymes . . . 8d. — Raye for mending Clock & Chymes . . . 7s. 1d. — for mending the Chimes to John Raye . . . 3s. 4d. — 1630 Pd to Ray the Smith for mending the clock & chymes . . . 8s. 6d.*

In 1633 Raye and Slatford both found work to do on the clock, viz., *Goodman Raye smith for mend' the clock . . . viijd. — Thos Slatford for work done to the Chymes . . .iis. viijd. — Pd him for work done to the clock . . . iijs. iid. — Pd Goodman Raye smith for work done to ye Clock & Chimes . . . iiis. iiijd.* In 1634 the annual contract was reintroduced, viz., *Pd Goodman Raye for keeping the clock & chymes for a yeare . . . xxs. — Pd the Clerk his quartridge 4 times xs. for keeping the clock & chymes.* This arrangement continued until 1641 when Thomas Ranklyn replaced Raye and drew the annual allowance of 20s. In this year is the first entry referring specifically to *mendying the Jacke . . . 4s. 6d.* In 1644 Ranklyn was paid 4s. for making two new spindles and mending the clock, and two years later 10s. 10d. was paid for 29 lbs. of bell ropes and for clock weights.

In April 1653 the City Council had once more to make a grant to the parish for clock expenses. £3 was allowed out of the City treasure towards setting up again the chimes of St. Martin's parish provided that the said parish agreed in future to keep them always in repair; this sum was paid to Ranklyn. By 1656 it was necessary to buy *a greate Bell rope & a chyme wyre waying 17 pounds . . . 9s. 6d.* and next year to pay *Rainkoll the smith old arreages . . . 6s. 0d. — More to him for worke don about the Clocke chimes & greate bell . . . 6s. 4d.* Soon afterwards Ranklyn died and a final payment was made to *Widow Ranckling for iron work about the clock and bells . . . 6s. 9d.* In 1660 it was the clerk who submitted *his bill for mending ye chimes & clock and Jackes . . . 5s. 2d.* Eventually John Shewell, an apprentice of Ranklyn, took over work on the clock. There were charges for clock ropes, 8s. 6d. in 1663, mending the clock, 11s. 10d. in 1667, and an extensive overhaul in 1669 costing £4 8s. 0d. Shewell drew an annual allowance of 10s. a year for looking to the clock which did not include major repairs.

In 1676 the 5 old bells were recast with added metal to make a ring of 6 by John Keene at a cost of over £89, and at the same time the layout of the chimes was rebuilt; the entry in the Accounts merely states, *William Young mending the clock and making the chimes . . . £16 10s. 0d. — 2 lines for the chimes . . . 10s. — a weight for the chimes . . . 3s. 6d. — Pd John Rixon for painting the Diall . . . £2.* From what one knows of Young's work on other church clocks it may be deduced that he scrapped the old chime-work that was in use in 1601 and 1626-8. The City Council had to grant the parish £40 towards recasting and hanging the 6 bells and setting up a new chime and repairing the clock. In 1680 there was purchased *A Sett of new Ropes & a Rope for the Clock weighing 55 pounds . . . £1 7s. 6d.* Small amounts were paid during the next 5 years for repairs, wire, oil, etc., to Young, Thomas Walker and William Cox.

The Accounts for 1686 record, *For writing an Agreement with Mr. Young about the clock and chimes . . . 1s. — Mr. Young for making a pendulum to the clock & work about the chimes & a new hand . . . £5 16s. 0d.* This agreement no longer exists but it must refer to conversion to an anchor escapement and long pendulum, and would make St. Martin's clock the third in Oxford to be so fitted, the pioneers being those of Wadham College and St. Mary the Virgin. It is possible that the clock of St. Mary Magdalene was converted about the same time. Until his death ca. 1695 Young continued to look after St. Martin's clock which needed only new ropes and wires. Shewell and Walker (q.v.) were the last of the smiths to be employed on this work at St. Martin's.

John Oakley was the first clockmaker to be employed. He mended the clock in 1704 for £1 10s. 0d. and the chimes in 1705 for £2 5s. 0d. The next clockmaker to be consulted was Edward Moore who was engaged at an annual wage of £2 from 1715 for cleaning, mending and new work to the clock and chimes; he probably did the winding also. His bill for 13 May 1726 is the first extra entered in the accounts and it was for *Mending ye hamers springs latches & for new keys to the chymes . . .15s.* and his last bill was for £1 2s. 9d. In 1727 are the entries, *Mr. Moor's Quarter for the clock and chime . . . 10s. — Paid Mr. Short for turning the wheels for the clock and chime . . . 2s. 6d. — Paid Nicholas Benwell 3 quarters salary for ye clock & chimes . . . £1 10s. 0d.*

Benwell was sexton at Christ Church and for the following ten years was responsible for winding the clock and chimes at £2 a year. Then a Mr. Mumford was engaged but he put in rather heavy bills for which there are no details. Again the parish was in financial difficulties, such that on 12 August 1748 the City Council resolved that *10 gns. be paid to the Churchwardens of St. Martin's towards repairing or new erecting of the clock & chimes at St. Martin's church and if the same shall be done in a decent & substantial manner the treasurer shall make up the sum to 12 gns.*

A satisfactory solution was found by getting Thomas Reynolds (q.v.) to complete the repairs for £13 8s. 0d. and by employing him to look after the clock, etc. for £1 16s. 0d., raised to £2 in 1763, a year. He held this post for 45 years and after the initial expenditure very little further attention was needed through his period of service, that is, to 1799.

Considerable repairs were done to the church in the last quarter of the 18th century and in the early 19th century. In 1819 it was decided to pull down and rebuild the church towards the cost of which the City Council contributed £600. The foundation stone of the new church was laid in October 1820 and it was opened in June 1822. An engraving by G. Hollis done in 1819 before the demolition of the church shows that the dial and jacks were at the east end of the south aisle as they had been for at least two centuries. In the new church a two-hand circular dial without any jacks was erected above the centre of the chancel window. The jacks and their bells were stored in the Town Hall. The new clock was put in by W. & J. Taylor of Oxford in 1848 and the old clock disappeared with its chimes. In 1896 the 75-year old church was demolished entirely except for the 14th century tower, known as Carfax Tower. In 1898 the clock was slightly reconstructed and Messrs. Payne & Son of Oxford, who then had the contract for winding and maintaining the clock, reported that it had sufficient power to work the Quarter Boys if they were replaced.

A new design was adopted for the dial and accessories when it was erected on Carfax Tower eastern wall. The dial is framed by scrolled sides and a gabled pediment and below are two fluted pillars on bracket supports joined above by an entablature inscribed FORTIS EST VERITAS. Accurate copies of the original jacks stand below and strike on two bells made by J. Taylor and Co., Loughborough, 1898 (Plate 2, fig. 3). The old jacks remained in the Town Hall.

In 1938-9 an electric master clock with a ½ minute pulse and electrically driven chiming and hour-striking mechanisms, which also operate the quarter jacks, was installed by Gent & Co., Ltd., Makers, Leicester. The Taylor clock of 1848 has disappeared and the 16th century jacks still rest in the Town Hall cellars. Chiming takes place on the 6 bells, a complete ring made by Richard Keene of Woodstock in 1676-78, preserved from St. Martin's church.

OXFORD — St. Mary Magdalene

The rolls of the Churchwardens' Accounts begin in 1561 but there is no mention of a clock until 1605 when there are two entries, *Item for Bords and Workmanshippe to cover the Clock from Duste . . . xid.,* and, *Item payd Hughe Corbate for kepinge the Clocke the wholle yeare . . . iiijs.* The previous year, 1604, is missing and may have recorded the purchase of the clock. Hugh Corbet (q.v.) was retained to look after the clock at 4s. a year until about 1616. The clerk also was paid 10s. a year for keeping, i.e. winding, the clock. From 1617 onwards John Raye (q.v.) was called in for minor repairs to the *clocke Irons* and for wire, oil, rope, etc., and a weight. In 1622 repairs to the clock cost 22s., a new weight was needed costing

4s. 6d. and the glazier had to repair the clock window, 2s. 6d. The last item is puzzling; it may mean a window in the clock-house, which was not on the ringing room floor as a ladder for the clock was bought in 1627 at a cost of 18d. The clerk continued to receive 10s. a year for winding and in 1648 he is named Goodman (John) Stubbs.

From 1650 to 1686 the accounts are missing so that there is no sure indication that a conversion to a long pendulum took place during that period; nor is there any evidence subsequently, although it is unlikely that it continued with a foliot throughout the 18th century. From 1686 to 1714 the clock was wound by the sexton for £1 and later £1 6s. 8d. a year; repairs were inexpensive and could not have covered a new escapement. When the clock-keeper, Edward Stockford, a cordwainer, died in 1714 Richard Robinson, a smith, was appointed keeper at £2 a year. It was not until 1722 that a clockmaker was consulted, and for some reason he was not an Oxford man but William Ball of Bicester (q.v.). The repairs to the clock which he did in 1722 cost £3 5s. 0d. and 2 pulleys, lines, wire and some timber, in all 6s. 6d., were supplied. Ball was retained at 5s. a year to keep an eye on the maintenance of the clock until 1736; the winding was done by Crosier, the sexton, for £1 a year.

First mention in the Accounts of an outside dial is in 1725 when a new dial board was made by Edward Riggins, carpenter, and 1s. was spent *on ale for taking down ye hand*. Riggins overcharged for this and other work. After considering his faults at two vestry meetings the Churchwardens presented him at the Bishop's Court in December 1726. This storm in a teacup cost the parish 8s., *spent at passing Mr. Riggins accounts*, and 3s. 6d., *Spent at a Parish Meeting about Mr. Riggins*, and 4s. 1d., *Paid upon Mr. Riggins account for Waiting upon Dr. Irish [of All Souls] and Measuring the hand of the Church clock;* in the end 9s. was recovered from him.

In 1737 a gunsmith, Thomas Pavier, succeeded Ball as repairs adviser, *a Greed with Mr. Pavey to look after ye Clock for a Crown a year by Consent of ye Parish and to keep ye Clock in Repares same as Baull did*. This arrangement lasted for ten years to 1746 but notwithstanding Pavier was paid extra for repairs. The sexton, Thomas Irons, wound the clock at an increased wage. Eventually the clockmaker Thomas Reynolds of Holywell Street (q.v.) was called in to overhaul the clock which he did for £5 10s. 0d. He was then retained at 5s. a year and remained in this capacity for 53 years. This proved to be a fortunate arrangement for the parish of St. Mary Magdalene as the clock thereafter functioned very satisfactorily; repairs were needed only in 1776 (£2 2s. 6d.), 1787 (£1 15s. 6d.) and 1794 (£1 6s. 0d.). A receipt for the year 1769 has been preserved, *The 18 of April 1769 Receiv'd of Mr. Bolton five shillings one years stipend for the Church clock and for new lines five shillings and taking down and puting up the Clock when the new flower was made 4 shillings in all foorteen shillings in full pd me Thos Reynolds*.

After Reynolds' death his partner, Thomas Earle (q.v.), continued from 1800 but he died 8 or 9 years later. The daily winding was done by the sexton. After Earle, the clockmaker George Rowell (q.v.) was employed and his duties included winding at £3 3s. 0d. p.a. In 1813 £9 18s. 0d. was paid to *Mr. Bliss for Painting and Guilding the Clock* and £5 10s. 0d. to Mr. Payne *for a new front Face and repairing the back Do. of the Clock*. This is the first reference in the Accounts to two dials one on the north and the other on the south wall of the tower. The only charge for clock repairs in 1813 is 12s. to Rowell; and a marginal note in the Accounts, *The Clock Painted Expenses £16 0s. 0d.*, adds no more information.

George Rowell was succeeded in 1834 by Richard R. Rowell who looked after the clock for the rest of our period. His annual wage was £3 3s. 0d. for winding and care with an extra £1 8s. 0d. for cleaning every few years. The present clock by Thwaites and Reed was installed by Rowell & Son, Oxford, in 1903.

OXFORD — St. Mary the Virgin

The history of the clocks in this church, the University Church, is to be found in the Accounts of the Proctors and of the Vice-Chancellor of the University and of the Churchwardens of the Parish. St. Mary's, being the University church from the earliest centuries, expenditure on new building, additions and repairs to the fabric and on maintenance of fittings, which included bells, organ and clock, was met from the University Chest. At later periods the Churchwardens were responsible for part of the maintenance and repairs to fittings out of the parish revenues but were also helped by occasional grants in aid from the University and one or two Colleges. In the case of the clock, therefore, one may find an entry in one set of accounts to which there is no corresponding reference in the other accounts — when both happen to have survived for the same financial year.

Most of the Proctors' Accounts for the period 1464 to 1496 have been preserved. Major repairs to the church clock, which was certainly not then new, took place in 1471 and 1477. Between 1469 and 1497 the parish clerk, who had many other duties, was paid by the University 3s. and 4s. a year for winding and time-setting the clock (Salter, 1921). The Churchwardens' Accounts for 1473-1474 reveal that the clock was sited in the Solarium and that it was driven by ropes in pulleys, *de viijd. solut p. ijbs asseribus ad solarium horologii — de iid. solut pro positsione cordule in ly poleys.* Two carpenters worked for a day on the solarium and closing up the windows in the bell tower. The end of the 15th century was marked by lack of repairs to the church and according to Anthony Wood it *was so ruinated in Henry VII raigne that it could scarce stand.* It was extensively rebuilt by the University about 1490-1503. Wood refers to *the clock house next the Church* which bore the University arms; these were pulled down *when the church was new modelled.* The clock house apparently had a slated roof as the Churchwardens' Accounts for 1491-1492 have an entry under the head, costs of slaters and laborers, Paid for a peck of slate pins and for nails used over the horloge, iiid.

The church records for the 16th century are largely incomplete. One of Wood's stray notes has *Horologium reparat 1510.* The Churchwardens' Accounts for the score or so years that survive have no reference to a clock but from the Vice-Chancellor's Accounts it is known that a new clock was erected in 1524 towards the cost of which fines on the scholars were allotted. In 1550 is an item, *to Tho. Masey for mendinge of St. Maryes Clock 25° Junii travellinge by the space of two Weekes thereon . . . £1 14s. 0d., and, Itm for a Lock to St. Maryes Clock . . . 10d.* The Churchwardens' Accounts for 1549-1550 merely state, *Paid for mendyng the house by the clock . . . vid.* And that unfortunately is all that is known about time-keeping at St. Mary the Virgin's during the 16th century, except that a sundial devised by Nicholas Kratzer was erected outside the church in 1520.

For the first 40 years of the 17th century the Accounts have nothing to say about the clock or a clock-keeper. Some disorganisation may have been due to the recasting of two bells in 1612, of three more in 1623 and to the cracking of the tenor bell in 1626. When the Great Tom bell of Christ Church tower was recast in 1626 Dr. Richard Corbet, who was Dean of Christ Church at that time, wrote a poem *To Yonge Tom* in which he refers to the

1523 clock at St. Mary's, . . . *thre Clock hanges dumbe in towre And knows not that foure quarters makes an howre,* and to the cracked tenor bell which *Runge like a quart pott to the congregation* (quoted by Sharpe, 1951). Not only was St. Mary's clock frequently out of action but in 1637 its keeper, the parish clerk John Dyer, was under suspicion of stealing £100 from the Danvers Chest. The Chancellor of the University, Archbishop Laud, who claimed the patronage of the clerkship of the church, then intervened. He wrote to the Vice-Chancellor on 26 May 1637, *I hold it very fit that the same man that is Clerk may not have the keeping of the Clock at St. Marys to shorten Hours at his Pleasure, especially in Lent, to the great hindrance of those Disputations and the disordering of the University in all Exercises. But I think it very fit that some honest man were taken into that service that would not be so easily found, nor perhaps so easily corrupted as the Clerks use to be.* (quoted by Jackson, 1897).

Eventually the clock was condemned and the Vice-Chancellor's Accounts for 1640-41 record, *to John Reyer for makeing ye new Clocke att St. Maries and quarter clocke . . . £22 — To Adams Tomlins Striplinge and Jeffes for their severall workes done about ye same and the clockhouse . . . £7 9s. 6d.,* (see John Raye, Thomas Adams, John Jeffs). Of the 1523 clock there was no information when Capt. W. H. Smyth made enquiries in 1848.

In 1643 the Churchwardens paid *Slatter for lyme and Slatts and his worke for slatting over the clocke House . . . 5s. 8d.,* and in 1647 *Item pd. Ed Wild for sclatting plastring over the clocke house & pointing ye windowes after glazing . . . 10s. 2d.* These entries seem to show that the clockhouse was not inside the church tower but possibly on the roof of the adjoining north chapel. Adams and Jeffs were frequently employed at the church and Adams was clock-keeper in 1650-2 if not longer and paid by the University. No charges for repairs are entered in either of the Accounts.

The curious division between the University and the Parish for expenditure on the clock is emphasised by a unilateral decision of the Churchwardens to add chiming mechanism to the clock in 1657. They employed William Young, the Oxford specialist on chimes, *It' pd for the Chimes . . . £9; for weight to them . . . £1 19s. 3d.; for the Rope . . . 5s. 6d.; to the Carpenters . . . £1 7s. 5d.; for sweeping ye Tower . . . 1s.; In All . . . £12 13s. 2d.* This is followed in 1658 by, *Item for a Safe for the Chimes . . . £1 18s. and 3d.; Spikes for them . . . 3d.; for a paire of hinges . . . 1s. 6d.; for a Lock & Key to the said Safe . . . 2s., £2 4s. 6d.; Item to Doctor Wilson . . . 10s. & Mr. Aires 5s. for setting the Tunes of the Chimes . . . 15s., Item to the Clarke for windeing upp the Chimes . . . 15s.* Clearly the "chimes", a carrillon barrel with its own weight drive, was in a safe in the tower and not in the clockhouse. Having bought the chimes the parish became perpetually responsible for winding and repairing them. The clerk, Richard Hart, was paid 15s. a year for winding but seems to have been careless and to have damaged the works. Consequently he was not kept on and William Young was given an annual contract for £2 to look after the chimes, which he repaired in 1660 for £7 6s. 0d. Young tended the chimes regularly from 1660 to 1672 his wage being raised to £2 10s. 0d. in 1668.

In 1670 an important horological event was recorded. The Annual Accounts of the Vice-Chancellor for the period 8 September 1669 to 16 September 1670 under the heading *Expensae extraordinariae* have the following entry, *Item to Mr. Knibb for altering ye Univ'sity clock to a Pendulum . . . £6 7s. 0d.* Clearly this innovation did not originate with the Churchwardens and one may speculate why the Vice-Chancellor was prepared to "modernise" a 28-year old clock that was functioning satisfactorily. Correct time-keeping was important for University functions as Archbishop Laud had stressed shortly before St. Mary's new clock was purchased, and there was Kratzer's sundial outside the church for correction with solar time (Beeson, 1961). It is most likely that the main stimulus to this horological venture was the advice and influence of Christopher Wren. Under Wren's direction Joseph Knibb had investigated the new pendulum for the Wadham College clock in 1669-70 (Beeson, 1957) and its prospects were apparently good enough to recommend it also for the important time-keeper at St. Mary's.

The conversion of Raye's clock by Knibb was completed early in 1670, the clock-house was repaired by John Dewe, a plasterer and both jobs were paid off before September 1670. The Wadham College clock still exists as evidence that it was made with an anchor escapement and a seconds pendulum, and therefore supports the inference that Knibb replaced the foliot of St. Mary's clock by a similar escapement and pendulum. If his alteration had comprised a crown wheel and Huygens pendulum one would expect a later conversion to an anchor escapement, but nothing in the Accounts of the Vice-Chancellor or Churchwardens can be interpreted as such.

William Warner was in charge of the chimes from 1673 to 1676. In 1677 the Vice-Chancellor paid £2 10s. 0d. *to the Parish of St. Maries for hurt done to their Chymes by the Workmen* engaged in repairing the steeple. Warner was succeeded by William Ferriman from 1677 to 1689 each of them at an annual wage of £2 10s. 0d. During this period of nearly 30 years no repairs of any importance were charged to either the Vice-Chancellor or the Parish. In 1689 are entries which may refer to Kratzer's sundial, as a clock dial has not previously been mentioned (see under Sundials).

Accounts for the next 16 years are missing and in 1705 the keeper of the chimes was the elder Pittaway, a smith, who was followed by his widow for the years 1713-18, and then by their son, Edward Pittaway, until 1741; the wage remained at £2 10s. 0d. At this period the University paid Nathaniel Day 13s. 4d. a year *for looking after St. Mary's clock*. In 1733 the smith's men were at work on the chimes and the Churchwardens' Accounts for 1734 state without details, *Paid the carpenters and smiths Bill for the Chimes ... £13 13s. 0d.*

By 1740 Raye's clock had reached the end of its useful life and the University ordered Thomas Paris of Warwick to supply a one-hand quarter-striking and chiming clock, viz., *It. pd. Mr. Paris for St. Mary's clock ... £100 — It Carriage of Ditto ... 27s. 6d. — Smiths Work to Do ... £5 10s. 0d. — It Carpenters Bill to Do ... £11 4s. 0d. — Painter to Do ... £16 9s. 0d. — Total £134 10s. 6d.* There was also a payment of £78 13s. 0d. to Townsend, the mason and next year to Green, the painter, for gilding at St. Mary's, £3 7s. 0d. It is possible that Townsend's bill included alterations to the parapet of open trefoil work round the base of the spire, which was broken into by the siting of two large circular clock dials in elaborate carved stone frames; prints of the period show the dials with one large counterpoised hand. A description of the Paris clock is given later (see Plate 3, fig. 4).

After 1741 the costs of winding and repairing the new clock were divided between the University and the Parish although the chiming train is an integral part of the Paris clock. John Oakley, clockmaker, (q.v.) was appointed by the Vice-Chancellor to wind the clock at £5 a year and the Churchwardens paid him £2 10s. 0d. a year to wind the chimes. The Vice-Chancellor's Accounts always refer only to " St. Mary's clock " and enter payments for its repair; the Churchwardens' Accounts refer only to winding, cleaning and mending " the chimes."

Oakley died in January 1749 and his duties were performed thereafter by John Herbert, clockmaker (q.v.) at £5 a year. For the University Herbert looked after the clock for 45 years until 1794, but he ceased to wind the chimes for the Churchwardens about 1757. The parish then reverted to the custom of giving the job to a smith, to James Boswell, paying him the standard wage of £2 10s. 0d.; it thereafter became a family perquisite, to Mary Boswell his widow and to Martin his son. During the 50 years ending 1794 the Vice-Chancellor paid for repairs on 22 occasions a total of £39 14s. 3d. The Churchwardens disbursed sums on wires, buffer springs and cleaning the chimes. Bills for 1761 give some idea of the division of liabilities. One by John Taylor, carpenter, shows that the parish was responsible for boxing in the wires and bell-hammers: *1761 July 22, for 4 days and a ½ work to ye Chimes . . . 9s. — for a new wheel to Do . . . 15s. — August 18, for 15 foot of oak 3 by 4 to ye partition to shut in ye Chimes . . . 3s. 1½d. — for 11 foot of Elm 3 by 4 to Do . . . 1s. 6¾d. — for 52 foot of ¾ board to Do . . . 6s. 6d. — for nail . . . 1s. 6d. — for 3 Days and a ½ work . . . 6s 6d.* And another by James Boswell: *1761 July 30 to ye chimes takeing to pieces & cleaning & mending & new wire . . . 15s. — August 15 to ye Chimes mending . . . 2s. 6d. — December 22 to ye Chimes winding up this year . . . £2 10s. 0d.* The University's contribution in 1761 was £2 15s. 0d. to Herbert for work done to the clock and 1s to Boswell " for mending the clock."

In 1794 Richard Pearson, clockmaker of Oxford (q.v.), was appointed keeper at £5 a year until 1803 (when the Account Book ends). During this period the University spent £25 on repairs on 8 occasions. Very little further information is available until 1851 when the tower was restored. According to the *Herald*, " the parapet of open trefoil work, round the base of the spire, has been restored and the clock [sic] which interrupted the continuation on two sides, has been removed and will be replaced with skeleton face and minute hands (the old one having only hour hands) by Mr. Taylor, who constructed the City Church clock." The cost of the work by William Taylor (q.v.) appears in the Vice-Chancellor's Accounts for 1851-1852 as, " Taylor, Clock, dials etc., balance of account £43 7s. 6d."

Plate 3, fig. 4, shows the clock when removed for overhaul by E. Dent and Co. in 1961.

The very large flat bar frame has corner-posts with some shallow moulding and U-shaped finials, which probably ended in brass knobs when supplied by Paris. Of the 5 pairs of vertical pivot bars two pairs end in fish-tail scrolls and are of different origin to the rest of the clock. As they support oblique bars carrying the indicator dial and extra gearing for the transmission to the new dials, they were probably so placed by Taylor. The going train in the middle compartment with a smaller endplate on its barrel, has a pinwheel escapement with semi-circular pins and offset Lepauté type pallets. The crutch is cranked so as to reach the cylindrical iron rod of the pendulum which is centrally mounted and suspended from a bracket fixed to the stone wall behind; it beats one and one-fifth seconds. The striking train on the left has an internally-toothed locking plate and a peg-stop on the arbor of the 2-vaned fly.

On the barrel arbor of the chiming train inside the frame is a narrow squirrel-cage iron barrel with pegs to operate 5 hammer-levers, and outside on the same arbor is the quarter locking-plate. The fly is 4-vaned. Striking levers are pivoted at the corner-posts of both ends of the frame. All wheels are of brass and all fastenings are by nuts.

Skeleton dials are mounted on the north and south walls of the tower and were made in 1851 by William Taylor. A good engraving showing the one-hand dial of 1741 is in J. Peshall's City of Oxford, 1773.

OXFORD — St. Michael

Except for an obscure item in 1470 there is no reference to a clock in the Churchwardens' Accounts for 1404 to 1562 (Salter, 1933). It is, *pro uno bigato lignorum traditorum cuidam nomine Loufgent pro donacione cuiusdam horologii . . . xxd.*, that is, for a cartload of timber from the bequest of Isabella Lovegent as a gift for some sort of clock; the word may be *dolacione*, i.e. cutting or hewing the wooden framework. No further documentary evidence is available but Skelton's engraving of Northgate and St. Michael's Church tower shows a diamond-shaped dial high up on one side of a small window.

OXFORD — Trinity College

In the tower of the west range of the Garden Quadrangle is a clock made by John Hawting of Oxford (q.v.). It is a finely finished 3-train movement in an iron bar frame, the top cross-members of which are of unusual design in that they are curved upwards at each side to brace the long middle vertical bars. The 8-day going train has a seconds pendulum with a dead-beat escapement of which the escape wheel has broad teeth placed horizontally to the rim and the anchor pallets work at the side of the wheel as in a pin-wheel escapement. Plate 7, fig. 11. The hour train has 8 pins on the 2nd wheel and a locking-plate. The quarter train has 10 pins and a snail on the 2nd wheel for the ting-tang quarter-striking. On the indicator dial is "JOHN HAWTING *Oxford* 1787." The clock serves two 4-foot diameter slate dials. In an open turret above the tower roof are 3 clock bells hung dead, the largest of which is dated 1787. The College Archives are not accessible for information on maintenance and there is no record of an earlier clock. An overhaul was undertaken by Thwaites and Reed, London, in 1955.

OXFORD — University College

A clock by A. & J. Thwaites, 1792, is in a small clock-house in the gable block of the south range of the Main Quadrangle. The firm's bill to Mr. Harrison dated July 1792 specifies it as:— " a new Eight day Turret Clock with 1 ft. 8 in. Great Wheels, the first Pinions of 18 Teeth & the rest of the Wheels & Pinions in proportion to shew One Outside Dial Hours & Hands & to strike Hours only with a Gridiron Pendulum of Zinc & steel fixed in the Wall, with a shifting Cock on front to adjust the hanging, a Strong Cast Iron block with flint the Centre of the Verge Pallets . . . A Copper Dial plate of 3 ft. Diameter painted Black with Gold figures & Moulding . . . To a Man fixing the above 43 Days with Expenses &c, & Expenses for self going to Oxford Sundry Times . . . To a Bell wt. 2c. 1 qr. 13 lbs. 0 oz."

The escapement is a recoil and the pendulum beats 2 seconds reaching through a floor into the room below. It is enclosed in a narrow shaft which conceals most of its length but reveals that the gridiron is a simple arrangement of compensating rods. The rack-striking train is not now used; its clock-bell is dead hung. On the setting dial is inscribed, *Made by Aynsth & John Thwaites Clerkenwell London. By the direction and under the inspection of Wm. Harrison A.D. 1792.* This latter is presumably William Harrison of Charlbury (q.v.). There is no record of an earlier clock.

OXFORD — Wadham College

The Bursarial disbursement accounts of Wadham College, which begin in 1660, have yielded a complete history of the maintenance of the College clock for 200 years from Michaelmas 1670, but the archives have no record of its acquisition. Traditionally it is supposed to have been presented by

Christopher Wren and to have been designed by him; it was always described as Wren's clock in guide books, etc., but is now, since 1947, in the History of Science Museum, Oxford. From the earliest entries in the Bursar's Accounts and from contemporaneous facts about the clock of the church of St. Mary the Virgin (Beeson, 1961) it is reasonably certain that the clock was made by Joseph Knibb.

The first entry in the Wadham Accounts states, *Knib for keeping ye Clock for one year ending at Mich. 1671 . . . £1;* so again in 1672; in 1673, *To John Nibs for minding ye clock . . . £1.* Thereafter John Knibb was paid £1 a year until 1721 — he died in 1722. Joseph Knibb, who matriculated as a University servant in 1667 and obtained the freedom of Oxford in 1668, departed to London in 1670 and John obtained his freedom in 1673, hence the mention of his name first in that year. Except for the renewal of clock lines in 1690, 1698, 1705 and 1709 the clock needed no repairs until 1716. In the Accounts for Xmas to Midsummer 1716 there was a charge *Mending ye Clock . . . £7 0s. 0d.* This was followed by supplementary charges for "mending" of 1s. in 1717 and 12s. in 1718. Nothing in the Accounts helps one to identify the details of the mending which from their cost must have been extensive. A dial is first mentioned in 1721, *To Mr. Cole for fixing and proving ye Dial agst Chap . . . 16s. 6d.;* there is no charge for connecting up with the movement, and this reference may be to one of the sundials on the Chapel wall buttress.

The custodian who succeeded John Knibb was John Free, formerly apprentice of John Knibb, who started business on his own in 1709. He was paid at the increased rate of £1 10s. 0d. a year to 1726, when he died. His widow, Penelope Free, carried on the business and was paid £2 10s. 0d. a year from 1730 to 1734 for the care of the Chapel clock and one other in the Warden's Lodgings. During this period only £1 17s. 6d. was spent on repairs, in 1733. This may refer to a brass middle wheel in the going train which from its distinctive crossing-out could have been supplied by Edward Hemins of Bicester (q.v.).

Between 1735 and 1750 the annual rate for "winding" or "looking after" both clocks was £2 10s. 0d. paid to James Oakley, clockmaker (q.v.). From 1751 to 1754 and half of 1755 Robert Denton, clockmaker of Cornmarket Street, was responsible for winding at £2 10s. 0d. a year; repairs cost 18s. In 1753 the Convention of the College decided that *Handsome Inward or Inside Doors be made to the Chaple and a better Way up to the Clock.* This work may have involved repositioning the movement and the weight pulleys, etc. At any rate, in the Accounts for Midsummer-Xmas 1755 are the items, *Newmaking the Clock pd. Reynolds £7 7s. 0d.,* and, *His Bill July 31 1755 £1 19s. 6d.* and also £1 for winding the clock for half a year.

Thomas Reynolds of Holywell Street (q.v.) repaired and maintained other turret clocks in Oxford from 1745 onwards. On the Wadham College clock he placed an indicator dial with his own name on it, and about the same time made a turret clock for Swalcliffe Church which has a frame of the same construction and finials of the same design as in the Wadham clock (see figure, p.25 for comparison of the finials). This might suggest that the "newmaking" by Reynolds included a completely new frame for the old movement of the Wadham clock but other evidence confirms that the frame is the original work of Joseph Knibb.

After 1755 the clock appears to have been wound by one of the College staff for about 40 years. At any rate only occasional payments to Reynolds were recorded for cleaning and small repairs varying from 2s. 6d. in 1776 to £1 3s. 6d. in 1765. His last bill in 1783 was £1 1s. 0d. for *cleaning and repairing the Clock and adding new Pullies.* In the same year and until

1794 the College employed for repairs John Hawting, clockmaker (q.v.); his bills amounted to £2 3s. 0d. But in 1795 Reynolds was back again on an annual contract of £1 1s. 0d. for care and cleaning which lasted until 1800.

From 1801 to 1810 this payment was continued annually to his partner, Thomas Earle (q.v.). He received also £3 18s. 6d. for repairs (1805-1807). In 1811 George Rowell of Broad Street (q.v.) was given an annual contract of £1 for care of the clock. This was taken over about 1835 by Richard R. Rowell of the same firm and continued unbroken with remarkably few small repairs until 1870.

In March 1870 the College decided that the clock was no longer serviceable and ordered a new one from Dent and Co., London, which cost £55 18s. 11d. Apparently the last straw was the breaking of the circular ratchet on the great wheel of the going train which put an end to the daily winding. The dial of the new clock was erected on the Chapel wall in 1873 in a carved stone surround surmounted by a winged cherub head under a gabled pediment. The arms of Wren and of the College are painted in shields on the wooden dial board.

After storage in the ante-chapel for about 70 years the clock was transferred to the History of Science Museum. In spite of the various unexplained expenditure on repairs the movement is not substantially altered in structure. (See Plate 5, figs. 7 & 8). The frame, 21 × 10 × 14 inches, is unusually small and well-finished for a turret clock of 1670. The narrow sides are each of one piece; the pairs of front and back long members are permanently welded to the sides and there are no cross-pieces; square screw-nuts are used to fasten the 6 vertical pivot-bars. The ribbon scroll finials are directed inwards (see text figure, p.25). The wheel work of the 2 side-by-side trains is brass, and the end plates of the wooden barrels are brass; and the great wheel ratchets are slender rings not plates. In the striking train the 2nd wheel is original, its locking-plate is external, the fly is 2-vaned and control is by peg and lever. In the going train the 2nd wheel is an 18th century replacement. The one-second pendulum is hung from a threaded rod and wing-nut in the typical Knibb fashion, with a small regulating dial on the cock. The anchor has a long neck and spans a wide arc of the escape wheel. It may be noted that one end of the anchor arbor is marked with incised rings as are also the ends of the barrel arbors. A large indicator dial is inscribed, THO REYNOLDS OXON.

This clock is the earliest example with an authentically dated anchor escapement and antedates that by William Clement made in 1671 (Beeson, 1957).

OXFORD — Worcester College

The clock over the entrance gateway, made by Thwaites and Reed in 1856 is an 8-day, hour, rack-striking movement with a dead beat escapement and a heavy one-second pendulum in a frame of cast iron, cylindrical, corner-pillars ending in ball finials signed on the going barrel plate and the setting dial, " Thwaites & Reed Clerkenwell 1856 ". It serves two circular dials by direct horizontal connecting rods. The drive and the rewind were electrified by Thwaites and Reed in 1961.

OVER WORTON — See Worton, Over

ROUSHAM HOUSE

In the Stables Building is a clock by John Davis, son of John Davis of Windsor (1678-89) who made the carillon clock in the Curfew Tower of Windsor Castle.

The Rousham clock of 2 side-by-side trains in a simple frame has a one-second anchor pendulum and locking-plate striking. The striking release lever is raised by a spiral cam or worm on a vertical shaft. It serves two dials in a cupola. The indicator plate is signed, I. DAVIS WINDSOR 1760.

SHELSWELL PARK
The clock turret on the stables has a bell dated 1757 (Sharpe, 1953).

SHIPTON UNDER WYCHWOOD — St. Mary
An old clock is still preserved. The tower has a slate dial.

SIBFORD GOWER — The Holy Trinity
The church was built in 1840 and has at the west end a convex circular dial with an open turret above containing one bell. The indicator dial on the clock is inscribed, " Rev. T. S. Morrell, Perpetual Curate, John Hitchcox Richard Wilkes, Churchwardens . . . T. Strange Banbury 1841 ". The provenance of the clock is not known and it has been almost entirely rebuilt in 1841. The old bar frame, 25 × 16 × 23 inches, has 4 finials recurved inwards and finished with brass flask-shaped knobs. The going train has a semi-dead-beat anchor escapement, a pendulum cock with a heavy screw device for location as a form of beat-screw. The pendulum beats $1\frac{1}{4}$ seconds and has a wooden rod of 2 × $2\frac{1}{4}$ inches thick which takes the crutch peg in a slot. In the striking train the great wheel has 12 hammer-pins flattened on one side; and the peg and lever warning gear has been redesigned and resited. Pinions integral of 7 and 8 leaves, and rope-driven barrels.

SOUTH NEWINGTON — See Newington, South

SPELSBURY — All Saints
The early Churchwardens' Accounts deal only with revenue but include an item in 1574 for *ye clock wherle . . . xxd*. It is not clear if this is a sale or a gift for purchase.

STANTON - ST. - JOHN — St. John the Baptist
The clock in the tower was given by Mrs. Jane Yerke in 1730 (V.C.H. vol. V). Churchwardens' Accounts are available only from 1788 to 1823 and record annual payments of £1 5s. 0d. up to 1813 and of £2 2s. 0d. from 1814 to 1823 for winding and keeping the clock in order. It disappeared in recent years. The present clock is by John Smith and Sons, Derby, 1925.

STEEPLE ASTON — See Aston, Steeple

STOKE LYNE — St. Peter
The Oxford Archdeaconry Papers for 1757 record orders to repair the clock.

STONESFIELD — St. James the Less
This church is stated to have acquired in November 1743 an old clock made in 1543, which was originally in a nearby manor house that was demolished. The clock taken from the church is now in Judd's Garage, Wootton, north of Woodstock, extensively rebuilt, in working order and serving a modern two-hand dial.

SWALCLIFFE — SS. Peter and Paul

In the tower is a disused clock the indicator dial of which is inscribed, THOMAS REYNOLDS OXFORD, and a plate on the frame reads, THE GIFT OF WM. WICKHAM Esq. The frame, 32 × 15 × 21 inches, consists of two end-sections each of one piece with the front and back horizontal bars welded to them as in the Wadham College clock; the 4 finials are similar snub-ended scrolls (see text-figure p.25). The going train has a circular iron ratchet-plate inside the great wheel with the spring and pawl fixed to the wheel spokes. All the wheels are of brass with 6 straight arms. The anchor has a short stalk and its crutch carries a peg. The pendulum cock is bolted to the vertical pivot bar and is without adjustment for the suspension of the pendulum which beats $1\frac{1}{4}$ seconds. Wheels of the striking train are of brass with crossing-out as in the going train. The brass outside locking-plate has $\frac{1}{4}$, $\frac{1}{2}$ and $\frac{3}{4}$ hour pins which serve a separate chiming train. Hour-striking control is by peg and lever, and free-wheel fan with large vanes.

The separate chiming unit is mounted on a wooden 4-legged frame and comprises a large wooden barrel of numerous staves and 4 stout spokes, 11 rows of iron pegs and 11 hammer-levers mounted on the frame. A train of 3 iron wheels with bifurcated and welded spokes is geared up to reduce the winding effort. The circular barrel-ratchet and curved spring-click are of the same pattern as in the clock. The fly arbor has at one end a long 4-armed free-wheel fan and at the other a short 2-armed subsidiary fan. An 8-pointed star forms part of the stop work. The Oxford Archdeaconry Papers for 1755 record orders to repair the chimes in Swalcliffe church. The chiming unit may therefore be a surviving part of an earlier clock which has been replaced.

Thomas Reynolds made this clock about the time he repaired the Wadham College clock in 1755 and evidently copied the unusual construction of its frame. A comparison of the details in the two movements shows that the Swalcliffe clock is otherwise entirely Reynolds' own design. Heavy stone weights used to drive this equipment are still preserved.

References to the clock in the Churchwardens' Accounts are available only for the period 1776 to 1814 and deal mainly with mending the wires of the chimes. But there are two items in 1809 for " fastening up the face of the Church Clock ", and " helping to put up the clock face ". This dial was replaced by a circular skeleton dial inscribed, " In Memoriam V.R.I", and the clock continued in use for a long time in the 20th century.

The present clock is by John Smith and Sons, Derby; a plate on the movement is inscribed, " The clock in the Tower was given to this church by Muriel Maud Norris of Swalcliffe Close A.D. 1940 ". The church has a ring of 6 bells.

THAME — St. Mary

The Churchwardens' Accounts for the New Town begin in 1442 (partly printed by Ellis, 1901). Many of the payments are qualified by the phrase " for our part " which means the portion of the total chargeable to the New Town of Thame, the balance being paid by the Old Town.

The sexton was clock-keeper and was paid at the rate of 6s. 8d. a year until 1468. Three years' arrears of pay due to him in 1443 show that the clock was working as early as 1440. The first entry for repairs is in the Accounts for 1442, viz. *for makyng of ye clocke . . . vid.* Sums received by the Churchwardens for several years *as for kepyng of ye kloke* appear to have been taxes on the citizens for maintenance of the clock. Fees were likewise charged for the use of the great bell. Up to about 1455 repairs were done by Thomas Smith and Robert Smith (q.v.), a minor job costing 2½d., *mendyng of ye sayll of ye clok & for a wyr to ye same . . . iiijd.*, and larger jobs costing up to 2s. 8d. From 1458 to about 1464 Thomas Cottiswold (q.v.) did the repairs.

After an item in the Accounts for 1465, viz. *Ité we payde a nernyst for mendyng of ye clocke . . . xld*, there are some years missing but an undated section may refer to the occurrence of an accident about this time. The clock-keeper's pay was stopped, *It' for Cristmas was non a lowyd hym for ye clocke stode stille;* and repairs to the clockroom were needed. *It we have payde to John Cowper for fotyng of ye clocke & for makyng of a flor by to save ye rode loft yif ye pese falle for ii dayes labur & di . . . vd. — It we payde for hys mete & drynke . . . iid. ob — It was payde to Thom's Smyth for gret naylys . . . ob qr — It we payde for other nayls . . . ob — It was payde for timb'r for ye clock fete . . . id ob — It we payde for ii new Gyistys for ye new flor . . . iid. — It we payde for plankes to ye flor . . . iiid.* The reference to an earnest and to meat and drink shows that John Cowper was not a local man. Possibly a new clock was acquired about this time as according to an item for the year 1474, *Ite we payde to Jhon smyth A for Crystmas for mendying of ye clok ye wyche ye smyth of tryng mad for owr part . . . xiid.* There is no entry of the full price of such a purchase, unless it was made about 1465, and there is no previous reference to the unnamed smith of Tring in Hertfordshire.

The Accounts for 1447-1480 *for iij hole yeres & nine weeks beyond* deal mainly with the installation of new organs and nothing is written about the clock. But in 1481 the clock-keeper's wages start again and the smith is paid 12d. *for wyrke about ye clokk.* Later on there was another interruption; the second bell was taken down in 1487 and sent to the bellfounder at Wokingham to be recast but it had to be sent back again the next year. When it was finally rehung the clock had to be set up again, so for the year 1488-9, *It' sol. p'r le settyng up horrilogii quando sc'da campana fuit le new hanged . . . iijd.* Thomas Powlen also was paid *for makyng ye armys to ye grete pece of ye clokke . . . iiijd.* It is evident that the 15th century clock was sited in the first or second storey of the tower and not at ground-level, and that the line of fall of the weights was on the east side above the rood-loft.

For the period 1489 to 1527 the Accounts no longer exist and only one reference to a clock is on record, as in 1502 William Smith of Bicester repaired the clock for 3s. 8d. in full payment of a bargain made with him.

The next period of the itemised Accounts that has been preserved is for 1527 to 1665, thereafter only totals of receipts and expenses have been entered for a few more years. In 1527 the clock was being kept by William Barret for 6s. 8d. a year and he continued to do so until 1539. During these twelve years only small repairs were needed. In 1529 is an item, *pd for mendyng of a Barell for the clocke & naylyse . . . xiiijd.*, and in 1530, *Itm pd to ye Clocke maker for mendyng of ye clocke . . . iis. viijd.;* this is an early use of the term "clockmaker", and indicates a craftsman called in from elsewhere. Repairs by local men were e.g., in 1535, *pd to the Smyth for mendyng of the clocke And for a boy for the same Days . . . xijd.*, and in 1537, *pd to Geffray ffounteyne for mendyng of the Clocke . . . xijd.* In 1539 the clockmaker was again employed and was paid 10s.

The next year John Grene became the clock-keeper at 6s. 8d. annually. In 1541 fifteen pence was paid *to the Smyth for naylys to mend the great bell & for mendyng of the hamr of the Clocke & the vice whele.* Thomas Peres or Perry replaced Grene in 1542 and was paid at the same wage until 1548. The clock, however, needed repairs costing 8s. and wire costing 8d.

From the Accounts for 1543 we learn that there was also a clock in the Moot Hall in Thame (see later under Moot Hall). In 1544 Brown, the smith, was paid 9d. *for Mendynge the hamer and the guge [sic] of the clok,* and in 1547 the clockmaker had to do repairs costing 3s. 4d.

In 1548, the second year of the reign of Edward VI, after an injunction to realise church plate and other valuables when needed to pay for keeping the fabric in repair, the churchwardens began to sell plate, jewels and bells. The great bell was removed from the tower, broken up and sold in London for over £36, together with plate worth £38 10s. 0d. As a result the clock which was striking on the great bell had to be shifted, thus, *Item payd to peter for ye Removyng of ye Clok frome ye great Bell to ye iiijth bell . . . vs.* — *Itm payd to John Browne for Makynge the Irons & for a staff x ffoot longe to ye same . . . viijd.* — *Itm payd to ye Cloke Maker for Mendynge the Whelles and ye fflee of ye clok . . . iijs. iiijd.* — *Itm for his Bordynge & bedd by ye space of ij Days . . . viijd.* Thomas Perry, the clock-keeper's wage was then raised to 10s.; he also had the duty of ringing the curfew, 2s. But in 1551 another large bell was sold for £34 and some more plate. How long this went on is uncertain but the Churchwardens for the years 1547 to 1552 were later accused of complicity. The clock escaped and repairs costing 10s. in 1553 and 20d. in 1554 were done by Robert Johnson, after which references cease and it may have remained out of use for some years.

In 1562 it seems to have been decided that a new clock was needed: *Itm pd to hughe fuller for goynge for the clockemaker to Dorchestre & Nettilbedd . . . xiiijd.* — *pd to the clokmaker in money . . . xxijs. viijd.* — *pd for a keye for the clok . . . iiijd.* — *payd more to the clokemaker . . . xvjs.* — *pd to Spryngold for Ropis for the clok . . . vis. iiijd.* — *pd to the plomber for Castinge of Payses for the cloke . . . xvjd.* — *payd to the Clokemaker for Makinge ye cloke . . . xxs.* — *payd to John Wared ye Carpendre for Work in the Church . . . ijs.* The low price of £3 8s. 8d. for the clock, etc., represents the half share paid by the Churchwardens of New Thame and may be compared with the expenditure in 1573. The maker, who remains unidentified, was employed again in 1565, *Payd to the clokmaker for Mendinge of bothe the Clokks . . . vijs. viijd.*, and, by Old Thame in 1566, *Pd to the clockemaker . . . vs. iiijd.*

During the next six years the only clock items are for mending in 1567, 12d., to Thomas Springold for a rope in 1568, 6s. 8d., for a clock line in 1570, 6d. and *for a bolte & a plate for the clok to Couwlys . . . viijd.* in 1571. Cowlys was a locksmith. For the year ending 8 April 1573 the Accounts of New Thame have one entry, *Itm payd to Robert heth for my portion for makyng the clok . . . xxxvis. viijd.,* and those for Old Thame and Priest End the following entries: *Item for tymber aboute the Clok . . . ixs.* — *Item for Di a hundred borde . . . iis. ijd.* — *Item for nayles . . . xd.* — *Item for lath and layth nales . . . xijd.* — *Item more for a borde about the dyall . . . xijd.* — *Item for carpenters wages about the place & the frame of the cloke x days & di & the whele & for mending of the seate . . . xs. vid.* — *Itm payd to the cutler for ernest of the cloke . . . iiijd.* — *Item mending the Lok & the Key to the clok house . . . iijd.* — *Item payd to heth xxxvjs. viijd.* Thus the total expenditure on the clock was about £4 18s. 0d. Next year New Thame Accounts recorded, *Item payd for the Dyall in the Church . . . vis. viijd.* and Old Thame accounted for, *Item payd to Thomas Springold for a rope for the Cloke . . . iijs. iijd.* A dial costing 6s. 8d. on a board costing 1s., if actually erected *in* the church along with a new movement confirms that there were two clocks in use. Previously in 1565 there was a reference to *bothe the clokkes,* possibly one of which was in the Moot Hall. Later in 1582 there is one called the town clock, viz. *Item for mending the towne clocke . . . iis.*

vjd., paid by New Thame, and, *Item a pece of elme to mend the clok ... vjd.* paid by Old Thame. And again in 1583 Old Thame paid, *Item to Couwlys for the church clok ... xvjd.*, and New Thame, *Item for wyer to mend the clok ... id.* It is therefore difficult to decide which clock is concerned in each of the items of earlier expenses.

In 1585 there was further trouble with the bells and the third was recast at Reading; this involved more work on the clock in the church. In 1586 Old Thame paid, *to Tho Heyley ffor mending the bell whele and putting up the clok house and mending the seats in the church ... ijs. vjd.*, and, *Item to Mr. Ayer for mending the clok ... vs.* New Town also paid Mr. Ayer for mending the clock, 10s., and at the same time 5s. to Walker for mending the clock; here, it would seem, two clocks are concerned. No more entries occur to clarify the allocations except for one in the Accounts of New Thame in the year ending 20 April 1589, viz., *It' payde for mendinge the little clock.*

From this year, 1589, until 1611 only summaries of the totals are accounted and then in 1613 details are again recorded. *Itm to Edward Herbert for mending ye Clock ... ijs.*, and, *Itm for the mending of ye Church clock ... ijs. vjd. — for a Lyne for ye clock xljd.* In 1614 again, *Itm for a Lyne for the Clocke ... iiijs. iiijd.* In 1616, *Itm for mending the Clock ... vs. ijd.* and similarly in 1617, 5s. No more items occur until 1625 when there are two, *To Willm Smith for mending the Clock ... vs, — for a Rope for the Clock ... iiijs.* William Smith mended the clock again in 1630 for 3s. 4d. Repairs again in 1632 for 6s. 8d. and in 1634 for 1s. 4d. and the purchase of a rope from Thomas Springall for 5s. In 1641, *It paid to Richard Smith for mending the Church Clock ... 8s. 6d.* and at Easter 1642, *Pd Thomas Harris for mending the cloke and for a locke for the Steple Doore ... 2s.*, and *Pd Richard Smith for mending the Clocke ... 3s.* In 1649 are two rather incompatible items, *Itm for mending ye clock ... £1 0s. 0d.* and, *Itm for mending ye clock ... 4d.* In 1652 the clock was again mended for 5s. Finally, the last entry in this period of the Accounts is for 1657, *paid to Wm Parker for mending ye church clock ... 2s.*

A new book of Accounts starts in 1772. Between that date and 1665, probably at the end of the 17th century, a new clock was acquired, the frame and striking train of which now form part of the existing church clock. This is confirmed by the first entry for 2 December 1772, *John Tomlinson To mending the Pendulum of the clock ... 1s. 6d.*, and in the following January, *To a spring to the Fliers of the Clock ... 1s. — To stud to the Locking Plate and mending the Wyre of Do ... 1s. 6d.* And in 1775, *John Tomlinson To Bridge to the lock of the Clock House ... 6d.* More substantial repairs were done by Tomlinson in 1782 costing £8 10s. 2½d. and he cleaned the clock in 1787 for 10s. 6d. and in 1788 for 14s.

Next year the church underwent thorough repairs including extensive alterations to the clock by William Buckland (q v), which cost £56 8s. 1d. What these were it is not possible to deduce in view of further modifications done in 1828. Between 1789 and 1819 several bills were paid to Buckland and to John Tomlinson and in 1819-1821 to Job Tomlinson also.

In 1828 and 1829 the considerable work that was done on the clock is attributed to Job Tomlinson (q.v.); it is covered by the following entries. *1828, Job Tomlinson Bill for Vane, £14 0s. 6d.; part of Bill of £65 9s. 0d., for new copper figures and repairing Clock to Church, Painting etc., £35 9s. 0d.*, and, *10 August 1829, Job Tomlinson Clockmaker remainder of his bill of £65 9s. 0d. paid £30.* This clock is now sited in the lofty ringing chamber in a boarded house surrounded by a narrow gallery supported on two pine beams between the north and south walls of the tower and accessible by a long ladder. The pendulum swings in a triangular boarded casing below the platform. This unusual construction probably dates from Buckland's time and the work on the going train and transmission to the dials is mainly due to Tomlinson. Plate 8, fig. 12.

The iron frame has corner-standards with top and bottom buttress mouldings and outwardly stalked finials ending in the square knob with a central nipple characteristic of late 17th century design. The sides are braced with X-shaped pieces fixed, as are the pivot-bars, with nuts. Of the two side-by-side trains the striking part has the original forged iron wheels and an internally toothed locking-plate to which has been added a slipping clutch and a pawl on its pinion. The going train barrel is original but with an added ratchet plate for a maintaining detent; the second wheel, the escape wheel and the anchor and the bevel wheels, etc., to the connecting rods are of 19th century brass. The pendulum beats two seconds with a wooden rod and a heavy bob. Its suspension is unusually elaborate.

A block gripping a pair of short springs has a horizontal threaded rod allowing lateral adjustment in the cock; the springs support a square iron frame through the base of which depends another screwed rod with a winged nut for vertical adjustment; the fork of the crutch embraces this rod and is also adjustable laterally and finally the top of the crutch is adjustable on the anchor arbor. The hand-setting ring is inscribed " Repaired by Job Tomlinson 1828 T. Hedges N. W. & T. Seymour Church Wardens ". The weights have recently been increased and hung on wire lines through a doubled pulley-system so that the clock now runs for a week.

Winding the clock from 1830 to 1840 cost about £2 a year rising to £2 2s. 0d. in 1842 (W. F. Miles), £2 10s. 0d. in 1847, and £3 in 1859. Job Tomlinson kept it in order until as late as 1865.

THAME — The Moot Hall

The Church Accounts for New Thame and Old Thame show that there was a clock in the Moot Hall early in the 16th century. An entry for 1543 is *Itm payde for mendynge of the Clook in the moote halle . . . vs.* In 1565 an item records payment of 7s. 8d. to the clockmaker for *Mendinge of bothe the clokks,* and in 1582 the New Town paid 2s. 6d. for mending *the towne cloke.* The entry in 1589, *It payde for mendynge the little clocke . . . vs.* must refer to the same one.

There is no subsequent history.

WARDINGTON — St. Mary Magdalene

The present clock is by John Smith and Sons, Derby, 4 July 1900. A few records occur of an earlier clock mainly in the 19th century; the older books are lost.

WATLINGTON — The Town Hall

The brick Market House or Town Hall was built by Thomas Stonor of Stonor in 1665. An indenture of 28 May 1664 states the building shall contain a room for a clock house for the public use and benefit of the inhabitants of the town of Watlington. There is no subsequent documentary history of this clock which still exists. It has an iron bar frame with two trains side-by-side. The corner-standards have late 17th century mouldings at the top and bottom, and end in scroll finials; decorated nuts are used for fastening all the bars. The wheels of both barrels and the locking-plate are of forged iron; the remaining wheels are of brass and not all of the same period. A seconds pendulum and a recoil anchor escapement enable the movement to run on daily winding. Hour striking is controlled by peg and lever. Plate 8, fig. 13.

The setting dial is inscribed, " IMPROVED by Wm. James Watlington " (he has not been identified) and may refer to an early 19th century addition of motion work leading to a two-hand convex dial on the outside wall; or to a commemoration of Queen Victoria's Silver Jubilee. The present copper dial, however, was fixed in 1929.

Originally sited in the centre of the room below a square wooden turret housing the bell, the clock may, according to local belief, have had a verge and crown wheel escapement; in this case it is an unusual example of a layout of side-by-side trains for a verge escapement. Two unused holes on the top cross-bar and a replanting of the pivots of the striking lever arbors may be evidence of a later conversion.

Rawlinson, writing about 1720, recorded, *The inscription upon the Markett Bell commonly called the Cross Bell hanging in a wooden turrett upon the Market House on which strikes the town clock: This Bell Belongeth to the Inhabitants of Watlington Ralph Walis Richard Gregory 1665.* It was recast in 1764 with the original lettering repeated and the addition of, *Thos Hine Ino Ewes Constables 1764.*

The photograph in Plate 8, fig. 13 was taken shortly before the clock was overhauled by Thwaites and Reed in 1961.

WIGGINTON — St. Giles

An unusual clock for which there is no documentary history but a local tradition that it came from Bloxham church about 1880.

The flat bar frame, 30 × 18 × 23 inches, is fastened by square nuts. The corner-standards have no true finials but plain tapered extensions. The two trains are end-to-end pivoted in the middle of the frame in separate bars. In the going train the iron main wheel has a horseshoe-shaped spring click working on its spokes. It has been converted to an anchor escapement which has needed 2nd and escape wheels of brass and integral pinions of 9 and 12 leaves. The pendulum cock has one square headed nut and the rod has a wide loop through which the winding square protrudes, and a lenticular bob with a winged nut for regulation. The beat is $1\frac{1}{4}$ seconds.

The striking train is largely if not entirely original with the barrel at the top and the fly at the bottom. The pinions of 6 and 8 leaves are integral with the almost circular tapered arbors. There is a check pin on the fly arbor instead of a hoop wheel, and the release lever pivots on the side of the frame being raised by a peg on the going train barrel arbor. The iron locking-plate with offset spokes is planted in the space between the two trains. The bell crank-lever is pivoted at the top of the frame in an extension of a vertical bar and an extra lateral bearing. Two large stone weights are hung on the driving ropes. There is no clock dial.

It is noticeable that the going train arbors are all slightly out of truth and oblique to the horizontal but the frame is not distorted. As the frame is not fastened by wedged tenons and the welded spokes of the iron wheels are set on their arbor with square locking nuts the clock has no features which might date it earlier than the mid-17th century in spite of its unusual construction.

WOODSTOCK — See Blenheim Palace

WOODSTOCK — St. Mary Magdalene

In 1792 John Briant, Hertford, supplied a striking clock.

WOOTTON — St. Mary

The earliest mention of the old clock is in the Churchwardens' Accounts for 1707. The old dial on the south wall of the tower was of the usual diamond shape. In 1876 the roof of the chancel was raised, the tower was repaired and the clock was sold as scrap for £1 in 1877. The present clock is by John Smith and Sons, Derby, installed in 1877.

WORTON, NETHER — St. James

A somewhat altered clock with a bar frame, 22 × 12 × 19 inches, and two trains and an anchor escapement for a one-second pendulum. It is said to have been originally at Heythrop House and thence removed to Sandford St. Martin whence it was rescued by a schoolmaster of Nether Worton. It is possibly the mid-18th century clock at one time in Sandford St. Martin church.

WORTON, OVER — The Holy Trinity

An undated sanction was given by Charles Talbot, B.L., Vicar-General from 1714 to 1733, *to empower ye Church Wardens of Overworton to sell ye said old useless bell and with the money arrising to provide a Church Clock applying ye remainder to ye use of Church.* The church has no tower and no clock.

WROXTON — All Saints

A clock made in 1846 was restored in 1876.

YARNTON — St. Bartholomew

According to the Churchwardens' Accounts (Stapleton, 1893) the local blacksmith and wheelwright, William Burbridge, was employed in 1611 on a half yearly wage of 6s. 8d. to keep and repair the clock and ring the bell. This clock was replaced 30 years later, from which fact one may conclude that it was made in the 16th century and quite possibly near the beginning. At this time there was only one bell, the sanctus, in the church and the tower was in course of extensive reconstruction at the expense of Sir Thomas Spencer, Lord of the Manor. By 1620 five new bells had been obtained and a completely new bell frame was made, which had to be altered soon after to house 6 bells. The work was not entirely satisfactory and frequent repairs were needed in the twenties and thirties.

In 1641 the new one-hand clock was bought for £5 18s. 0d. plus the scrap value of the old clock. Seven days work on the clock house was needed when it was installed. Repairs between 1648 and 1658 were done by Thomas Ranklyn, smith of Oxford (q.v.) for £2 7s. 0d. Again in 1651 the smith and his boy came to Yarnton to mend the clock, but unfortunately the Churchwarden had to pay almost immediately, *For mending him again when the Rogue pulled him in peeces,* and, *For carrying the wheeles on my back to Oxford three times to mend,* which, however, cost only 4s.

In 1665 repairs to the clock, which involved taking down the hand and setting it up again, cost £1, after which it functioned well for 15 years. Then in 1680 the clerk paid 1s. 2d. *for mending the thing that winds up the clock.* Was this the handle or part of the barrel ? However, a year later a serious breakdown occurred. The Churchwardens called in " the Witney clockmaker " and " the Oxford clockmaker " and " the painter Hamilton " to inspect and estimate for the job, but they gave it to George Harris of Fritwell (q.v.). The hand was again taken down and the clock was put on a horse and sent to Fritwell with a man in charge (cost 4s.). Harris put the clock in order in 1682 for £1 10s. 0d. The dial was painted by Christopher Matthews of Oxford (q.v.) for £1 10s. 6d., boards for it cost 9s. and the fixing of the dial on the tower cost 9s. 10d., which included the workmen's bills at the Six Bells inn.

In 1685 the pendulum was mended for 1s.; this first mention of a pendulum suggests that the crown wheel and foliot escapement of 1641 was converted to anchor by Harris in 1682, and that the opinions of the Witney and Oxford clockmakers had been taken on the question of conversion.

In 1703 John Knibb of Oxford did repairs costing 13s. and in 1716 a new hand was fitted at a cost of £1 16s. 0d., probably not a minute hand as nothing is mentioned about altering the dial.

The final entry in the Accounts is for repairs in 1730 done by Aress (sic) of Kidlington, and there is no further information for the 18th century.

SUNDIALS & SAND-GLASSES
SUNDIALS

SUNDIALS were widely used throughout the 16th, 17th and 18th centuries, but, although needed to check the time-keeping of turret-clocks, references to their use in the financial accounts of Oxfordshire churches and colleges are not numerous. They concern mostly mural dials. Further information about those that have disappeared from Oxford colleges can be gleaned from 17th and 18th century engravings. R. T. Gunther (1923) has assembled many pictorial representations from these sources, particularly from David Loggan's prints of 1675 and those of W. Williams in 1733.

At the beginning of the 16th century the problem of equating apparent solar time with equal hours attracted attention. Nicholas Kratzer, astronomer to Henry VIII, was then a Fellow of Corpus Christi College as Professor of Mathematics, astronomy, geography and kindred subjects. While there he devised two elaborate pedestal or column dials, one for his college and the other for St. Mary the Virgin (see below).

ADDERBURY — St. Mary

Churchwardens' Accounts: 1824, Gt. Cakebread for a Sun diall . . . £3 3s. 0d. A print of 1797 shows a square dial set on the west side of the south transept window. The new one was centred in the gable above the window and survives.

BANBURY — The old church of St. Mary

Beneath a parapet on the south wall of the tower was a dial inscribed, *Proprio sumpti Caroli North ann. 1704.* This was destroyed with the church about 1791.

BLENHEIM PALACE — Woodstock

When Langley Bradley supplied a turret clock for Townsend's tower in 1710 he specified four horizontal dials which were set up on stone pedestals in the palace gardens. They were made by John Rowley, the instrument maker, three costing £60 each and the fourth £50, plus £5 for Rowley's journey to Woodstock to set up the dials and take levels. The pedestals, each about 4ft. high, were carved by Henry Banks who charged £19 7s. 3d. All the dials are different, one showing an equation table, another a moon dial, the third a projection of the sphere and the fourth diagonal minutes, azimuths and difference of meridians. The gnomons are pierced and enriched with the Marlborough arms and coronet, and trophies are engraved on the dial plates (Green, 1951).

BODICOTE — The Weeping Cross

A dial was erected on the mutilated pedestal of the Weeping Cross in 1730 and inscribed, *Given by Mr. Richard Wise Clockmaker in London Anno Domini 1730.* Nothing now remains (Beesley, 1842).

BURFORD — St. John the Baptist

Churchwardens' Accounts: *1670, Pd. Christopher Kempster for a diall sett over the Church porch . . . £1.* Christopher Kempster, whose monument is in the Leggare Chapel of Burford church, was employed as master mason in building St. Paul's Cathedral and in the erection of several churches in London after the Great Fire of 1665. He owned Kitt's Quarries southwest of Burford which supplied stone for some of Wren's churches (Monk, 1891).

CHINNOR — St. Andrew

Churchwardens' Accounts: *1664, Item paid for erecting a new Diall over the Church Doore . . . 6s.* — *1687, Item paid for ye stile of ye Church Diall . . . 3s.* The church has no clock.

CLAYDON — St. James the Great

Churchwardens' Accounts: *1746, pd. for panting ye diall . . . 4s.* — *1761 July ye 4, paid for painting the Sun Diall . . . 5s.* The clock has no dial.

GREAT HASELEY — St. Peter

Chuchwardens' Accounts: *1667, Paid for flourishing the Sundyall . . . 5s.* — *1675, Paid for flourishing the dialls . . . 2s. 6d.* — *1689, A part for ye Sun Diall . . . 6d.*

OXFORD — All Souls College

The elaborate mural dial, showing the minutes, formerly on the south wall of the Chapel and now on the south wall of the Codrington Library, has a cartouche of arms of the college, festoons and a segmental cornice, incorporating a winged cherub head above and a motto below, *Pereunt et imputantur*. The colouring has been skilfully restored. It was made in 1658-9 by William Bird, the stonecarver, of Holywell Street, Oxford, at a cost of £54. As Christopher Wren was then Bursar of the College it is surmised that he designed the dial.

Loggan's print of 1675 shows a small dial on a chimney, since disappeared.

OXFORD — Balliol College

Three dials pictured by Loggan in 1675 have disappeared, two south-facing mural dials and one globe dial standing on a column in a walled garden.

OXFORD — Botanic (Physic) Garden

In the spandrels between the heads of the lower niches on the Danby Gateway are two incised dials facing southwest. They may have been made by Nicholas Stone in 1632 or later ca. 1694.

OXFORD — Brasenose College

The large restored mural dial on the south west front of the Old Quadrangle dates from 1719. The arms of the college are on a shield in the upper central area. It originally cost £9 and the painting and gilding cost £7 7s. 0d.

OXFORD — Christ Church

There was a mural dial on the southern square turret at the east end of the Cathedral in 1705. That on the end of the west wing of Peckwater Quadrangle, marked for the hours, VII-IV, dates from about 1750. Two mural dials on the Cathedral in 1673 have disappeared.

OXFORD — City Council

In 1672 a handsome dial was set up at South Bridge, Oxford, at the City's expense. It was carved by Richard Wood, stone cutter, for £1 15s. 0d.

OXFORD — Corpus Christi College

Nicholas Kratzer (q.v.) designed an elaborate polyhedral dial of twelve faces surmounted by a sphere, which was erected ca. 1520 in the garden of Corpus Christi. It has disappeared but drawings and descriptions of it are included in Robert Hegge's Treatise of Dials and Dialling written ca. 1625-1630. Hegge's drawing is reproduced by Gunther (1923) and another drawing of it is in a MS Book probably written about 1630, now in the History of Science Museum; the two pictures differ only in minor details. Hegge describes the dials as Aequinoctial, Polar, two-faced Vertical, Convex and Concave and adds, *In other dials neighbouring Clocks betray their errours but in this consort of Dials informed with one Soul of Art they move all with one motion.*

Charles Turnbull (q.v.) designed an elaborate column sundial for his college. It stands now in the centre of the Front Quadrangle and bears two dates, 1581 and 1605, the former being that of its construction and the latter that of subsequent additions. In 1698 Townsend, the mason, repaired it for £12, and the column was further restored and repainted in 1876, 1907 and 1937. It now consists of a late 17th century pedestal with defaced carvings, an octagonal shaft with moulded base and enriched capital supporting a rectangular block with a cornice and pyramidal capping; near the top of the column on the south is a gnomon and painted on the shaft is a perpetual calendar with the inscription, *Horae omnes complector* — all added in 1605. The block above has carved cartouches of the arms of Elizabeth I, the University, the founder Richard Fox and Bishop Oldham. The lower part of the scroll-work frame round each shield acts as a gnomon to the dial face engraved below it. On the cornice are four mottos, *Posui Deum Adjutorem meum, Est reposita Justitiae corona, Gratia Dei mecum, 1581. Est Deo gratia.* The capping has two dials and is finished with a pelican standing on a sphere (Oxford Roy. Comm., 1939 and Gunther, 1923). A copy of Turnbull's dial is at Princeton University, U.S.A.

OXFORD — Exeter College

A mural dial on the south wall of the tower and another facing west in the Front Quadrangle were figured by Loggan in 1675.

OXFORD — Magdalen College

A west-facing dial on the Library showing hours II-VI was figured by Loggan in 1675. Another on the battlements over the east side of the Library, ca. 1635, probably survived until 1824.

OXFORD — Merton College

The mural dial on the east side of the last buttress against the north wall of the Chapel was probably set up between 1622 and 1650. It comprises four sets of lines, (a) diagonal lines measuring the time, (b) hyperbolic curves measuring the sun's declination, (c) straight lines showing the hours from sunrise, and (d) vertical lines measuring time. The first three sets require a gnomon as index the fourth uses the edge of the adjoining buttress as index. The dial functions only between 6 a.m. and 9 a.m. in the summer months.

Two mural dials, one facing east and the other facing west in the Fellows' Quadrangle, possibly dating from 1610, were figured by Loggan and Williams.

OXFORD — New College

A south-facing mural dial against the end of the Hall was depicted by Williams in 1730.

OXFORD — Oriel College

A south-facing mural dial under the middle gable on the north side of the Front Quadrange was figured by Williams in 1733.

OXFORD — The Queen's College

There was a south-facing mural dial over the low window of the Provost's Lodgings, ca. 1673, and a globe dial on a column in a garden opposite St. Edmund Hall, ca. 1675.

OXFORD — St. Edmund Hall

The wooden mural dial on the south front of the quadrangle dates from ca. 1741.

OXFORD — St. Cross, Holywell

The church has no clock. References to the dial on the tower which still exists are: *1667, Payde for makeing the South Diall . . . £1 15s. 0d. — Payde for an Iron to Mend the gnomon of the South Diall . . . 1s.*

OXFORD — St. John's College

In 1643 a dial in the Grove made by John Jackson, mason, cost £12 11s. 6d.; some indication of it is given in Loggan's view of 1675. In 1661 two dials in the Quadrangle which were set up three years earlier by John Dale, Fellow, one on the east side and the other on the north side, were blown down in a storm. The south-facing dial on the Chapel wall was shown by Loggan in 1675 and by Williams in 1733. A small dial marked WS on the west side of a chimney stack and showing the hours I-IIII has disappeared since 1675.

OXFORD — St. Giles

A square mural dial over the south porch door is illustrated in Rawlinson's engraving of 1754 in Peshall, 1773.

OXFORD — St. Mary Magdalene

Churchwardens' Accounts: *1620, Item dressinge the Diall . . . viijs. vjd. — 1688, Paid Mr. Taylor for new dooen ye Sundyall . . . 6s. — for setting hime up againe . . . 1s. 8d. Tho. Crosier for Carrying & fetching ye Diall . . . 6d.* John Taylor, portrait painter, became Mayor of Oxford in 1695.

OXFORD — St. Mary the Virgin

The pedestal dial by Nicholas Kratzer (q.v.) erected in the churchyard in 1520 is recorded in Kratzer's Latin Ms., *De Horolgiis,* preserved in the Library of Corpus Christi College. A manuscript copy of the note is inserted in the Bodleian Library's copy of Tycho Brahe's *Astronomiae instauratae mechanica,* 1598, and *Stellarum octavi orbis inerrantium accurata restitutio,* 1598. It states, *Anno 1520 Ego Nicolaus Krasterus . . . illo tempore erexi columnam seu cylindrum ante ecclesiam Divae Virginis cum lapicido Wilhelmo Easte servo Regis.* The stonework was carved by the master mason William Este (q.v.) and consisted of a cylindrical shaft engraved with mathematical lines and supporting a cubical block with 4 dials, above which was a pyramidal top surmounted by a globe and cross. The dial on the east showed the old

German or Babylonian hours by green lines, the south dial the diurnal hours, the west dial the Italian hours by blue lines, the north dial the times and oblique motions of the sun and moon. It was set up near the east end of the south side of St. Mary's church.

Because it was a good place to attract public attention a condemnation of Luther's doctrines was fastened to this dial by Kratzer himself in 1521, as he recorded, *Eo tempore Lutherus fuit ab universitate condemnatus cuius testimonium ego Nicolaus Krasterus in columno manu propria scripta posui.*

The sundial was kept painted and gilded. The University, for example, gave £1 to the Churchwardens in 1672 *towards beautifying their Sundiall* and Corpus Christi College gave 10s. in 1673. The Churchwardens' Accounts for 1688-89 include, *Payd the Carpenter for Scafolding about the dial . . . 4s. 6d. & for Masons work there . . . 10d. and to the plumber . . . 1s. 3d.=6s. 7d.—Item payd to Xstofer Matthews for paynting and guilding the Diall . . . 16s. & to Mr. Prujeon for drawings . . . 8s.=£1 4s. 0d.—Item payd Mr. Young for a bolt to the Tower & a lock and key to the Vestrie d* [page torn] *to the Dial . . . 4s. 8d.*

A view of the top of it appears in Loggan's drawing of 1675. The wall round the church and the sundial with it were demolished in 1744.

OXFORD — St. Peter in the East

A dial about 3ft. square in a shallow recess over the south door was figured by Ackermann in 1814. It is mentioned in the Church Accounts for 1665, *For beautifying Queen Elizabeths monument and renewing the dyall at ye Church door . . . £1 6s. 0d.* And in 1685, *To Rich Tipping for a cock for the Dial and work about the bells . . . 6s. 6d. — To Lyonell Broughton for a scaffold for the Dyal . . . 3s. 6d. — For playstering ye Dyall . . . 1s. 6d.* And in 1689, *Pd. Mr. Peesly & a labourer for work and the use of the stuff for the scaffold for new doing the Dial . . . 2s. — Mem that Mr. John Prijon and Mr. Henry Wyldgoose nue drawed and painted the dial gratis.*

OXFORD — Trinity College

A mural dial made by Francis Potter (1595-1678) for the College disappeared about 1670. Potter probably also made a globe dial with gnomons for the President's Garden and a pillar dial carved as a pile of three geometrical solids which stood in another small garden. Three more mural dials have also disappeared, one in the Front Quadrangle figured by Loggan in 1675, another in the north Court existing in 1733 and demolished in 1802, and a west-facing dial in the First Court illustrated by Williams in 1733.

OXFORD — Wadham College

The remains of two mural dials still exist on the south side of a buttress of the Chapel, the lower probably dating from 1612 and the other possibly added in the time of Warden Wilkins (1648-1659). The dial fixed and proved by Cole in 1721 may be one of these. A globe on the shoulders of Atlas probably designed by Wilkins about 1653, figured by Loggan in 1688 and by Williams in 1733, was destroyed by a storm in 1753. A horizontal dial on a column standing near the north wall of the Fellows' Garden was made ca. 1730.

STEANE PARK

On two adjoining walls of Sir Thomas Crewe's chapel is a double sundial which was probably added after ca. 1634. The design of the carved swags of the dial show an acquaintance with Inigo Jones' ideas and no parallel can be cited earlier than Nicholas Stone's Gateway in the Botanic Garden.

SAND-GLASSES

SAND-GLASSES were often used in parish churches in the 16th century and much more commonly in the following century. The earliest mention of one in Oxfordshire church accounts is for 1582. One bought for the pulpit of St. Helen's church, Abingdon, in 1591 cost 4d. During the 17th century the average price of an hour glass round about Oxford was 8d. but the stand or frame for it, of wood or wrought iron, was relatively costly. Towards the end of the 18th century the price had risen to 1s. 6d. and before the middle of the 19th century half a crown was usual.

For a history of sand-glasses and the composition of the sand used in them see Drover et. al., 1960.

BECKLEY — St. Mary

According to the Oxfordshire Archaeological Society's Report for 1933, p.15, the Jacobean pulpit has an hour-glass and stand.

CHINNOR — St. Andrew

Churchwarden's Accounts: *1660, Item paid for mending the hower glasse . . . 1d*

HEYFORD, UPPER — St. Mary

According to Blomfield, 1884, the pulpit dated 1618 had an hour glass in a frame; it was pulled down when the present church was built in 1867.

MIXBURY — All Saints

A memorandum of church goods in 1662, *It A Beer and an hower glasse.*

OXFORD — City Council

In 1667 the Council purchased an hour glass for 8d.

OXFORD — St. Mary the Virgin

Churchwardens' Accounts: *1646, Thomas Adams made, a frame for an houreglasse & for an houre glasse . . . 12s. 10d. — 1650, Item pd. to Tho. Adams for an Iron ffname to the ministers pue to putt an Howerglasse in . . . 4s. 2d. — Item pd. for an Howerglasse . . . 8d.*

Proctors' Accounts: *1652, It for a ffname for ye Houre glasse belongyng to ye Pulpitt . . . 10s.*

OXFORD — St. Peter in the East

Churchwardens' Accounts: *1636, ffor the yron worke for the houre glasse . . . 4s.—ffor the houre glasse . . . 8d.* There was no clock in the church from 1600 to 1640.

THAME — St. Mary

Churchwardens' Accounts: *1582, Item the oure glasse & the yron to fit it in . . . ijs. — 1625, to Barthyollmew White for making the frame of the howre glasse . . . iis. vjd. — 1629, Item for an howre glasse . . . xd. — 1633, Item paid for one hower glasse . . . 8d.*

PART TWO
MAINLY BIOGRAPHICAL

BIOGRAPHICAL DICTIONARY

THE NAMES of some 320 apprentices, clockmakers, smiths and others concerned with the making and maintenance of clocks and watches are arranged alphabetically together with their places of abode or business. The dates added to each name in brackets are those for which documentary or factual evidence is available; they may represent the years of birth and of death, or of any period in between — not necessarily the whole period of active work.

No individuals are included who were not in business before 1850.

These names are reclassified by places in the Topographical List which follows the Dictionary.

ABBREVIATIONS

C. & W.=clock and watchmaker, and includes individuals who may have been only retailers or repairers.

C. and W., fig.=the numbers used for spandrel-corners by H. Cescinsky and M. Webster (1914), Pl. I-VII, pp. 92-98.

For C. and W., fig. 3 see Plates 22 and 23
         23 „ Plate 12, fig. 18
         24 „ Plate 14, fig. 23
         26 „ Plates 11, fig. 16, and 13, fig. 21
         33 „ Plate 15, fig. 25
         37 „ Plates 13, fig. 20, and 15, fig. 24.

ca.=circa, approximate date.
q.v.=quod vide, see the Biographical Dictionary, etc.

LANTERN CLOCK HAND, George Harris, Fritwell, ca. 1670, natural size

ADAMS, THOMAS — OXFORD (1621-1664)

Smith, locksmith, turret clock. Apprenticed to John Bates, blacksmith of Oxford (free October 1613). On 16 August 1621 having certified to the City Council that he had served his late master for full 7 years he was made a freeman. Lived in the parish of St. Mary the Virgin. Was employed by the University from 1630 to 1663 on various works connected with the Bodleian Library, the New Schools, the Convocation House and also by the Churchwardens at the church of St. Mary the Virgin. In 1640 he assisted John Raye to install a new quarter-clock in St. Mary's and was employed as its clock-keeper in 1650-1652. Made for the Churchwardens of St. Mary's an iron frame for an hour glass in 1646, another for the hour glass at the minister's pew in 1650 and a third for the University for the hour glass belonging to the pulpit.

His will is dated 4 October 1664. A. Wood, however, states that he died in September 1664.

ALDWORTH, SAMUEL—OXFORD, LONDON, CHILDREY (1673-?1730)

C. & W. Son of John Aldworth, yeoman of Childrey, Berks. Apprenticed to John Knibb, clockmaker of Oxford, on 27 May 1673 for 7 years; stayed as assistant to his master for about 15 years. On 27 June 1689 was made a freeman of Oxford and set up business on his own in that city but in 1697 became a brother of the London Clockmakers' Company and moved to an address in The Strand, London. Late in life he retired to his birthplace, Childrey, and continued to make clocks there.

Samuel Aldworth was closely associated with John and Joseph Knibb. While a journeyman in 1683 he was signatory with John Knibb to a bond for the administration of the estate of Richard Knibb in Claydon, Oxon. When Joseph Knibb sold up his London business in 1697 Aldworth moved to London taking with him some of the clocks he had made in Oxford and very probably acquired some of Knibb's stock at the Suffolk Street sale. Consequently there is considerable similarity between his clocks and those of the Knibbs.

In 1705 and 1709 he attended the clocks and watches of Lady Diana Fielding in Duke Street, St. James, previously maintained by Joseph Knibb (Lloyd, 1957). About 1720 he made an arch-dial bracket clock (now in the Victoria and Albert Museum) which he supplied to John Knibb. And about this time, at the age of 65, he retired to Childrey where he made an unusual long-case movement (see below).

Oxford: Lantern clock signed *Saml Aldworth Oxon,* ca. 1690-95.

Bracket clock, 8-day timepiece, repeater on 2 bells, back-plate signed *Sam Aldworth Oxoniae Fecit* over which a tablet inscribed *Londini* is added; below the chapter ring signed *Sam Aldworth Londini Fecit,* and therefore may be dated ca. 1696, Plate 24, fig. 47. (Coll. W. Phillipson).

London: Several 8-day bracket clocks in ebonised cases, verge escapements, repeater on 2 to 5 bells, e.g.

Bracket clock, ebonised case 12½ inches high, 8-day verge movement, time-piece, pull repeater on 2 bells, latched plates and dial legs, 5 baluster pillars, back plate bordered, engraved with leaf scrolls, a fish and signed *Sam Aldworth Strand London;* chapter ring similarly signed, ca. 1700 (coll. Beeson).

Bracket clock, ebonised, 8-day, striking and chiming, and alarm, arch dial, with an oval cartouche inscribed *John Knibb* OXON concealing *Sam Aldworth in the Strand, London,* ca. 1720 (V. and A. Museum).

Childrey: Longcase, month, arch-dial with subsidiary dials for the months and signs of the zodiac, days of the week and diurnal signs, 31-day calendar ring and seconds dial in the arch, signed on the dial *Sam Aldworth at Childrey From London.* The anchor and escape wheel are mounted in vertical extensions of the plates and the escape wheel is driven by a contrate wheel at the top of an 8-inch long shaft driven from the centre wheel. The spur wheels of the month and week dials are driven by trains of 3 wheels. Very few movements are known with a seconds hand so placed, ca. 1725 (coll. H. I. Wilkes).

ALMOND, WILLIAM — OXFORD (1750)

Son of Thomas Almond, carman of Oxford, apprenticed to Thomas Reynolds, clockmaker and whitesmith of Oxford, on 23 April 1750 for 7 years.

ANSLEY, RICHARD — WITNEY (1823)

C. & W. In Witney in 1823, not in Directory of 1853.

ARCHER, JOHN — HENLEY ON THAMES (1532-1541)

Smith, turret clock. Probably the son of John Archer of Henley (died 1506). Mended the chimes of the 15th century clock in Henley church in September 1532 and again 1535-1541. He was not a burgess.

ARSBORN, THOMAS — BLOXHAM (1853)

C. & W. In Bloxham in 1853.

ATKINS, JOSHUA — CHIPPING NORTON, LONDON (1780-1823)

C. & W. According to Baillie joined the London Clockmakers' Company in the late 18th century; was in business in Chipping Norton in 1823.

Long-case, 30-hour, 11-inch square dial painted with flowers, plain oak case, hood with lateral scrolls and central ball finial, signed *Jos Atkin Chipping Norton,* ca. 1800.

Watchpaper: *"Josh Atkins* Watch & clock MAKER *Chipping Norton;* To make the Watch go Slow (etc.)"; leaf and flower border.

ATKINS, WILLIAM — CHIPPING NORTON, LONDON (1764-1778)

C. & W. A Quaker, son of William and Johanna Atkins of Norwich. Probably apprenticed in London. Married Sarah, daughter of Joseph and Mary King at Adderbury on 18 September 1764. Took as apprentice Richard Coles who was registered at the P.R.O., London, on 5 July 1771. Advertisements in his name dated 1 November 1764 and 7 February 1778.

Watchpaper: *"William Atkins in* Chipping Norton Clock & Watch Maker from *LONDON";* border of 12 scrolled lobes.

ATTE LEE, JOHN — HENLEY ON THAMES (1412-1448)

Smith. Also Joh. at le Smyth and Joh. Lee. Admitted burgess of Henley on 23 March 1412; elected constable on 13 September 1420 in which post he served at least 18 years. Died ca. 1448. In 1443 he had devised tenements to supply 5 poor men with coats.

He is the only smith recorded in Henley at this period and must therefore have undertaken repairs of the church clock, the first recorded payment for which was in 1410.

AULKIN, RICHARD — OXFORD (1770)

Son of Richard Aulkin labourer of Oxford, apprenticed to Thomas Reynolds, whitesmith of Oxford on 11 June 1770 for 7 years.

BALL, WILLIAM (1) — BICESTER (1705-1740)

C. & W. Probably born late 17th century. Married Catherine Jagger of Bicester on 1 June 1735. Probably not the William Ball, Will and Administration in 1786.

The following letters were written by Henry Purefoy of Shalstone, Bucks, to William Ball at Bicester (Eland, 1931).

" Mar. 8 1735 — When you was here ye 17th of Jan. last you promised mee to bring me an Allarum of a guinea price in a month or six weeks time & to take ye Clock Allarum agen and give mee a guinea and an half in money. The Clock Allarum has not gone since you was here neither can I make it go & I am unwilling to let anybody else meddle with it, as thinking it not a credit for it be known that you make a thing not to perform as it should do. I desire you'll let me have an answer by the Bister cutler directed for mee to be left at Mr. James Paine's, a baker at Brackley. Whether you will make me this guinea allarum or no & what day you will bring it over, now the roads are so good I hope I shall see you soon w'th it ".

"July 28 1736 — The Allarum I had of you would not go. I was forced to send for a man of your Businesse to look at it, for w'ch he had a shilling, it failed again in a week & I was forced to send again to have it sett in order. I am credibly informed 'tis a mean piece of work & will not perform, so 'tis good for nothing to mee. Pray come over here & make it good to mee, w'ch will save mee the Trouble of giving you further trouble, for I am resolved not to be imposed on in this manner".

Long-case, $11\frac{3}{4}$-inch square dial, bordered; 8-day movement, 5 pillars, outside locking plate, signed on chapter ring *Wm Ball of Bister fecit*, in panelled oak case, with lift-up hood, twist pilasters, ca. 1705-10 (coll. Beeson).

Long-case, 8-day, unusual arch dial, centre seconds, age and phases of the moon, day of the month, day of the week, date, signed *Wm Ball Bisceter*, in lacquered case, ca. 1720, illustrated in Britten 6th ed., fig. 796; now in Queen Anne walnut case in ownership of David B. Miller, U.S.A.

Long-cases, 30-hour, 10 inch square brass dials, one or two hands, engraved with birds in flight on matte ground signed *William Ball Bicester*, 4-post or plated movements in plain oak cases, ca. 1750-1765. Some of these may be by William Ball (2) below.

Alarm wall clock, arch dial 5 × 7 inches, centre matte, applied chapter ring and spandrel pieces, one hand. Movement with 5-inch, turned pillars and 3-inch square plates, verge escapement. Signed, *Wm Ball Bicester*, on a disc in the arch, ca. 1740 (coll. J. W. McDonald).

Retained from 1722-1736 on the clock of St. Mary Magdalene, Oxford.

Lost watch signed *Wm Ball* advertised 11 November 1758.

BALL, WILLIAM (2) — BICESTER (1738-1823)

C. & W. ? son of William Ball (1) above. Born 1738; married Ann Castle of Bicester on 9 May 1776. Buried at Bicester 30 January 1823, aged 85. Recorded as watchmaker in 1823.

Long case, 8-day, arch-dial 12 × $16\frac{1}{2}$ inches, centre silvered, engraved, signed below, *Willm. Ball* BICESTER, recessed seconds dial, central date hand, chapter ring without marks between the hours, corner-pieces C. and W. fig. 30, strike-silent in arch. Movement rack-striking. Case painted (? mahogany), hood with bell-section pediment and 3 ball and spike finials, about 8 ft. high, ca. 1770 (C. J. Coggins).

Probably made some of the 30-hour long case clocks mentioned under William Ball (1).

Maintained Merton church clock annually from 1779 to about 1800.

BARTON, JOHN — HENLEY ON THAMES (1659-1685)

C. & W. Married, aged about 26, Mary Day, widow of Henley, at Bix, Oxon on 23 October 1685.

BARRITT, GEORGE — WITNEY (1800)

Apprenticed to William Harris, watchmaker of Witney on 14 February 1800.

BASSETT, JOSEPH — CHIPPING NORTON (ca. 1780)

Longcase 8-day silvered and engraved arch dial, chapter ring without marks between the hour numerals, minutes not numbered, large seconds ring numbered at 15, 30, 45, 60, counterpoised hand; calendar lunette; the two hands with openwork stems and heads; corners engraved with shamrock and thistles alternately; arch engraved with bird, its wings outstretched, perched on a ribbon inscribed TEMPUS FUGIT; signed across the centre *Joseph*

Bassett Chipping Norton. Mahogany case, the waist with fluted columns, the hood with Chippendale scrolls, ca. 1780.

BECK, I. — HENLEY ON THAMES (early 19th cent.)

C. & W. Not in 1853 Directory.

Watchpaper: printed on orange and also on violet paper, *" I Beck WATCH & CLOCK Maker HENLEY Oxon";* 9-lobed border.

BEDFORD, WILLIAM — HOOK NORTON (1842)

C. & W. In Hook Norton in 1842.

BELCHER, JOHN — OXFORD (1723)

Son of Jonathan Belcher, carpenter of Abingdon, Berks. Apprenticed to Humphrey Brickland, watchmaker of Oxford, on 22 August 1723 for 7 years (registered at P.R.O. as Jonathan Belcher).

BIGNELL, THOMAS — OXFORD (1764-1812)

C. & W. Son of John Bignell of the University of Oxford, yeoman. Apprenticed to John Hawting, whitesmith and clockmaker of Oxford, on 24 May 1764 for 7 years, at a premium of £16 of which £10 was paid by the University and £6 by his father. Freedom of Oxford on 16 October 1812.

BIRD, MICHAEL (1) — OXFORD (1648-1689)

C. & W. Eldest son of John Bird, mercer of Oxford (Mayor 1615). Apprenticed to Thomas Taylor, London Clockmakers' Company, on 11 October 1648 for 7 years, and turned over to Edward Gilpin, i.e. to 1655. Freedom of Oxford on 8 September 1654. In business at 19 Cornmarket Street, east side, St. Michael's parish. Paid tax on 2 hearths in 1665 and Poll Tax on wife and 6 children, one female servant and one apprentice in 1667. Elected in 1668 to have a loan of £25 from Sir Thomas White's fund for helping young honest tradesmen.

In September 1678 he was elected to the Common Council and paid 3s. 4d. for not serving as Head Constable and £2 13s. 4d. in lieu of entertainment. Thereupon he was chosen Mayor's Chamberlain, an office which required him to pay a further fee of 40s. and £4 in lieu of entertainment. This office of Senior Chamberlain was held for a year after which his name appears in the list of Chamberlains until 1687. On 28 September 1685 he was appointed with others " to take a view of all the encroachments in the City and suburbs and to take notice of what shops are kept by any foreigners and which foreigners follow trades within the City and to report from time to time to the Council during the next six months."

In accordance with his policy of conciliating the nonconformists James II sent an order to the City Council on 21 January 1688 requiring the election of various officers including Michael Bird to be a Bailiff " without administering unto them any oath or oaths which His Majesty is pleased to dispense on that behalf." On 16 February the Council duly dismissed those office holders objected to by the King and elected those nominated. But Michael Bird and 4 others " being of this house and not present were removed pursuant to the King's letter."

E. T. Leeds expresses the opinion that Bird by persuasion must have belonged to the strongly Protestant faction of his day. Anthony Wood records that on the occasion of an anti-popery riot in Oxford in April 1683 a Will (sic) Bird, clockmaker, was one of the chief rioters in an assault on

Oxford undergraduates and was arrested by the Proctors but rescued by the crowd.

Michael Bird's Apprentices: Robert Heyten on 29 May 1656 for 8 years; Anthony Hodges on 28 March 1664, 8 years; John Harris on 5 July 1668, 9 years; Michael Bird junior on 15 April 1672, 7 years; Richard Saunders on 14 December 1674, 7 years; Wright Lane on 8 August 1676, 7 years; Nathaniel Bird on 6 August 1678, 7 years; Wright Bird on 5 April 1682, 7 years; Greenaway Curtice on 28 February 1688, 7 years. The last named apprenticeship ended in October 1689.

Bird issued a token in 1668 which is one of two of larger size than the rest of the Oxford issues; obverse, *MICHAEL · BIRD · HIS · HALFPENY, a cock in centre — reverse, *OXFORD · WATCH MAKER · · ·, (in centre) · · · M · B · 1668 · · · .

He died in 1689. His widow, Sarah, renounced probate and letters of administration were granted to his son Michael on 12 February 1690.

One longcase, 30-hour, one hand signed *Michael Bird Oxon* and two watches.

BIRD, MICHAEL (2) — OXFORD, LONDON (1672-1713)

C. & W. Son of Michael Bird (1), watchmaker of Oxford. Apprenticed to his father on 15 April 1672 for 7 years. Freedom of the London Clockmakers' Company in 1682. Administrator of his father's will in 1690. Baillie gives the dates CC 1682-1713.

Bracket clock signed *Michael Bird London.*

BIRD, NATHANIEL — OXFORD, LONDON (1678-1693)

C. & W. Son of Michael Bird (1) watchmaker of Oxford. Apprenticed to his father on 7 August 1678 for 7 years. Freedom of the London Clockmakers' Company in 1693.

A longcase clock.

BIRD, WRIGHT — OXFORD (1682-1686)

Son of Michael Bird (1) watchmaker of Oxford. Apprenticed to his father on 5 April 1682 for 7 years. On 10 June 1686 his indenture was cancelled.

BLUNDELL, JOHN — HORLEY (?1678-1700)

C. & W. Possibly apprenticed to George Nau, clockmaker of London, on 2 December 1678 for 8 years and (according to Baillie) later became free of the London Clockmakers' Company

Long-case, 30-hour, 11-inch square brass dial, two hands, calendar, centre engraved with scrolls, chapter ring signed *Ino Blundell Horly 1700;* 4-post brass frame in oak case; found in a village near Horley, Oxon (coll. Beeson).

BOWLES, EDWARD — THAME (1783)

C. & W. Married Hannah Pearson, both of Thame, on 1 March 1783.

BRICKLAND, HUMPHRY — OXFORD (1710-1750)

C. & W. Apprenticed to John Knibb, watchmaker of Oxford, about 1710 but was not registered in the Hanasters Book. Freedom of Oxford on 14 July 1723. Took Jonathan Belcher as apprentice on 22 August 1723.

Witnessed John Knibb's will in February 1721, remaining as journeyman to his master until 1722. Died in 1750.

Succeeded John Knibb as keeper of the clock of St. John's College from 1723 to 1750.

BROADWATER, HUGH — OXFORD, LONDON (1680-1697)

C. & W. Son of John Broadwater also Griffin, fisherman of Iffley, Oxon. Apprenticed to John Harris, watchmaker of Oxford, on 20 December 1680 for 7 years. Freedom of Oxford on 9 April 1688; apparently moved to London in 1689, becoming free of the London Clockmakers' Company in 1692. Signed the Association Oath Roll of Members of the L.C.C. in 1697.

A watch.

BROOKS, RICHARD — OXFORD (1765)

Son of Richard Brooks, decd. shoemaker of Iffley, Oxon. Apprenticed to Thomas Reynolds, whitesmith of Oxford, on 5 April 1765 for 7 years.

BROWN, JAMES — OXFORD (1852)

C. & W., jeweller. At 5 Park Place, St. Giles, Oxford in 1852.

Watchpaper: printed on yellow paper, " J. BROWN Watch & Clock Maker *Jeweller 5 Park Place* OXFORD ", below the Royal coat of arms on a shield with lion and unicorn supporters.

BROWNE, JAMES — OXFORD (1663-?1696)

Son of James Browne, printer of Oxford. Apprenticed to Richard Quelch, watchmaker of Oxford, on 18 May 1663 for 8 years. There is no record of his obtaining the freedom of Oxford but the Window Tax of 1696 lists a J. Browne in the house in Turl Street previously occupied by Quelch.

BUCKLAND, WILLIAM — THAME (1786-1821)

C. & W., smith. Worked for the Churchwardens of Thame from 1786 to 1821. Married Elizabeth Ross of Haddenham, Bucks, on 15 March 1794. Was Overseer of the Poor, Thame, in 1787; looked after the town fire engine for £1 a year from 1788 to 1814.

Act of Parliament clock, circular dial 2ft. 4ins. in diameter, yellow with black figures, case door lacquered with Chinese scene, remainder black and gold, 15-day movement, dead beat escapement signed *Willm Buckland THAME*, ca. 1780 (coll. Beeson).

Long-case, 8-day in mahogany case, ca. 1790.

Long-case, 30-hour, 12 inch square painted dial, oak case, ca. 1800.

Overhauled the clock of St. Mary's church, Thame, in 1788.

Repaired the clock of St. Peter's church, Great Haseley, in 1794 and 1814.

BULL, LIONEL — OXFORD (1761)

Son of Jonas Bull, carpenter of 28 Castle Street, Oxford. Apprenticed to John Hawting, clockmaker and whitesmith of Oxford, on 11 August 1761 for 7 years.

BULL, THOMAS — OXFORD (1592)

C. & W., smith. In the Oxford Council Acts for 24 November 1592, *Hit is agreed at this counsell that Thomas Bull clockmaker shalbe from hence-*

forth free of this citie so that he put in suche sufficient securitie before the feast of the purificacion of our lady next, as the more parte of the thirteen associats shall like of, for the repairinge and kepinge of the clocke jacks chyme and dyall at Carfoxe during his life and doe also before the same tyme com in and be sworne and pay iiijs. vjd. for the officers fees according to the custome of this citie, he having suche allowance of St. Martin's parishe as they shall agree and thinke good. No agreement was made between Bull and the Churchwardens of St. Martin's; on the contrary they preferred to appoint a smith, Hugh Corbet, to the duty of keeping the clock and chimes at 20s. a year.

Nothing more is known of Thomas Bull unless he was the man who supplied ironwork for the east window of Jesus College Chapel in 1636.

BULLER, ROBERT — BANBURY, ?ABINGDON (1705-1719)

C. & W. A Quaker, son of John Buller, baker of Neithrop, Banbury; born 7 March 1705. His mother, Mary Buller, was a most active Friend until the mid-18th century. Apprenticed to William Gunn, a Quaker clockmaker of Wallingford, Berks, on 19 March 1719. Possibly the Buller of Abingdon, Berks, maker of a lost watch advertised in 1759.

BURDITT, J. W. — BANBURY (mid-19th century)

C. & W. Long-case and regulator clocks with 8-day and month movements.

Cleaned the Adderbury church clock in 1868.

CAKEBREAD, RICHARD — OXFORD (1566-1592)

Blacksmith. Admitted freeman of Oxford on 11 September 1566 (Ricd Tayler alias Cackebrede). Apprentices: Thomas Rolewright, freedom of Oxford 17 July 1584; Thomas Rysson, free 28 July 1587; John Slatford, free 4 August 1592. Described as deceased in 1592.

Repaired Christ Church Cathedral clock in 1583.

Worked for St. Martin's church in 1579.

CAMOZZI, CHARLES — BICESTER (1832-1850)

C. & W., jeweller, ironmonger and general hardware dealer, in Market Place. Married Eleanor — children born 1832-1840. Business carried on by Eleanor Camozzi in 1852.

CARPENTER, WILLIAM — BANBURY (1834-1853)

C. & W., silversmith, jeweller, musical instruments, in Bridge Street. His tradesman's card and watchpaper printed by Cheney, Banbury in 1844 and illustrated in "John Cheney and his descendants," used the London Clockmaker's Company coat of arms.

CARTER, HENRY — OXFORD (1852)

C. & W., silversmith, jeweller in 19 High Street.

Lever watch, fusee, back wind, signed, "*Carter High Street OXFORD* No. 15090." (coll. Beeson).

CARTER, WILLIAM — OXFORD (1779)

Son of William Carter, labourer of Church Cowley, Oxford was apprenticed to Thomas Reynolds, whitesmith and clockmaker, Oxford on 14 August 1779 for 7 years.

CARTWRIGHT, WILLIAM — OXFORD, LONDON (1705-1759)

C. & W., turret clocks. Apprenticed to Alexander Warfield, clockmaker of London, on 2 July 1705 for 7 years; free of London Clockmakers' Company 17 January 1714. Went to Oxford in 1759 and set up business opposite the Black Horse in St. Clement's. An advertisement of 15 September 1759 states he was a maker of church and turret clocks.

"Long walnut case clock about 1713 signed Benjamin Cartwright junior London" (Britten) ? error.

CHAPMAN — OXFORD (later 19th century)

C. & W., turret clock. At 15 Broad Street, not in Directory for 1852.

Long-case, 8-day regulator, no. 1172 in carved oak case, ca. 1860.

Bracket clock, 8-day, no. 170 in mahogany case of mid-18th century design signed " Chapman Oxford 170."

An unusual ting-tang-quarter turret clock for St. Edward's School, Oxford ca. 1885.

CLARKE, JAMES — FINMERE (mid-19th century)

M.D., surgeon, living at Finmere House in the mid-19th century, made several long-case clocks with the assistance of William Bayliss, carpenter of Finmere. Most of the movements are small turret clocks with a variety of escapements including gravity types and bell chimes. Also repaired and modernised the Finmere church clock in 1858. His collection was dispersed in 1953 from Waterstock House.

Long-case, 8-day, turret clock movement, maintaining power, reversed seconds dial, dead beat escapement, long cylindrical pendulum bob, in pine wood case by W. Bayliss (coll. Beeson).

CLEMENTS, JOHN — OXFORD (1808-1853)

C. & W., silversmith, copper plate engraver. Set up business at 44 High Street in August 1808; married 19 September 1818; same address in 1853.

Watchpaper: " CLEMENTS *WATCH & CLOCK* MAKER *High Street* Oxford "; above, an eagle with watch suspended from beak and Tempus Fugit; "Plate & Jewellery Repaired. Engraving neatly executed "; plain outer border.

COLES, RICHARD — CHIPPING NORTON (1771)

Apprenticed to William Atkins, clockmaker of Chipping Norton on 5 July 1771. Possibly set up in Buckingham.

COOPER, JOHN — GREAT HASELEY (1818)

Painter. Painted and gilded the dial of the church clock at Great Haseley in 1818.

CORBET, HUGH — OXFORD (1589-1616)

Locksmith, gunsmith, turret clocks. Freedom of Oxford on 15 April 1589 paying 50s. and 4s. 6d. Appointed Constable on 2 October 1592. His apprentice, Thomas Lane, free on 16 April 1604, another apprentice free on 4 May 1614.

Repaired the clock and chimes of St. Martin's church, ca. 1592 to 1595, and thereafter to 1600 was employed on an annual wage of 20s.; repaired the chimes " being greatlie decayed " in 1601.

Keeper of St. John's College clock from 1598 to 1603. Keeper of St. Mary Magdalene clock from 1604 to 1616.

COSTER, CHARLES — HENLEY ON THAMES (1853)

C. & W. silversmith, jeweller, in Hart Street in 1853.

COSTER, JAMES — HENLEY, GREAT MARLOW (1800-1823)

C. & W. jeweller, in Hart Street, Henley on Thames in 1823. Took William Whitern as apprentice on 19 August 1806.

Watchpapers: (1) "*Jas Coster Watch & Clock Maker* HENLEY, Plate Jewellery Wedding Rings &c"; borders of pearls, jewel stones and leaves — (2) *Jas Coster Watch & Clock Maker* HENLEY and at Great Marlow"; same border.

COTTISWOLD, THOMAS — THAME (1458-1464)

Smith. Repaired the clock in St. Mary's church, Thame, on several occasions from 1458 onwards. In 1465 earnest money of 40 pence was paid for clock repairs but the Accounts are missing for the completion of the work.

CREED, THOMAS — OXFORD, ?LONDON (1657-1699)

C. & W. Son of Henry Creed, husbandman of Gloucestershire; apprenticed to Richard Quelch, watchmaker of Oxford on 1 September 1657 for 8 years. Not a freeman of Oxford, possibly admitted as brother of the London Clockmaker's Company in 1668, CC. to 1699.

CURTICE, GREENAWAY — OXFORD, LONDON (1688-1702)

C. & W. Son of Thomas Curtis, joiner of Oxford. Apprenticed to Michael Bird, watchmaker of Oxford, on 28 February 1688 for 7 years but on 4 October 1689 this was cancelled on his master's death and he seems to have been transferred to Wright Lane of Oxford. On 3 December 1694 he was bound to Joseph Foster, clockmaker in London and should have remained until 1701. After the death of his father in April 1699 he returned to Oxford and was made a freeman of Oxford on 15 May 1699 — the year of John Knibb's mayoralty. On 30 September 1700 he was given a Chamberlain's place as Mayor's child and paid a fine for not serving as Constable. On 30 May 1701 he obtained a bailiff's place for a fee of £3 3s. 0d. In April 1702 he died and was buried on the 7th at his parish church, All Saints.

A clock in the London Museum.

Lantern clock signed, *Greenaway Curtice Oxon* in the Virginia Museum, Richmond, U.S.A.; this is figured and dated 1685 in the Museum's 1937 Catalogue of the Henry P. Strause collection; clearly it should be dated ca. 1700.

DENTON, ROBERT — OXFORD (1730-1769)

C. & W. Matriculated at Oxford University on 31 March 1730 as watch- and clockmaker privilegiatus. Had a shop in High Street near Carfax from at least as early as 1740; this house was advertised as to be mortgaged or sold on 11 February 1754, but it was still in his occupation on 29 November 1766. He was buried on 22 February 1769 at All Saints Church, Oxford. (Baillie's date of 1790 is an error).

Long-case ca. 1750; lost watch advertised 1730.

Maintained Wadham College clock from 1751 to 1755, and St. John's College clock from 1751 to 1753.

DENTON, SAMUEL (1) — OXFORD (1756-1795)

C. & W. Matriculated at Oxford University on 6 June 1774 as watchmaker privilegiatus. Younger son of Robert Denton (above) and brother of William (below) with whom for a time he was in partnership. Carried on the business in the house in Cornmarket Street previously occupied by William. In 1774 married Mary Piercy (died 1779). He was buried on 30 December 1795 at All Saints Church, Oxford.

Long-case, 30-hour, 11-inch square brass dial signed *Saml Denton* OXON, in painted oak case, ca. 1775.

Verge watch in silver case H.M. 1787/8, maker IH, signed, Samuel DENTON Oxford No. 369.

DENTON, SAMUEL (2) — OXFORD, LONDON (1776-1827)

C. & W., silversmith. Son of Samuel (1) and Mary Denton, baptised 21 June 1776 at All Saints Church, Oxford. Apprenticed to John Wontner, clockmaker, Minories, London on 9 September 1790, £42 paid. Married Elizabeth Gardiner on 15 November 1802 at All Saints church. Freeman of Oxford by Act of Council on 8 November 1802 and resigned his freedom of Oxford in 1827 in a joint petition with several others. In High Street opposite King Edward Street from about 1820. Was succeeded in business by Joseph Steele.

Long-case, 8-day, white arch-dial, 12 × 17 inches, signed, Saml Denton Oxford, in tall varnished oak case.

Watchpapers: (1) "*Saml Denton* Watch & Clock Maker *High Street* Oxford *Sells all sorts of Gold & Silver plate &c as cheap as in London*"; border of scrolling and leaves.

(2) "DENTON WATCH & CLOCK *Maker High Street OXFORD*," on a shield surmounted by a clock dial; below, Gold Rings &c; simple border of leaves.

DENTON, WILLIAM — OXFORD (1756-?1774)

C. & W. Elder son of Robert Denton (above). Matriculated at Oxford University on 21 June 1756 as watchmaker privilegiatus. Owned a fourth share of a tenement later numbered 62-64 Cornmarket Street which he eventually sold to Isaac Board, a sadler; the property consisted of 2 shops with a frontage of 21 feet, in front of the Sun Inn. According to advertisements he was in occupation as late as 1769, and a survey of the area in 1772 records a Mr. Denton in occupation. For some years he was in partnership with his younger brother Samuel (1) (above). He may have died before 1774.

Long-case, 8-day, arch dial, seconds hand, date aperture, applied chapter ring and spandrels, signed on silvered disc in arch, *Willm. Denton Oxford*. Case oak, hood pilasters brass capped, height 7 ft., ca. 1760 (L. Robinson).

Long-case, 8-day, arch-dial in a tall inlaid walnut case, ca. 1770. Watch ca. 1764.

DENTON, WILLIAM & SAMUEL — OXFORD (1756-1774)

C. & W. Brothers and sons of Robert Denton (above) were in partnership in a house near the Sun Inn, Cornmarket Street from 1756 to about 1774 (advt. of 30 October 1756).

DRURY, CHARLES WILLIAM — BANBURY (early 19th cent.)

C. & W., jeweller. In Market Place in 1823. The business was taken over by Thomas Strange, probably about 1830. A Charles William Drury of North Bar Street voted in 1847.

Long-case, 30-hour, square painted dial signed, *C. W. Drury Banbury,* in oak case with scroll top hood. Probably some of the painted dial long case clocks signed Drury, Banbury (see below) should be ascribed to him.

DRURY, JAMES — BANBURY (1780-1800)

C. & W. Britten gives the dates 1780-1800; the evidence has not been traced. Long-case, 30-hour, $10\frac{1}{4}$ inch, square silvered dial signed *Drury BANBURY,* engraved chapter ring and corners, one hand, calendar, plated movement; back of the dial engraved J; ca. 1780 (coll. Beeson).

DRURY, WILLIAM — BANBURY (1773-?1810)

C. & W., jeweller. In Market Place until early 19th century. On 25 February 1773, described as bachelor, watch and clockmaker, he married Mary Fairfax, both of Banbury. On 5 February 1785, then a widower, he married Elizabeth Savage, both of Banbury. Served as Overseer of the Poor in 1788-89.

Long-case, 30 hour, 10-inch square brass dial, 2 hands, corners engraved, signed Wm. Drury Banbury, in tall oak case.

Long-case clocks, 30-hour, 11-12 inch square painted dials, signed, DRURY *Banbury;* 8-day, 12-inch square painted dial in panelled inlaid oak case; 8-day, white arch dial in panelled inlaid oak case; 8-day, 3-train movement playing 7 tunes on 8 bells; 8-day, regulator movement with a 12-inch circular silvered dial and centre seconds hand in mahogany case, 6ft. 8ins. high; all signed, Drury Banbury.

Repaired the Bodicote church clock in 1801.

Watchpaper: In centre, "*Plate* Rings Jewels & Watches *Chime Repeating and Plain Clocks* Made & Mended by *Wm Drury* Banbury "; monthly calendar border.

DRURY — OXFORD (?1790)

C. & W. Baillie gives the date ca. 1790 CC.

Watchpaper: Printed in blue, "*Drury* WATCHMAKER *OXFORD*"; above, a draped cloth from which two watches hang; below, Father Time recumbent with scythe and hour-glass; simple line edging.

DURRAN, EUSTACE — BANBURY, BRACKLEY (mid 19th cent.)

C. & W., silversmith, jeweller; at 90 High Street, Banbury.

Watchpaper: "EUSTACE DURRAN Watchmaker SILVERSMITH & JEWELLER 90 HIGH STREET (on circular strap) BANBURY (on central shield) RESPECTFULLY SOLICITS YOUR PATRONAGE & RECOMMENDATION (around strap)"; outer border of flowers and foliage surmounted by Royal Arms with lion and unicorn.

Another watchpaper of similar design is by DURRAN & SMITH — BRACKLEY, but the borders surmounted by the arms, crest and motto of Brackley (Northants). A third has in the centre "EUSTACE DURRAN Watch Maker JEWELLER & ENGRAVER BRACKLEY, surrounded by a circular strap and a serrate hatched outer border.

DURRAN, JAMES HOPKINS — BANBURY (1832-1854)

C. & W., silversmith, jeweller. Succeeded Charles Saunders in Parsons Lane in 1832; voted in Polls of 1837 and 1847; in High Street in 1853. According to "Shoemaker's Window" Durran used to draw the wires which plush-makers use for cutting the pile of the plush. His advertisement of 1854 mentions, thermometers, barometers, looking glasses, umbrellas, carpet bags, walking sticks, etc., on sale.

Repaired the Claydon church clock in 1834 and 1852-54.

DUTTON, T. — BANBURY (1841-1842)
C. & W. In Bridge Street in 1841 and 1842.

EARLE, THOMAS — OXFORD (1789-1812)
C. & W. Apprenticed to Thomas Reynolds of Oxford. Freedom of Oxford 23 May 1796. Was in partnership with Thomas Reynolds of Holywell Street from at least as early as 1797 and carried on Reynolds' contracts after his death. James White was apprenticed to him before 1812.

Looked after Wadham College clock from 1801 to 1810.

EDWARDS, THOMAS — DEDDINGTON (1773)
Apprenticed to John Fardon (2) clockmaker of Deddington but left his master in 1773.

ENOCK, EZRA — SIBFORD GOWER, LONDON (1799-1860)
C. & W. A Quaker, son of John and Elizabeth Enock of Sibford, born 10 July 1799 at Sibford Gower. Married, 1st, Eliza Harris at Sibford Gower on 6 April 1827, who died next year in London; 2nd, Ann Prophett at Adderbury on 22 May 1832; children born 1833 and 1834. Was in business in New Road, Whitechapel, Middlesex, from 1827. Returned to Sibford Gower in 1832 and died there April 1860, aged 60.

His son John (1834-1883) worked as a repairer in adjoining villages.

Long-case, 30-hour, painted dial wheatsheafs in corners, balloon below XII, signed, E. ENOCK, Sibford, in oak case (compare Joseph Williams).

ESTE, WILLIAM — OXFORD, BURFORD, ABINGDON (1505-1526)
Master mason. In Abingdon in 1505, in Burford in 1522; worked in Oxford 1505 to 1522 for Corpus Christi, Balliol and Magdalen Colleges. Died ca. 1526. Together with Louis Foose and Martin Williamson contracted in 1505 to make a striking turret clock for Magdalen College for £10. Made the column sundials designed by Nicholas Kratzer for Corpus Christi College and for the church of St. Mary the Virgin, Oxford, in 1520 (see under Sundials).

FAIRBROTHER, JAMES — WITNEY, CHARLBURY (1823)
C. & W. In Witney in 1823. " Fairbrother, a clockmaker and farmer and viol player at Charlbury and Finstock " (Kibble, 1928).

FARDON, JOHN (1) — DEDDINGTON (1700-1743)
C. & W. A Quaker, probably the son of Thomas Fardon, yeoman of North Newington, and Hannah, born about 1700. Apprenticed to Thomas Gilkes, clockmaker of Sibford Gower. Settled in Deddington about 1723. Married, 1st, Elizabeth daughter of John Pottinger, blacksmith of Adderbury West, on 2 August 1731, and, 2nd, Mary Cox daughter of a yeoman of Milton near Adderbury, on 9 August 1735. One son, John, born of the second marriage in 1736. His will dated 8 April 1743 shows that he was possessed of a house in Market Street and three cottages in Hoose Lane, Deddington and closes known as Deep Slade in North Newington. He also had a mortgage of £100 on a quarter land belonging to George Pottinger, his brother-in-law

PLATE 9
Fig. 14, JOHN HAWTING of Oxford, Clock in the Radcliffe Infirmary, Oxford, ca. 1775.

PLATE 10
Fig. 15, RICHARD GILKES of Adderbury, Movement of a 30-hour, one hand wall clock, ca. 174

Fig. 16, above, THOMAS GILKES of Charlbury, Dial 11-inch square of a 30-hour clock.
Fig. 17, below, THOMAS HARRIS of Deddington, Dial 10-inch square of a 30-hour clock.

PLATE 11

Fig. 18, above, RICHARD GILKES of Adderbury, Dial, 9-inch square of a 30-hour clock.
Fig. 19, below, RICHARD GILKES of Adderbury, Dial, 10¼-inch square of a 30-hour clock.

PLATE 13

Fig. 20, above, RICHARD GILKES of Adderbury, Dial, 11-inch square.
(Photo by courtesy of Mrs. C. A. Allitt).
Fig. 21, below, RICHARD GILKES of Adderbury, Dial, 11-inch square of a rack striking clock.

PLATE 14

Fig. 22, FRANCIS WEBB of Watlington, Dial, 10-inch square of a 30-hour clock. Fig. 23, JOHN MAY of Witney, Dial, 10-inch square of a 30-hour clock.

PLATE 15

Fig. 24, THOMAS GILKES of Charlbury, Dial, 10-inch square of a 30-hour clock.

Fig. 25, JOHN FARDON of Deddington, Dial, 10-inch square of a 30-hour clock.

PLATE 16

Fig. 26, THOMAS FARDON of Deddington, Act Fig. 27, WILLIAM PEACOCK of Banbury, Act of

in Adderbury West. He left bequests to 15 people amounting to £133, and the residue to his wife Mary, his son being a minor (Beeson, 1958).

Long-case, 30-hour, 10 inch square dial, one hand, centre with 3 zones of rings and very close zig zags, corner-pieces C. and W., fig. 23; movement 4-posted, iron top and bottom plates, loop and spikes; oak case, flat-topped hood, door of unusual design with 4 roundels and a cross-strip forming a sinuate upper edge, sides veneered; signed on chapter ring, *Ino: Fardon Dedington,* ca. 1740 (coll. N. S. Savage).

FARDON, JOHN (2) — DEDDINGTON, LONDON (1736-1786)

C. & W. A Quaker, only son of John Fardon (1), clockmaker of Deddington and Mary, born 11 July 1736. Only 10 years old when his father died he seems to have been apprenticed in London, and later to have returned to Deddington, probably after 1760. On 9 May 1772 he advertised for a journeyman and on 11 October 1773 notified that he had lost his apprentice, Thomas Edwards. Died 6 December 1786 and was buried at Adderbury West. (Beeson, 1958, figs. 7 and 9).

Long-case, 30-hour, 10-inch square dial, 2 hands of filed steel, centre with calendar lunette, engraved scrolling, and signed *John Fardon Dedington,* corner-pieces C. and W., fig. 33; movement plated, outside locking-plate, ca. 1770 (coll. Beeson). Plate 15, fig. 25.

Long-case, 30-hour, 10-inch square brass dial, 2 hands, centre with radiant sun flanked by basket and leafage, lych gate in wall, signed below centre *Ino Fardon Dedington,* corner-pieces of univalve shells in fan scrolling; movement plated, outside locking plate, ca. 1760 (coll. J. A. K. Fergie).

Long-case, 30-hour, 10-inch square brass dial, 2 hands, chapter ring with ¼ hour and unnumbered minute divisions, centre signed INO FARDON DEDDINGTON on a curved strap below a bird with outstretched wings; corner-pieces rococo scrolls; plated movement; case mahogany, probably later; ca. 1770 (coll. C. S. Launchbury).

Watchpaper: " From London *John Fardon* AT DEDDINGTON *Makes & Mends all sorts* of Repeating Horizontal & *plain Clocks* and Watches." Outer border with " Sun Faster · Sun Slower."

FARDON, JOHN (3) — DEDDINGTON, WOODSTOCK (1791-1853)

C. & W. A Quaker. Possibly son of John Fardon (2) above. In partnership with Thomas Fardon (below). Recorded as a watchmaker in Deddington in 1791 (Universal Directory). Repair scribings of John Fardon dated 1801-1830. Baillie gives " John Fardon Woodstock 1780 from Deddington."

A watch and clockmaker of this name in Woodstock in 1853 (Oxfordshire Directory). There may be two individuals of this name.

FARDON, THOMAS — DEDDINGTON, ADDERBURY (?1787-1838)

C. & W. A Quaker. Possibly son of John Fardon (2) above. In partnership with John Fardon (3) above. Described as watchmaker and ironmonger in 1791 (Universal Directory). In Deddington in 1833, married Lydia — who died 4 September 1836, aged 68, then residing at Adderbury East and was buried at the Meeting House. Thomas died in December 1838.

Act of Parliament clock, case of well-figured mahogany, circular dial, 2ft. diameter, signed on dial, THOS FARDON DEDDINGTON (coll. Beeson). Plate 16, fig. 26.

Act of Parliament clock, case of black and gold with door lacquered in the Chinese style, signed below the circular dial.

Verge watch, signed, *Thos Fardon* DEDDINGTON No. 4077, Arabic hour numbers, back wind, copper coloured cock (coll. Beeson). Other lost watches advertised in 1810 and 1821.

Installed the clock in Kidlington church in 1805, and in Deddington church in 1833.

Long-case, 30-hour, 10-inch square painted dial, signed, THOS FARDON ADDERBURY, one-hand lantern type movement in 4-posted frame with brass plates, ca. 1787. This clock probably represents a short interlude after the death of Richard Gilkes in 1787 and before the establishment of Joseph Williams, then 24 years old. (coll. Beeson).

T. and J. Fardon: Long-case, 30-hour, 12-inch square painted dial, Arabic numerals, calendar, signed, *T. & J. Fardon* Deddington, two-hand plated movement in oak case with rosewood inlay, ca. 1810, (coll. Lady Fremantle).

Verge watch, silver pair case HM 1801, signed, *Thos & Jno Fardon's* DEDDINGTON 549, (Beeson, 1958, fig. 11). Plate 19, fig. 36.

FARDON, THOMAS — WOODSTOCK (1823)

C. & W. A Quaker, possibly the same as the foregoing. In Woodstock in 1823.

Watchpaper: "*Fardon* WATCH & *Clock Maker* Woodstock Sells Gold Rings &c," on a large scroll held by standing Father Time with scythe and hour-glass; on left, an urn-clock on a pillar; above, " Tempus fugit"; rope border.

A bill of Aynsworth Thwaites, London, to the Rev. Reading of Woodstock " Cleaned the Watch Name Fardon filed the ballance wheel & set it deeper stopt up the Cock holes put a New banking Pin & smoothed the Pivots ... 6s."

FAULKNER, JOHN — BURFORD (1790)

Apprenticed to Richard Warner, clockmaker of Burford, on 16 November 1790, (P.R.O. registration).

FIELD, THOMAS — BICESTER (1787)

Apprenticed to William Musselwhite, clockmaker of Bicester, on 5 February 1787. (P.R.O. registration).

FLOWERS, JOHN — BANBURY (1797-1814)

C. & W. Married Ann Grant, both of Banbury, on 19 July 1802. As widower, married Esther Leatherbarrow, both of Banbury, on 11 September 1814. Attended vestry meeting, April 1797.

FORD, JOHN — OXFORD, ? AYLESBURY (1682-1725)

C. & W. Son of William Ford, Cleric late of Heyford, Oxon. Apprenticed to John Knibb, clockmaker of Oxford, on 28 April 1682 for 8 years. Freedom of Oxford on 10 July 1691. According to the Window Tax of 1696 a John Ford had a large house in High Street, Oxford. In April 1708 he was loaned £25 by the City Council. Probably moved to Aylesbury, Bucks, where he was recorded as insolvent in 1725. [Britten states " Jno Forde Oxon 1750 ", a date much too late].

Longcase, 8-day, 10-inch square dial, winged cherub corner-pieces, matte centre, calendar, fleur-de-lys on chapter ring as used by John Knibb, signed,

John Ford Bucks; movement with outside locking-plate. Case of oak veneered in front with walnut; slide up hood with twist pillars and portico top, ca. 1710.

Installed the clock in Finmere church in 1697.

FOWLER, WILLIAM — OXFORD (1667)

Son of William Fowler, mason of Oxford. Apprenticed to Richard Quelch, watchmaker of Oxford, on 25 June 1667 for 8 years.

FRANCKLYN, THOMAS see RANKLYN, THOMAS

FREE, JOHN — OXFORD (1696-1726)

C. & W. Son of John Free, gardener of Veldin (sic), Essex. Apprenticed to John Knibb, clockmaker of Oxford, on 7 July 1696 for 9 years, and lodged in Knibb's house (Poll Tax of 1702). Freedom of Oxford on 28 March 1709. Lived in a part of nos. 98 and 99 High Street in St. Mary's parish. Married Penelope Colegrave in 1704, children born 1711-1717. His elder son John, born 31 May 1711, became a D.D. and a freeman of Oxford on 30 July 1753. Thomas Hearne writes of him in his Diary, "*9 Feb 1725/6. At about seven of the Clock at night last Monday, died suddenly in his Chair, as 'tis said, one Mr. John Free aged about fifty, of St. Peter's Parish in the East, Oxford. He was a Watchmaker & was look'd upon as the best Workman in that sort of business in Oxford; but he lost many of his Customers upon account of his being a very great Whig. He was commonly call'd Skinny Free, because some time ago he is reported to have said that he wish'd all the Tories throughout the World were flea'd, and their Skins hung up on Trees, and their heads upon Pinnacles. He was a sober Man and saving & his Wife (who is living with three Children she had by him) and he lived lovingly together. He was a Man that read much & would talk well enough of History. He also busied himself much in finding out the Longitude. The first time he discovered himself to be so very Violent a Whig was some Years since, when Oxford was so much pester'd with Souldiers, soon after the business at Preston. He was buried in St. Peter's Church in the East at 4 o'clock afternoon on Thursday following.*"

His wife continued the business.

Verge watch, small silver pair case, silver champlevé dial with gilt edging, and gilt hands, signed on the dial, FREE OXON, and on the movement, *Ino Free Oxon.* (Ashmolean Museum).

Maintained the Wadham College clock from 1722 to 1726.

FREE, PENELOPE — OXFORD (1710-1734)

C. & W. Wife of John Free above, continued the business after his death. Maintained the Wadham College clock and one other in the Warden's Lodgings (1726-1734).

GAINSBOROUGH, HUMPHREY — HENLEY-ON-THAMES (1718-1776)

Congregational Minister, engineer. Son of John Gainsborough, crape and shroud maker of Sudbury, Suffolk; born 1718, educated at Sudbury Grammar School; pastoral minister at Newport Pagnell, Bucks in 1743; congregational minister at Henley in 1748 until his sudden death on 23 August 1796, aged 57. He was buried at his Meeting House. As engineer invented a tide mill and hydraulic machines, a wagon weighing machine and a steam engine with a condensing chamber. He made a rolling ball clock which was presented to the British Museum on 6 December 1788 but has since disappeared, as has also another timepiece based on a sundial deposited in the Museum in 1784. His portrait painted by his brother, Sir Thomas Gainsborough, is illustrated by Silvio A. Bedini (1959).

GARDNER, WILLIAM — OXFORD (1755-1758)

Son of William Gardner yeoman of Taploe, Berks., apprenticed to Thomas Reynolds, clockmaker of Oxford on 5 September 1755 for 7 years; in May 1758 Reynolds advertised that Wm. Gardner had left him.

GIBBS, JOSHUA — DEDDINGTON, SOULDERN (1805-1855)

C. & W. Worked for or apprenticed to the Fardons of Deddington as is indicated by a repair scribing on a John Fardon clock dial, " J. Gibbs Jan. 1805 ". (Beeson, 1958). He started business in Souldern and eventually succeeded the Fardons in Deddington until ca. 1855.

Longcases, 30-hour, 11 and 12-inch square painted dials, with arabic hour numerals, calendar, gilt openwork hands, signed, JOSHUA GIBBS SOULDERN; the hour ring is often painted green with gilt numerals. Cases oak plain or inlaid with mahogany; hood with Chippendale scrolls (W. J. Smith).

Watchpaper: (repairs dated 1850-1855). On a shield and its plinth, " GIBBS *CLOCK & WATCH* Maker DEDDINGTON & SOULDERN ", surmounted by a clock and flanked by Father Time with scythe and a man in frock coat and breeches using a sextant; border with, " Repeating Horizontal & Patent Lever Watches Repaired + LICENCED DEALER IN SILVER PLATE & GOLD WEDDING RINGS + ".

GIBBS, RICHARD — SIBFORD GOWER (1852)

C. & W. In Sibford Gower in 1852.

GILKES — Quaker Dials and 4-post Movements

Dials: A design for square brass dials used — apparently exclusively — by Quaker clockmakers of North Oxfordshire comprises, (a) in the central area a series of rings or narrow bands on a ground of radial strips of zigzags. The concentric rings were cut first and the blank interspaces filled in thereafter. The number of rings varies from 2 to 6, generally proportioned to the area of the dial plate; in the simplest workmanship they are plain and superficial, in the more finished forms they are deeply cut to a convex section with marginal circles. The zigzags vary in width from about 3 to 6 mm. and the strips may be overlapping or close to or fairly distant from the adjoining ones. These were produced by a graving tool working in a guide which ensured even width and a radial direction, and which could give more or less curvature to each line and a deeper impression at the apex of the angle. Consequently the space within the chapter ring could be decorated mechanically without the artistic ability needed to draw the symmetrical scrolling, leafage or floral designs often used on other dials of this period.

(b) The chapter ring has a diamond between the hour figures and a dot ● as a fourth quarter-hour mark.

(c) The spandrel pieces are similar to C. and W., figs. 7, 8, 23 (the most frequent), 24 and 26.

(d) The hand, for single hand movements, is carved in thick steel in simple designs of one or two pairs of loops usually with ears or cross-bars and a tail; the length varies from about $3\frac{1}{2}$ to 5 inches. For two-hand movements the finish of both hands is good and the curves and loops in the head of the hour hand are accurately drawn and chased.

Such dials, 7½ to 11 inches square, are termed Ring and Zigzag dials and were used by the Quakers John Fardon (I), Richard Gilkes, John Nethercot, William Green and also John Wilkes. See Plate 12, figs. 18 and 19, Plate 13, fig. 20.

The quality of execution of the engraving varies too much to be the work of one person, e.g. an itinerant dial engraver, and suggests that the dials were made in each clockmaker's own workshop, some of the cruder work probably by his apprentices. Nevertheless the Quaker makers undoubtedly helped each other out by supply or exchange of finished material; some of Gilkes' dials are indistinguishable from anonymous examples and from those signed by other men. Spandrel pieces, particularly C. and W., fig. 23 which were cast from progressively degenerating moulds until almost shapeless and never filed or trimmed, indicate a common source of supply.

Movements: Plate 10, fig. 15 shows a 30-hour movement by Richard Gilkes. In similar examples the 4 iron posts of rectangular section vary from about 5 to 6½ inches high by 5 to 8 sixteenths wide and are rivetted to the iron top and bottom plates. These vary from 4 to 5½ inches, shortest and longest sides, and brass lantern-style pivot-bars are slotted into them; the top plate carries a sinuous iron stirrup and the bases of the back pair of posts are bent outwards into spikes. The dial is pinned by 2 iron lugs to the top plate and by one to the bottom plate. The discs of the chain pulleys are one of iron and one of brass, more rarely both of brass. The locking-plate is more often of iron than of brass.

Variants of the wholly iron frame may have both plates and the 4 posts of brass. The latest 4-post movement to which a date, 1787, can safely be assigned (Thomas Fardon) has brass plates, locking-plate and pulley discs but the posts are of iron and there are no loop and spikes for hanging.

GILKES, JOHN — BURFORD (ca. 1800)

C. & W. Probably late 18th and early 19th centuries. Baillie mentions a watch signed, Gilkes, Burford, before 1768.

Verge watch, silver plated case, maker's mark DC, signed, " Ino Gilkes BURFORD ", back wind, ca. 1800 (T. Williams).

GILKES, JOHN — SHIPSTON-ON-STOUR, Warwicks (1740-1766)

C. & W. Probably a Quaker. Baillie states, an. 1758 84, Watch, Shipton (sic). Longcase, 10-inch arch-dial with oscillating crescent moon in the arch and " Tempus fugit " on a band encircling the opening, central area with arabesques, chapter ring with diamonds, spandrel pieces classical head in scrolls. Very tall narrow green lacquered case with Chinese motifs.

Longcase, 9-inch square dial, central area with arabesques, signed, *Jn Gilkes Shipston,* on a 2-inch diameter medallion below XII, spandrel pieces crossed univalve shells in scrolling; 4-posted movement; oak case. Longcase, 30-hour, 9¾-inch square dial, one hand, centre matted in vertical strips, 2 rings; movement 4-posted, brass top and bottom plates, loop and spikes, signed on chapter ring, *I Gilkes Shipston,* ca. 1750 (coll. Beeson).

Verge watch, signed *Jno Gilkes Shipston* on the plate and *Mary Gilkes* on the dial, silver case HM, 1766.

Mended Tredington church clock in 1740 and 1758.

GILKES, RICHARD — ADDERBURY (1715-1787)

C. & W. A Quaker, son of Thomas Gilkes, clockmaker of Sibford Gower and Anne his wife, born 18 April 1715. Apprenticed to his father. Started work in Adderbury East about 1736. There was then a small Quaker colony in the village and a Meeting House at Adderbury West built by Bray D'Oyley in 1675. His wife Grace was the daughter of another Sibford Gower family, William and Anne Gilkes, and he married her on 23 March 1744 at Sibford Gower Meeting House. Six children were born between 1746 and 1760; his wife died in January 1763. His first son Thomas died in 1757 aged 11 and his second son Thomas died in 1764 aged 6. Richard died without a male heir on 14 February 1787 aged 72 and was buried in the Friends' cemetery at Adderbury West. Although a prolific maker of 30-hour clocks in the earlier stages of his business, he does not seem to have made many clocks in the last 10-15 years of his life. After his death Thomas Fardon, clock maker of Deddington, continued the Adderbury business for a year until Joseph Williams became established as the Adderbury clockmaker.

Thomas Gilkes, minister and clockmaker of Charlbury, was his elder brother.

Richard Gilkes may be the originator of the ring and zigzag clock dial, described above, which became the Quaker trade mark until shortly after the middle of the 18th century. It is likely that he supplied engraved dial plates, chapter rings and spandrel castings to his Friends. From a survey of about 30 of his longcase clocks an idea may be formed of the progression of styles in a village clockmaker's workshop. The types examined include clocks owned by Mrs. C. A. Allitt, G. S. Evens, J. A. K. Fergie, C. L. Goggs, R. A. Johnson, M. Kelley, Mrs. F. Lester, S. M. Robins, Mrs. T. Tims and the author.

Four-posted frames: The clocks in groups (a) to (e) all have 30-hour movements in 4-posted frames as described above (Plate 10, fig. 15) with chain drive and one-second pendulums.

(a) Six with one-hand dials, 9, $9\frac{1}{2}$, 10, $10\frac{3}{4}$ and 11-inch square, quarter-hour divisions marked with a large dot, and between the hour figures a diamond, 4 or 5 rings and zigzags. The spandrel pieces include 3 with female bust and eagles, C. and W., 23, 2 with cherubs supporting a crown, C. and W., 8, and one with a vase of flowers in scrolls, C. and W., 24. None with a calendar. All signed *Gilkes Adderbury* on the chapter ring. Plate 12, fig. 18.

(b) One example with a two-hand dial, $10\frac{1}{4}$-inch square, no $\frac{1}{4}$-hour marks, diamond between the hours and a large dot below each 5-minute figure, 4 rings and zigzags. Spandrel pieces a pair of crossed univalve shells in scrolling. No calendar. Signed *Gilkes Adderbury* on the chapter ring. Plate 12, fig. 19.

(c) Four examples with two-hand dials, 10- and 11-inch square, signed *Richd Gilkes Adderbury* on a polished plaque below XII, no $\frac{1}{4}$-hour marks, two with a diamond between the hours and a large dot below each 5-minute figure, 3 or 4 rings and zigzags. Spandrel pieces female bust and eagles. The third is similar with a cut-out calendar lunette. The fourth has the central area of the dial matte and a sinuous scroll engraved below; the chapter ring with a diamond and $\frac{1}{4}$-hour divisions; the spandrel pieces cherubs supporting a crown.

(d) Two examples with two-hand dials 11-inch square signed on the chapter ring. One without $\frac{1}{4}$-hour marks, diamond between the hours and a large dot below the 5-minute figures, 5 rings and zigzags and a calendar lunette. Spandrel pieces rococco scrolls. Signed, *Rd Gilkes Adderbury*. The other without $\frac{1}{4}$-hour marks, fleur-de-lys between the hours, no dot below the minute figures, 6 rings and zigzags, and a calendar lunette. Spandrel pieces cherub supporting a crown. Signed, *Richd Gilkes* ADDERBURY, Plate 13, fig. 20.

(e) Three unique dials. One with a 9-inch, two-hand dial signed, *Gilkes Adderbury*, has the central area occupied by a compass-like pattern and spandrel pieces of the female bust and eagles.

The second with a 10$\frac{1}{8}$-inch square dial, one hand, signed on the chapter ring, *Gilkes Adderbury*, which has the usual diamonds and dots; the central area with a compass-like pattern of radial lancets engraved along their lengths with chevrons, and shorter lancets opposite the half hours; spandrel pieces as in the first.

The third, a 10-inch square, two-hand dial signed in the upper part of central area, RICHD GILKES ADDERBURY, above an unusual design of a tree with spreading tendrils bearing grape-like clusters, two birds standing on either side of the base of the stem and one bird below the signature; chapter ring without $\frac{1}{4}$-hour marks, a simplified fleur-de-lys between hours and a large dot below the 5-minutes; no calendar.

Plated frames: The clocks in groups (f) and (g) have movements in rectangular brass plates.

(f) Two-hand, 11-inch dial, the central area uniformly matte with a calendar lunette; chapter ring without $\frac{1}{4}$-hour marks, fleur-de-lys between the hours and no dot below the minutes, signed *Richd Gilkes ADDERBURY*. Spandrel pieces an urn in scrolling, C. and W., 26. Rack-striking 30-hour movement. Plate 13, fig. 21.

(g) Two-hand, arch-dial 11$\frac{3}{4}$ × 16 inches, central area with an engraved seconds dial and a calendar lunette, signed below the centre, *Richard Gilkes ADDERBURY*, and decorated with fine engraving of leafy sprays and an urn, in the style sometimes termed Nankin. Chapter ring with no $\frac{1}{4}$-hour marks, fleur-de-lys, and triangular marks below the 5-minutes. Spandrel pieces roccoco scrolling, C. and W., 36, Strike-Silent dial and pointer in the arch. An 8-day rack-striking movement, attached to the dial with square section posts.

As to the dating of these clocks those with ring and zigzag dials and one hand were first made about 1736, and the design was commonly used until about 1750 and less so thereafter. The earliest two-hand dials with large minute figures can be assigned to 1740-1750; those with a calendar lunette may represent a transition stage when the ring and zigzag design was being abandoned shortly after 1750. A plaque in the upper part of the dial may have been adopted about the same time.

Rack-striking movements in plates, such as in fig. 21, with corner pieces used in London up to 1770 may be dated ca. 1765. Finally, the 8-day arch-dial types represent the latest phase of R. Gilkes' products. No doubt 30-hour clocks continued to be supplied also but none are known with painted dials.

Turret clock: maintained the Adderbury church clock from 1747 to 1786.

GILKES, RICHARD — SIBFORD (ca. 1800)

C. & W. Parentage not traced.

Longcases, 30-hour, square painted dials signed, *Gilkes Sibford*, and *Richard Gilkes, Sibford*, in plain oak cases (Mrs. R. Hall).

GILKES, THOMAS (1) — CHARLBURY (1704-1757)

C. & W. A Quaker, eldest son of Thomas Gilkes, clockmaker of Sibford Gower and Anne his wife, born 8 January 1704; married Mary (died 1768) daughter of George Barrett on 17 April 1735; buried at Sibford Gower on

25 March 1757. He was the elder brother of Richard Gilkes, above, and was also apprenticed to his father. He is described as an eminent Minister. Two of his sons became clockmakers, Thomas (2) below, and Richard (1745-1822) who was in business in Devizes, Wilts.

Longcase, 30-hour, 11-inch square dial signed below leaf sprays, *G ∗ CHARLB:Y*, two carved mid-18th century hands, chapter ring without ¼-hour marks, nothing between hours, small dots marking the minutes, large figures, corner pieces urn in scrolling, C. and W., 26, Plate 11, fig. 16. (L. Milton).

GILKES, THOMAS (2) — CHARLBURY (1740-?1775)

C. & W. A Quaker, son of Thomas (1) above and Mary Gilkes of Charlbury and Sibford, born ca. 1740; married at Banbury on 27 August 1764 Sarah daughter of Richard and Mary Fardon of North Newington.

The date of his death is unknown; his widow died aged 80 on 20 May 1820 at Sibford.

Longcase, 30-hour, 10-inch square two-hand dial, signed above centre of dial, *GILKES Charl bury*, in leafy sprays, calendar lunette, no ¼-hour marks, nothing between hours, dots marking the minutes, large figures; corner pieces crossed univalve shells in scrolling. Movement 4-posted in painted deal case. Plate 15, fig. 24 (Beeson, 1958, fig. 12, Mrs. Pettifer).

Longcase, 30-hour, 9¾-inch square dial, two-hand, signed above leafy sprays, *T Gilkes Charlbury;* corner pieces crossed univalve shells in scrolling. Movement 4-posted in plain oak case, ca. 1775 (H. Cullimore).

GILKES, THOMAS — CHIPPING NORTON (1758-1770)

C. & W. Married Mary Sparkes by licence at Chipping Norton on 18 July 1758. Query, the same as Thomas Gilkes of Chipping Norton who married Alice Lord, also of Chipping Norton, by licence at St. Mary the Virgin, Oxford on 3 June 1745.

Longcase, 30-hour, 11-inch square, two-hand dial, no ¼-hour marks, nothing between the hours, minutes marked by lines, large figures; central area matte with polished scrolling and a bird in flight; corner pieces a female head in aureole and scrolling signed on chapter ring, *Thomas Gilkes Chipping Norton.* Plated movement, ca. 1770 (coll. J. Brice).

GILKES, THOMAS — SHIPSTON ON STOUR (2nd half 18th century)

C. & W. Possibly connected with one of the Thomas Gilkes above. Longcase, 30-hour, 10¾-inch square dial, two hands, calendar, spandrels engraved, signed, *Thomas Gilkes* SHIPSTON; Plain oak case 6ft. 10ins. high (coll. A. E. Shepard).

Longcase, 30-hour, 10-inch square dial, one hand; spandrels engraved and central area engraved with sprays of leaves and flowers, signed, *Gilkes* above and SHIPSTON below; oak case 6ft. 4ins. high (coll. C. Bryan).

Longcase, 30-hour, square dial, central area engraved with a few scrolls, signed in upper part, *Thomas Gilkes Shipston,* calendar ring engraved; chapter ring without marks between the hour numerals, minutes as dots with large figures in fives; cast corner-pieces, a head in reticulation, C. and W., 20. The two hands may not be original. Case dark oak with a gable top to the hood, ca. 1760.

GILKES, THOMAS — SIBFORD GOWER (?1665-1743)

C. & W. A Quaker, son of Thomas and Mary Gilkes of Sibford Gower, born ? 1665, married Anne, children born 1702 to 1715. Alive in 1743 according to the Will of John Fardon (1). Uncertain where he learnt his trade but, as a Quaker, he should have been apprenticed to a Quaker of whom the most likely is Richard Gilkes of the London Clockmakers' Company in 1686. His father was one of the group of Friends who bought land in 1681 for the establishment of the Sibford Gower Meeting House.

He trained two of his sons as clockmakers, Thomas born in 1704, who set up in Charlbury and Richard born in 1715, who went to Adderbury. John Fardon (1) of Deddington was apprenticed to him. He pioneered the clockmaking industry in north Oxfordshire villages which his brother Quakers developed so successfully that in the some villages they held the monopoly of this craft for over 150 years (Beeson, 1958).

Longcase, 30-hour, brass dial, one hand, signed, Thomas GILKES SIBFORD; 4-posted movement rope driven.

GILKS, TOBIAS — CHIPPING NORTON (ca. 1770)

C. & W. Baillie states " an. 1774 Watch ".

Longcase, 30-hour, dial for one hand fitted with two hands, central area matte; chapter-ring with $\frac{1}{4}$-hour marks and fleur-de-lys between the hours spandrel pieces, a large flower with expanded petals in roccoco scrolls and trefoils, signed on chapter ring, *Tobias Gilks Chipping Norton;* 4-posted movement with brass plates (coll. J. A. K. Fergie, Beeson, 1958, fig. 13).

Longcase, 30-hour, $8\frac{1}{4}$-inch dial, one hand, central area matte with 2 birds in flight; chapter ring with $\frac{1}{4}$-hour marks, fleur-de-lys between the hours; spandrels with a female head in scrolling, C. and W., 18; signed *Tobias Gilks Chippingnorton* on chapter ring. Movement iron 4-posted without spikes and loop, in a stained pine hood on a large bracket (coll. R. A. Johnson).

Longcase, 30 hour, one hand, central area matte with a symmetrical pattern of curved leaf-bearing lianes and hatched areas, chapter ring with fleur-de-lys signed, *Tobias Gilks Chipping norton,* spandrel pieces, head in scrolling C. and W., 21 (coll. Dr. P. Latcham).

GILLETT, THOMAS — OXFORD (1698-1702)

Son of George Gillett, late of Clanfield, Berks, apprenticed to John Knibb, clockmaker of Oxford, on 29 June 1698 for 7 years. According to the Property and Poll Taxes for 1702 he was then living in John Knibb's house in Holywell Street, described as a servant.

GODFREY, HENRY — OXFORD, LONDON (1676-1707)

C. & W. Son of George Godfrey, yeoman of Hillend, Berks, apprenticed to John Quelch, watchmaker of Oxford, on 29 June 1676 for 8 years. Free of the London Clockmakers' Company in 1685. Baillie gives 1707 as later date.

Longcases, 8-day, marquetry case (Victoria and Albert Museum); walnut case (coll. Beeson).

GOMES, WILLIAM — OXFORD (1550-?1585)

Clock-keeper at St. Martin's Church, Oxford from about 1550 at a wage of 8s. a year. From 1557 to 1585 the keeper of the clock was called the clerk. In 1586 John Jennings became the clerk.

GOWETH, JOHN — OXFORD (1669-1694)

C. & W. Son of John Goweth gunsmith of Oxford, baptised 14 November 1669 at All Saints Church; apprenticed to John Knibb, clockmaker of Oxford on 11 October 1686 for 7 years. Freeman of Oxford on 5 October 1694. His father was chosen Common Councillor of the City in September 1690 but was let off with a fine of £5 because he and his family were labouring under very great affliction. John senior died in 1706 and his executors paid rent until 1734 for the property he occupied in High Street, All Saints' parish. Presumably John, the clockmaker, occupied this house.

A watch, ante 1701 (Baillie).

GRAYSON, JOHN — HENLEY-ON-THAMES (1823-?1850)

C. & W., silversmith. In Bell Street, Henley in 1823 (not in 1853 Directory). Maintained Henley church clock from 1830 onwards at £5 5s. 0d. p.a.

Watchpaper: "GRAYSON *Watch & Clock* Maker & *ENGRAVER Henley*" in circle of pearls; " Plate and Jewelery (sic) Wedding & Mourning Rings ", in outer border.

GRAYSON, WILLIAM — HENLEY-ON-THAMES (1780-?1860)

C. & W., silversmith, jeweller. Apprenticed in 1780 (Baillie). In Bell Street, Henley in 1853. Married Jane Watts of Henley in 1839. A Mr. Grayson maintained the Henley church clock to 1860 at £5 5s. 0d. a year. Possibly two men of the same name.

Watchpaper: " WILLM GRAYSON Watch & Clock Maker HENLEY ON THAMES *Silversmith Jeweller & ENGRAVER* "; in upper half a shield surmounted by a plumed helmet and flanked by winged Time and a crowned man, on motto TEMPUS RERUM IMPERATOR (? Clockmakers' Company Arms); at sides, Horizontal Patent Lever & Repeating Watches; scalloped outer border.

GREEN, GEORGE SMITH — LEICESTER, CIRENCESTER, OXFORD (1750-1762)

C. & W. Started business in Leicester whence he moved to Cirencester trading near The Bell in Cricklade Street in 1750 and opposite the King's Head in 1751. Later he worked at Ross on Wye (Daniell, 1952) and thence went to Oxford. In 1758 he advertised, *At the Automaton Laboratory Confronting the Portal of All Souls College in Oxford, are fabricated and renovated Trochiliack Horloges, portable or permanent, linguaculous or taciturnal: whose circumgyrations are performed by internal spiral Elastics, or external pendulous Plumbages: Diminutives simple or compound in Aurum, or Argent Intiguments* (Oxford Journal).

He was the author of " Oliver Cromwell, an Historical Play " advertised in the Gloucester Journal on 24 December 1751 and sold at his shop in Cirencester (Buckley, 1929). He died at Oxford in May 1762.

Longcase, 11-inch square brass dial engraved with 2 birds, signed GEO: GREEN LEICESTER on a silvered arc, ca. 1750 (Leicester Museum).

GREEN, JOHN — OXFORD (1794-1823)

C. & W. In St. Aldate's parish in 1823. Freedom by Act of Council on 3 November 1794. Married Mary Smith of Oxford in 1802.

GREEN, SAMUEL — WITNEY, LONDON (1754-1823)

C. & W. Son of Henry Greene, distiller of Oxford, apprenticed to George Pyke, London Clockmakers' Company, on 21 September 1754 for 7 years, premium £42. Free of the London Clockmakers' Company 29 September 1772, l.c.c. 1787-1817. In Witney in 1823.

GREEN, WILLIAM — MILTON UNDER WYCHWOOD (1722-1770)

C. & W. Possibly the son of Isaac and Joan Green of Tadmarton, born 30 July 1722. Kibble, 1928, states "William Green of Milton under Wychwood had a clock club into which so much per week was paid to get a clock." He may possibly be the same as William Green working in 1780 at Milton south of Abingdon, Berks, who later moved to Newbury.

Longcases, 30-hour, single hand, with brass dials 7 to 11 inches square and 4-posted movement with wall loop and spikes. The dials are engraved with the Quaker design of concentric rings and radial zigzags using spandrel pieces and chapter rings similar to those of Richard Gilkes, signed on the chapter ring, *Wm Green Milton;* 2nd quarter 18th century (coll. Beeson, M. Margerisson, T. Williams).

Longcase, 30-hour, 10-inch square dial, 2-hand central area engraved with thin leaf sprays; chapter ring with a large dot below the 5-minute figures; spandrel pieces closely reticulate. Movement iron 4-posted, ca. 1770.

GRIFFIN, RICHARD — BURFORD (1853)

C. & W., jeweller. In High Street, Burford, in 1853.

GUNN, WILLIAM — WALLINGFORD (1714-1740)

C. & W. A Quaker, married Mary Fowler at Shutford, Oxon, on 19 May 1714. Took as apprentice on 19 March 1719 Robert Buller, son of Mary Buller, widow of Banbury.

Longcase, 30-hour, 10-inch square dial, one hand, central area matte; chapter ring with fleur-de-lys, signed, *William Gunn Wallingford;* spandrel pieces pair of cherubs supporting crown. Movement 4-posted, brass top and bottom plates. Case varnished pine, flat topped hood, waist door with glass pane, ca. 1740.

GURDEN, JOSEPH — OXFORD (1755-1772)

C. & W. Married Martha Woodbridge at St. Mary the Virgin, Oxford on 23 December 1755. Lived at No. 4 Turl Street. Buried 26 November 1772 at All Saints Church, Oxford.

A watch before 1762 (Baillie).

HAINES, ROBERT — OXFORD (1777-1800)

C. & W. Matriculated at Oxford University on 11 June 1777 as horologiorum fabricator privilegiatus. On 14 March 1778 he opened a shop at No. 133 High Street, then leased to John Green, carpenter; on 1 July 1789 the City transferred the lease to Haines with a fine of £5. In 1795 he advertised that he had moved from opposite New Market to opposite All Souls Church, i.e. to No. 117 High Street, which the City leased to him on 20 April 1799, fine £15; he ceased occupation before 1813.

Act of Parliament clock, black and gilt, Chinese lacquered door, irregularly 6-sided dial, signed, Haines Oxford, ca. 1780.

Watchpaper: "*Robert Haines* Watch *and Clock Maker* Oxford"; border of clock dials, scrolls and a circle enclosing R.H.

HARINGTON, THOMAS — HENLEY-ON-THAMES (1494-1515)

Clock-keeper. Also Thomas Clerke alias Harington. The Records of the Henley Borough Assembly for 12 September 1494 state, "*Et eodem die electus est et admissus pro aquebaiulo Tho Harington . . . et quod 'dictus Tho custodiat horecudium et campanas et habebit pro labore suo xxs. per annum et alia auantagia prouentia de campanis predictis ad anniuersaria ibidem custodienda"*. An *aquebaiulus* was a water-bailiff and *horecudium* is the term for a clock striking the hours. It is more likely that Harington was an *aquaebajularius* or holy water clerk. This Henley clock was in existence in 1410. As custodian of the bells (campana) Harington was responsible for ringing at anniversaries, burials and other ceremonies and for collecting the charges on the same out of which he received fees in addition to his stipend of 20s. a year. He was not a burgess.

When he submitted his account for bell-ringing from 9 May 1495 to 3 June 1496 he put in a bill for repairs which was disallowed as being his liability. In December 1499 John Balam was appointed second water-bailiff or holy water clerk to share with Harington half the profits, . . . *cum eo capiendo dimidium valoris et proficuorum proventium ex officio predicto preter de le clocke et chyme*. In December 1509 Thomas Lew (or Lewes) was appointed second aquebaiulus in Balam's vacancy; Lewes drew extra emoluments as a bridge-keeper (pontinarius) amounting to 10s. per annum or a house of like value. In August 1510 Harington was granted profits from the sale of wax for the Light of the Blessed Virgin Mary. The last entry for Harington is dated 20 April 1515 and runs, *in stipendium Tho. Harington clerici pro campanis et le clok et chyme* . . . *xs*.

HARRIS, GEORGE — FRITWELL (1614-1694)

C. & W., blacksmith, turret clock maker. Son of Jeffrey Harris of Fritwell, baptised 19 January 1614. By his first marriage children were born from March 1643 to May 1651. Second marriage to Betteris Toms of Glympton at Chipping Norton on 16 October 1654; children born from September 1655 to December 1666. In the Hearth Tax of 1665 he paid for one hearth, his forge being exempt. His Will dated 12 March 1693, proved 4 June 1694, shows that he owned a house and two workshops, a close, a quarter land and livestock. The estate excluding land and buildings was valued at £75 8s. 0d. He was succeeded in business by his son Nicholas (1657-1738).

The inventory of his effects made in May 1694, includes, Wearing apparel and money in purse, £20: 2 cows and a calf, £7 10s. 0d.: 1 hog, £1 3s. 0d.: crops growing in the field land, £5: malt in the house, £1 10s. 0d.: in parlor 1 round table and 6 chairs, £1 10s. 0d.: in kitchen 1 table, 1 jack, 2 spits and other lumber, £1: in the buttery 4 tubs and stand and other things, £1 5s. 0d.: 4 kettles, 2 brass pots, 1 brass pan, 3 possets and other brass, £2 10s. 0d.: pewter all sorts, £1: 10 pairs sheets, table cloth and napkins, £5: in the new chamber a bed, bedstead and other furniture, £3: in the middle chamber a bed, bedstead, trunks and boxes, etc., £2 10s. 0d.: in the chamber over the shop other furniture, £1 10s. 0d.: tools in the shop, £15 — Total, £75 8s. 0d.

Lantern clocks with an original unique feature. The great wheel in each train is provided with a barrel grooved for a gut line, and a second toothed wheel which engages with a large pinion mounted on an arbor pivoted in cocks on the base plate. The arbors protrude through winding holes in the chapter ring at V and VII.

(a) Centre of dial engraved with a Tudor rose and tulips, signed above, *George Harris in Frittwell fecit,* carved iron hand with tail, dolphin frets, verge escapement and bob pendulum, loop and spikes (coll. J. M. Surman) Plate 20, figs. 37, 38, and text figure on p.84.

(b) Similar but with an alarm dial and the chapter ring inlaid with black and red wax, alarm work on the back plate; mounted on a panelled oak pillar case (coll. J. A. K. Fergie).

(c) Centre of dial engraved with rose and tulips, signed above, *George Harris of Frittwell 1668*, carved hand with arrow-head tail, dolphin frets; top plate with a pair of eyelets instead of a loop for suspension; height 13$\frac{1}{8}$ inches, diameter of chapter ring 6$\frac{1}{4}$ inches. Movement with normal pull-up cord winding, originally with crown wheel escapement; the hammer is mounted on the top plate on a shaft pivoted horizontally, with its buffer spring; it is linked by a rod to the hour-strike lever pivoted in the bearing bars in the normal way (Cothele House, Cornwall).

This is the earliest *dated* lantern clock made in the County.

Turret clock with crown wheel and verge escapement made in 1671 for Hanwell Church. Plate 1, fig. 1.

Repaired the South Newington church clock in 1669 and Yarnton church clock in 1682.

HARRIS, JAMES — WITNEY (mid 18th century)

C. & W. In Market Place in 1852. A James Harris of Witney married Mary Poles of Oxford in 1795. Longcase, 30-hour, painted arch-dial signed, *In Harris Witney*, in varnished case.

HARRIS, JOHN — BURFORD (?1631)

Baillie records " John Harris Burford 1631 "; no further information is available unless he is the man who joined the London Clockmakers' Company in 1632.

HARRIS, JOHN — OXFORD (1668-?1696)

C. & W. Son of John Harris, chandler late of Oxford, apprenticed to Michael Bird, watchmaker of Oxford, on 5 July 1668 for 9 years. Freeman of Oxford on 10 February 1678. He probably lived in a house in High Street, St. Mary's parish, taxed in 1696 on 10 windows, but there were several men of his name in Oxford at this period. Hugh Broadwater was apprenticed to him in 1680, and Robert Veasey in 1682.

Verge watch in silver case, with tortoiseshell and silver pique outer case; silver champlevé dial with gilt border, signed, HARRIS OXON; black tulip hour hand; movement with early balance spring and cock foot with an irregular edge, tulip pillars, back wind, signed, *John Harris Oxon*, ca. 1680, Plate 18, figs. 30, 31 (coll. Beeson).

HARRIS, JOSHUA — WITNEY (1823)

C. & W. In Witney in 1823. Sold " Sheraton " banjo-cased barometers. Longcase, 30-hour, 10 inch square painted dial signed, *Joshua Harris Witney*, in pinewood case, 5ft. 8ins. high.

HARRIS, NICHOLAS — FRITWELL (1657-1738)

C. & W., whitesmith, turret clock maker. Third son of George Harris, blacksmith and clockmaker of Fritwell, and Betteris his wife. Baptised 17 April 1657. In 1694 his father's Will bequeathed to him " all those my working tools belonging to my trade in both shops "; in the inventory these are valued at £15. He also inherited " a quarter land ". Married Elizabeth on 11 April 1695. Died 20 May 1738 at Fritwell.

Lantern clock, two hands, signed in centre of the dial, anchor escapement (coll. R. H. Wade).

Mended South Newington church clock in 1674.

Made the Great Milton church clock in 1699 (Plate 6, fig. 9).

HARRIS, THOMAS — DEDDINGTON (1732-1797)

C. & W. A Quaker, son of Joseph and Mary Harris of Sibford Ferris, born 1732. On 20 September 1762 married Mary Fardon, daughter of Richard and Mary Fardon of North Newington, at Banbury. Succeeded John Fardon (1) clockmaker of Deddington during the minority of John Fardon (2). Described as a clockmaker of Deddington in 1762 he may have been joined by or displaced by John Fardon (2) about this time. Subsequent history untraced until his death on 1 August 1797, aged 65, at Milton and was buried at Adderbury West Meeting House. His wife died 20 April 1786, aged 57, at Bloxham and was buried at Adderbury West.

Wall clock, 30-hour, brass dial with cast spandrel pieces, one hand, iron 4-posted movement, chain drive; on oak bracket in a $\frac{3}{4}$-inch framed hood with glass side windows.

Longcase 10-inch square dial, 30-hour, one hand pierced in several loops, central area of dial above with a winged cherub head and scrolls, at the sides with what seems to be a Rosicrucian lamp in a circle of magical signs and a pot of flowers and a tree, and below centre the signature, *Tho Harris Dedington*. Chapter ring with diamond half hour marks and nothing between the hour figures; corner pieces intricately scrolled (coll. A. J. Fortescue, Beeson, 1958, fig. 8). Plate 11, fig. 17.

Longcase $10\frac{3}{4}$-inch square dial, 30-hour, central area matte, no calendar, 2 flat pierced hands, chapter ring without $\frac{1}{4}$-hour marks, nothing between the hours, large minute figures; spandrel pieces open strap scrolls, C. and W., 33, signed, Thos Harris DEDINGTON. Movement 4-posted, no loop or spikes, in heavy oak case with Greek key border below cornice, ca. 1765 (coll. A. J. Fortescue).

HARRIS, WILLIAM — THAME, ?LONDON (1756-?1763)

Son of John Harris, flaxdresser late of Thame; apprenticed to Boyer Glover, London Clockmakers' Company on 2 February 1756 for 7 years, premium £25. Subsequent history possibly London.

HARRIS, WILLIAM — WITNEY (1794-1823)

C. & W. In Witney 1794 to 1823. Took George Barritt as apprentice on 14 February 1800 (P.R.O.). Married Elizabeth Hinton of Witney in 1794, and, as widower, in 1803 married Dinah Hambridge of Witney.

Longcase, 30-hour, painted 10-inch square dial, signed *William Harris Whitney* (sic), in pinewood case.

HARRISON, WILLAM — CHARLBURY (1792)

C. & W. A Quaker.

Longcase, brass dial with silvered chapter ring, no movement, in Charlbury Society Museum.

Installed the turret clock made by Aynsworth and John Thwaites for University College, Oxford, in 1792.

HARVEY, ROBERT — OXFORD (1588-?1600)

C. & W. Freeman of Oxford on 19 September 1588 paying 40s. and 4s. 6d. fees, described as a clockmaker, the first freeman of Oxford so described contemporaneously in the Hannasters' Register. Britten and Baillie do not mention a Robert Harvey at this period but a clockmaker of this name accompanied the blacksmith organ-maker, Thomas Dallam, on his voyage to Constantinople to deliver his Great Musical Clock to the Sultan of Turkey

(Hillary, E., 1952). As Dallam made organs for Colleges in Oxford, an association of the two men is not improbable. An unusual balance-wheel wall clock with two dials by a Robertus Harvie of early 17th century design is illustrated in the Anthology of C. Fox (1947). The engraving of the dial is signed G.D.

HAWKINS, RICHARD — OXFORD (1637-1681)

Painter. Took apprentices from 1651 to 1663, including Christopher Mathews and John Rixon. Worked on the porch of St. Mary the Virgin in 1637, the south dial of the Schools in 1641, Divinity Schools in 1669, Old Ashmolean in 1681, etc.

HAWTING, JOHN — OXFORD (1745-1791)

C. & W., whitesmith, turret clock maker. Son of John Hawting, gardener of Oxford. Apprenticed to Thomas Reynolds, whitesmith of Oxford, on 17 May 1745 for 7 years, £10 fee paid by the University Charity. Freedom of Oxford on 16 January 1756. Lived in Holywell Street from as early as 1764, and according to the survey of 1772 at no. 35 on the north side, Broad Street end, in a house with a frontage of 37 feet. In 1754 he married Elisha Mathews of Oxford. On 3 January 1765 he married Mary Law at All Saints Church, Oxford, being described as " of Holywell in this City widower ". In 1767 he was elected to the City Council and became Chamberlain in 1772. Apprentices: John Heath in 1756, Lionel Bull in 1761, Thomas Bignell in 1764 and William Hawting in 1770. Died in 1791.

Longcase regulator, 12-inch square silvered dial, central minute hand, lunette for hours, shuttered winding hole; movement plates with 5 pillars, maintaining power, dead-beat escapement, gridiron pendulum; oak case. Made for the Radcliffe Observatory, now in the History of Science Museum.

Act of Parliament clock in a tall black and gilt case (Plate 9, fig. 14) white circular dial with black figures, waist door with a lacquered scene in the Chinese style representing Dr. Radcliffe examining a sick person brought in a carrying chair; to the right of the Doctor is Aesculapius with his serpent staff and behind is a Chinese building representing the Infirmary. Movement of 2 trains with long narrow weights, pendulum beating 56 to the minute, gridiron compensation; signed below dial, *John Hawting* OXFORD, ca. 1775. In the Board Room of the Radcliffe Infirmary, Plate 9, fig. 14.

Turret clock made for Trinity College in 1787. Plate 7, fig. 11.

Maintained the Wadham College clock in 1783 and 1784.

Watchpaper: "*Jno Hawting* Watch & Clock Maker *Holywell* OXFORD ". Sun Slower Sun Faster; lobed calendar border.

HAWTING, WILLIAM — OXFORD (1770)

Son of John Hawting, clockmaker (above). Apprenticed to his father on 5 April 1770 for 7 years.

HEAD, JAMES — BURFORD (ante 1757)

A lost watch advertised on 25 May 1757.

HEATH, JOHN — OXFORD (1756)

Son of Elizabeth Heath, widow of Burford. Apprenticed to John Hawting, whitesmith of Holywell, Oxford, on 23 February 1756 for 7 years.

HEMINS, EDWARD (1) — BICESTER (late 17th century)

Turret clockmaker. Not traceable in parish registers; described as "senior" in Victoria County History. Said to have made the clock in Islip church (q.v.).

HEMINS, EDWARD (2) — BICESTER (1720-1744)

C. & W., turret clocks, bellfounder. Said to be son of Edward Hemins (1) of Bicester (above). Had a bellfoundry in Bell Lane, Bicester, and made bells between 1729 and 1743 for numerous churches in villages in Buckinghamshire, Northamptonshire and Oxfordshire, also some Oxford colleges and private mansions (Sharpe. 1953). The foundry closed down after casting a bell for Ambrosden church in May 1743. Died in 1744. An obituary notice appeared in the *Reading Mercury* for 4 June 1744 — *Whereas the ingenious Edward Hemmins, Clockmaker in Bicester, Oxon, is lately deceas'd, and has left several very curious Pieces of Work, some of which are unfinished: This is therefore to acquaint any Person that is a very good Hand, and can come well recommended, that he may meet with good Encouragement by applying to his Executors, John Walker, Richard Walls, and Joseph Hemmins, who live in the same town. N.B. They will be secure from being pres'd during their being employ'd by the above Persons.* It is not known if or how the business was continued.

His will is dated 17 April 1739 and was proved 13 April 1745. He left to his brother Benjamin "*all my Working Tools Instruments and Utensills that are made use of and belong to the Trade of a Gunn Smith*" also, "*the Gunns Barrells Locks & Stocks which I shall have and be possessed of att the time of my Decease*". To his brothers Joseph and Benjamin and the above mentioned executors the house "*wherein I now dwell and all that my Workhouse or Shop called or known by the Name of the ffoundering Shop scituate & being in Bisseter aforesaid with the appurtenances thereto belonging*". The residue was left on trust to be sold for division to his mother and his four children, Martha, Edward, John and Richard. His wife Elizabeth received clothes, furniture and five shillings.

Lantern clock, chapter ring $6\frac{1}{4}$-inches diameter, line fleur-de-lys between hours, one carved steel hand, central area with leafy scrolls signed, *Edw Hemins of Bister,* dolphin frets. Movement with verge escapement, outside locking plate marked EH, bell cast by Hemins with his mark, EH above a fleur de lys (coll. Beeson).

Lantern clock, chapter ring $6\frac{1}{2}$-inches diameter, fleur-de-lys between hours, one hand, central area with leafy scrolls signed, Edd Hemins Bister, frets leafy scrolls; movement originally with verge escapement, converted to anchor, ca. 1720 (coll. Beeson). Another similar in the Painted Room, Oxford.

Repeating bracket clock (Baillie).

Longcase, 30-hour, 10-inch square brass dial, one-hand, central area matte, lozenge between the hours, corner-pieces head in scrolling C. and W., 21; oak case, ca. 1730.

Longcase, 8-day, unusual arch-dial, the chapter ring forming the arch and supported by winged Father Time seated on a rectangular plate engraved, Edward Hemins Bisiter Fecit. The main 2-hand dial has a seconds ring and calendar aperture but no spandrels. Two subsidiary dials and two small winged cherub heads to left and right of the Father Time. Case of oak with a massive flat-topped hood, a middle section with an arch-top door and a glass pane, and a separable basal section of stained deal. The pendulum has two brass bobs, the upper swinging opposite the glass pane. Formerly at the Manor House, Souldern.

Longcase, 8-day, arch-dial 16½ × 12 inches, chapter ring with chamfered inner edge bearing ¼-hour marks, fleur-de-lys, between hours, signed on an applied silvered ribbon, Edward Hemins Bisiter, corner pieces a head in close reticulation; strike-silent ring in the arch with dolphin spandrel pieces. Movement with 5 pillars latched to the plates. Case oak 7ft. 10ins. high, flat topped hood, broken arch topped door (coll. Beeson).

Turret clock from Bicester signed, *Edward Hemins* FECIT, ca. 1735, Plate 4, fig. 5 (coll. Beeson) (see description).

Turret clock in Aynho church, Northants dated 1740 (see description).

Turret clock in a cupola of an outbuilding at Ditchley Park.

Turret clock in Steeple Aston church, ca. 1725 (q.v.).

HEMMINS, JOSEPH — BANBURY, BICESTER (1741-1744)

Turret clock. Executor and probably relative of Edward Hemins (2) above.

Made the clock for Banbury old church in 1741 which is now at South Newington (q.v.).

HERBERT, JOHN — OXFORD (1742-1794)

C. & W. Served a regular apprenticeship in Oxford and was admitted free by Act of Council on 12 January 1749 for £5 5s. 0d. and officers' fees. At Michaelmas 1743 he was licenced to put up a sign, " A Diall ", at his shop in St. Peter in the East. According to the survey of 1772 this was a house of 13 feet frontage at 52 High Street. In 1756 he was elected Chamberlain and in 1774 Bailiff. In 1742 married Sarah Hawting; a son John was baptised at St. Peter in the East on 3 June 1759. Robert Tawney was apprenticed to him in 1755.

A watch before 1755 (Baillie).

Looked after the clock and chimes of St. Mary the Virgin from 1749 to 1794.

HETH, ROBERT, THAME (1573)

Made a turret clock for St. Mary's church, Thame, in 1573.

HEWITT, OWEN — WATLINGTON (1823)

C. & W. In Watlington in 1823.

HEYTEN, ROBERT — OXFORD (1656)

Younger son of Robert Heyten of Chadlington, Oxon. Apprenticed to Michael Bird, watchmaker of Oxford, on 29 May 1656 for 8 years.

HICKMAN, EDWARD — OXFORD (1818-1826)

C. & W. In High Street in 1823. Elected common Councillor in 1826.

HIDE, CHARLES — BANBURY (1757-1773)

C. & W. Signed Borough Vestry Book in 1757 and onwards. Mayor in 1769. Died at Banbury 1773, bond dated 12 July 1773.

A lost watch advertised in 1759.

HINE, JOHN — WATLINGTON (1777)
C. & W. Sale of stock advertised in 1777.

HITCHCOCK, WILLIAM — OXFORD (1675)
Son of Thomas Hitchcock, husbandman of Ratley, Warwicks. Apprenticed to John Knibb, watchmaker of Oxford, on 4 September 1675 for 8 years.

HOBDELL, HENRY BASHARD — OXFORD (1832-1844)
C. & W., silversmith. At 128 High Street. Married Emily in 1839. Business carried on in 1852 by Emily Hobdell.

Longcase regulator in mahogany case.

HODGES, ANTHONY — OXFORD (1664-1672)
C. & W. Son of Anthony Hodges (also Hedges) cleric of Wytham, Berks. Apprenticed to Michael Bird, watchmaker of Oxford on 28 March 1664 for 8 years, and lived with him at 19 Cornmarket Street (Poll Tax 1667). Freedom of Oxford on 13 January 1672.

HOLLOWAY, EDWARD — BANBURY (1833-1842)
Cabinet-maker, joiner. Apprenticed to Charles Hollowell, cabinet-maker of Banbury. In Church Lane and Calthorpe Lane. Made cases for grandfather clocks to take the movements of J. G. Walford.

HOLLOWAY, THOMAS — GREAT HASELEY (1736-1764)
C. & W., turret clock. A freeholder of Great Haseley in the poll of 1754. Died or retired in 1764 and was succeeded by Thomas Stockford.

Repaired the clocks in the church of Great Haseley from 1736 to 1761. Made a new clock and dial for Great Haseley for £8 15s. 0d. in 1760.

HOWSE, CHARLES — BANBURY, LONDON (1750-1804)
C. & W. Son of William Howse, shopkeeper, Mayor of Banbury in 1760. Apprenticed to Anthony Marsh, clockmaker of London on 10 September 1750 for 7 years, fee £16 16s. 0d. Free of the London Clockmakers' Company on 8 September 1761. On the same date Richard Stone of Thame was apprenticed to him. Master of the L.C.C. 1787-1804. According to his watchpaper and bill-head he was in Great Tower Street, London.

HUMPHRIS, ?WILLIAM — THAME (1850)
C. & W. A William Humphries was a joiner and carpenter in Derrick's Lane, Thame in 1850. Baillie lists a mid 18th century watch by William Humphreys without locality.

Longcase, 30-hour, 12-inch square painted dial, calendar, signed, *Humphris Thame*; case oak 7ft. 10ins. high, hood with scrolled top; early 19th century (A. Seymour).

HUNT, JOHN — OXFORD (1795-1823)
C. & W. In High Street in 1823. Baillie gives the date 1795. Married Jane Castle of Oxford in 1813.

Longcase, 30-hour, 11-inch square painted dial, seconds ring, no calendar; oak case, inlaid with mahogany, scrolled top to hood; signed, *John Hunt, Oxford*.

Longcase, 8-day, painted arch-dial; stained case; signed *Hunt Oxford*.

HUNT, THOMAS — OXFORD (1777-1800)

C. & W. Matriculated at Oxford on 10 May 1777 as horologiorum fabricator privilegiatus. In business in Cornmarket Street.

Watchpaper: Finely engraved, " T HUNT *CLOCK & WATCH* Maker *Corn Market* OXFORD *Ladies Ears pierced*", in a landscape with on left a blindfold woman holding scales of justice in one hand and a watch in the other; inner border with " Mourning Rings on the Shortest notice Plate and Jewellery Sold & Repaired "; outer border serrate.

Watch, early 19th century (Baillie).

HYDE, THOMAS — HENLEY-ON-THAMES (1464-1499)

Smith. Admitted burgess on 7 September 1464. In 1487 he was one of the persons supplying armour to the Borough Assembly, viz. *vnum le trussynge doublet vnam galeam pretio xs*. In 1499 his tenement was occupied by another and his name was not included in the list of burgesses, presumably dead. He is the only smith recorded in Henley between the death of John Atte Lee and the appointment of Thomas Harington to look after the church clock and chimes.

JACKSON, JOHN — HENLEY-ON-THAMES (1783-1786)

C. & W. In Bell Street. Advertised for a journeyman on 18 February 1786. Baillie gives the date 1783 for a watch. A John Jackson of Henley married Mary Mellet of Henley in 1750.

Longcase, 8-day, 10-inch square dial; bird and flower marquetry case, 6ft. 4ins. high.

Watchpaper: " *JACKSON* Watch & Clock MAKER *Bell Street HENLEY upon THAMES* "; wide border with swags of flowers and leaves.

JEEVES, ANTHONY — OXFORD (1770-1774)

C. & W. A longcase, arch-dial, inscribed, " Anthony Jeeves, musical clockmaker from Oxford ", and " Daniel Davidson " in the arch; in Gillespies School, Edinburgh (Britten), ca. 1770 (Baillie 1774).

JEFFREY, JAMES — WITNEY (1853)

C. & W. In West End, Witney in 1853.

JEFFS, JOHN — OXFORD (1640-1650)

Bellhanger. Made the bellframes for some village churches in Buckinghamshire (Sharpe, 1951). Worked on the bells of St. Martin's Oxford from 1640-1648 and at St. Mary the Virgin, Oxford in 1650. Assisted John Raye to install the turret clock in St. Mary the Virgin in 1640.

JENNINGS, THOMAS — FRITWELL (1722-1773)

C. & W., whitesmith. Son of Robert Jennings of Fritwell, born 1722. Married Philippa Taylor of Fritwell on 1 February 1756. Buried at Fritwell on 29 August 1773 and was succeeded by his brother William Jennings (below).

A lost watch advertised 5 July 1776.

JENNINGS, WILLIAM — FRITWELL (1716-1780)

C. & W. Son of Robert Jennings of Fritwell, born 1716. Succeeded Thomas Jennings (above) by advertisement dated 11 September 1773. Buried at Fritwell on 12 October 1780.

A lost watch advertised 14 June 1823.

Watchpaper: "*Jennings* Fritwell", under a dial numbered in fives to 30, Slower Faster; outer border of monthly calendar lobes.

JONES, JOHN — OXFORD (1645-1709)

Son of Matthew Jones of Pentyrch, Glamorganshire. Matriculated at Oxford in 1662, scholar and Fellow of Jesus College, D.C.L. 1677; licentiate of the College of Physicians 1687; Chancellor of Llandaff. Buried at Llandaff 1709. Invented a clock driven by air from bellows operated by weights, described in Plot's *Natural History of Oxfordshire,* 1676, p.230.

JORDAN, JAMES — STADHAMPTON (1751-1776)

C. & W. Born ca. 1751. Married Mary Billings, both of Newington, Oxon, on 26 September 1776. Worked for some years at Stadhampton with Thomas Jordan (below). Baillie gives the date, early 19th century.

Longcase, brass dial, signed, Jas. Jordan Stadhampton.

JORDAN, THOMAS — STADHAMPTON (1770-1790)

C. & W. Repaired the church clock of Great Haseley several times 1770-1790.

Watchpaper: "*Thos Jordan* CLOCK *and Watch Maker* AT *Stadhampton OXON*"; monthly calendar border.

KALABERGO, JOHN — BANBURY (1812-1852)

C. & W., jeweller. Born in Lombardy. In Market Place and in High Street, Banbury, 1832 to 1852. Mainly known for numerous barometers in Sheraton banjo cases. His nephew came to him as an assistant in October 1851 and as they were returning from one of their business journeys in January 1852, the nephew shot him in the head on Williams-cote Hill $3\frac{1}{2}$ miles from Banbury. The murderer was tried and executed at Oxford on 22 March 1852.

Longcase, 8-day, painted arch-dial; oak case veneered and inlaid with mahogany, boxwood, etc., hood with Chippendale scrolls.

KENDALL, LARKUM — CHARLBURY, LONDON (1721-1795)

C. & W. Born at Charlbury in 1721. Apprenticed to John Jeffreys, London Clockmakers' Company on 1 April 1735 for 7 years. A famous chronometer maker at various addresses in London. Died in 1795.

KENNING, WILLIAM — BANBURY, LONDON (1648-1687)

C. & W. Son of Alis and Martin Kenning, freemason of Banbury, baptised 24 October 1648. Probably went to London after 1675 and may be the William Kenning who was threatened with prosecution in 1682 by the London Clockmakers' Company for exercising the art not being admitted, but who obtained the freedom in 1684.

In the Banbury Borough Accounts dealing with loans from Chantry Funds is an entry, " 5 March 1687, Hardings money . . . One other bond of John Barnes att ye Poleax for £6 15s. 0d. bound with him John Kenning & William Kenning which did not seale ".

Lantern clock, chapter ring $6\frac{1}{4}$ inches diameter, dial centre with a Tudor rose and tulips, movement in plates $5\frac{3}{4} \times 5\frac{5}{8}$ inches, semicircular iron loop and spikes from the back feet, crown wheel and verge escapement, vertical pivot bars wedged against a pair of fixed pins, endless chain drive. Signed at top of dial, " William Kenning Banbury fecit 1674 " (coll. R. Rowntree).

KENNINGTON, JOHN — OXFORD (1772)

C. & W. ? Son of Mrs. Kennington of High Street between Rose Lane and the Botanical Garden. Succeeded the late T. Rose in St. Clement parish, Oxford, in August 1772.

KEYTE, RICHARD — WITNEY (1770)

C. & W. In Witney in 1770.

KNIBB, JOHN — OXFORD (1650-1722)

C. & W., turret clocks. Sixth son of Thomas Knibb, yeoman of Claydon and Elizabeth (née Wise), born 21 January 1650. Joined his elder brother, Joseph, about 1664 in Oxford as apprentice or assistant, and lived with him first in St. Clements and a year or so later in Holywell Street, Oxford.

In 1670 Joseph Knibb became a member of the London Clockmakers' Company, so that some of the responsibility for the Oxford business fell early into young John's hands. But it was not until 13 September, 1672, when he was 22 years old, that John applied for the freedom of Oxford. Although he had none of the required qualifications the City Council agreed to a proposal of the Mayor, William Cornish, on the following 27 September to admit him on payment of £30. This fine John considered excessive so he enlisted the support of Brome Whorwood, who was one of the Members of Parliament for the City and also D.C.L. of Trinity College, which had agreed in 1667 to take Joseph as a privileged tradesman of the University. On 22 March, 1673, the Council " at the earnest request of B. Whorwood, Esq., and out of respect for him " reduced John's fine from £30 to 20 marks and he became freeman by Act of Council on 11 April, 1673. Although there exists a very small alarm lantern clock inscribed " Johannes Knibb Oxon fecit 1669 ", it is clear from this evidence that he was not entitled to sign his clocks before 1673, and that the products of the Holywell workshop must have been issued with the signature " Joseph Knibb " until as late as the end of 1672.

Domestic and civic life:

John and his wife Elizabeth attended St. Cross Church, Holywell, throughout their lives in Oxford and were buried there; all their children were baptised there. They had three sons and five daughters between 1679 and 1695; three of the daughters died young and a fourth, Mary, died at the age of 22 (see page 119). His eldest child, Elizabeth, and his three sons, John George and Joseph survived him.

After Joseph's departure to London John continued to live in the Twentyseventh House, which in 1678 was split up by the University into six separate leases. John was tenant in occupation of one tenement leased to a non-resident landlord, Thomas Woodward, a mercer. This tenement was part of the four-storeyed gabled house with a backyard, a pump and a " house of office " shared with two neighbours, a shoemaker and a tailor.

A few doors to the westward was another block of three tenements known as the Twentysixth House, leased by Merton College to the City authorities and occupied up to 1681 by a barber, a stationer and a butler. Some time between 1682 and 1692 John Knibb moved into one of the three tenements in number 26. By now he was fairly prosperous and a citizen of repute. In September 1697 the City subleased to him two of the tenements in number 26 for thirty-one years with option of renewal. These had a frontage to the highway of 102 feet and a depth of 40 to 61 feet. The site was originally part of the extra-mural moat or City Ditch near Smith Gate, which had been filled in and gradually built over during the 16th century; nevertheless, conveyances for a hundred years consistently described the site as " heretofore a dung hill ". The rent amounted to £20 a year plus " one couple of good fatt capons " to be given to the Mayor each Michaelmas, and a covenant to grind all corn and grain used in the household at the City's Castle Mills. Some idea of the capacity of Knibb's house may be gleaned from the Poll Tax of 1702 when it accommodated himself, his wife, two sons, two daughters, a female servant, and two apprentices as well as the workshop. He let the smaller house next door to a coffeeman and a barber.

John Knibb's rise in civic affairs began in September, 1686, when he was appointed one of the twenty-four members of the City Council. On 23 October, 1688, he was elected Bailiff. This was the occasion when James II, in a final attempt to regain the confidence of his people, issued a proclamation restoring to corporations their charters and privileges, which he had earlier cancelled. The election took place publicly with full ceremony and the ringing of bells at Carfax. Twelve days later William of Orange landed at Torbay. At the coronation of William and Mary on 11 April of the following year the City of Oxford exercised its privilege of serving in the butlery. John Knibb was one of the six persons in attendance on the Mayor at Westminster, which service gained a knighthood for the Mayor, Robert Harrison. Knibb was Bailiff again from 1690 to 1696. Bailiffs were rent collectors and treasurers and also had the custody of offenders; on appointment each paid a fee of £5 which entitled him to certain profits of the Northgate Hundred.

On 11 October, 1697, John was chosen one of the eight assistants to the Mayor for which he had to pay £5 and £8 for entertainment. A year later on 19 September, 1698, he was elected Mayor with more fees to be paid. In 1700 he was Keykeeper and assistant member of the Mayor's Council. (He was not mayor as incorrectly stated by several authors). A Mayor could not be re-elected until after an interval of six years and John Knibb did not serve again as Mayor until 1710. As Keykeeper thereafter he continued to be active in civic affairs. In 1716 he was chosen Alderman.

Among his friends were the diarists and historians, Anthony à Wood and Thomas Hearne. In his diary for November, 1693, Wood notes that he borrowed a brass watch from Knibb; two years later Wood died but history does not record the fate of the watch. The antiquary Hearne, uncomplimentary as was his wont, wrote in November 1716, *" This Nibb is a man of so little understanding that he was never known to laugh "*. But it is Hearne who gives us the correct date of John Knibb's death; his diary records: *" 1722 July 19. Last night, about 8 Clock, died suddenly Mr. Alderman Knibb of Oxford, an old, quiet, harmless Man abt. 4 score years of Age. He lived by Smith Gate in Holywell Parish. ' Tis said he eat his supper heartily, went round New Parks with his Wife sate himself down in his chair and died. A few days since I talked with him about Antiquities, when he told me he hath seen Anslap Spire in Bucks from Brill he having, it seems, some Estate or Fortune at Anslapp."* The estate referred to by Hearne was that inherited from his brother Joseph in 1712. After paying survivors' legacies and an annuity to Joseph's wife John received the residual personal estate but Joseph's real estate in Hanslope, Buckinghamshire and in Farnborough, Warwickshire was held by him in trust for his own eldest son, John.

> Near this Place lyeth Interr'd the
> Daughters of IOHN & ELIZ: KNIBB.
> of this Parish.
> Hannah. HANNAH & Jane. & MARY.
> as also
> IOHN KNIBB Alderman OF ẙ CITY
> Died Iuly 22. 1722. Aged 72.
> ELIZ. the Wife of Joseph
> KNIBB.
> who Died Dec. 5. 1726. aged. 84.
> Mrs ELIZ.^th KNIBB Wife of the above
> Alder.^n KNIBB died Decr 23^d 1740.
> OHN KNIBB Junr Alder.^n of this CITY
> died Feb: 14th – 1754.
> Also DEB.^H his Wife died: A^it g^fh 8th: 1751.

John Knibb was buried in St. Cross Church on 22 July, 1722, according to the register. Some years later a memorial stone tablet to the Knibb family was placed on the north wall of the aisle of St. Cross Church and this incorrectly records the date of his death as the 22nd instead of the 18th (see adjoining text-figure). By his will John left all his personal and leasehold estate to his wife as executrix except for one Tenement to John, junior, from which an annuity was to be paid to her. He stated that the reason no other provision was made for his children was that their uncle Joseph had amply provided for them.

John Knibb, Clockmaker.

His early progress in business on his own coincided with the remarkable attraction of Oxford for young men who wished to be apprenticed to the 5 master clockmakers in the City. This boom is referred in the Historical Review, p.21. Between 1673 and 1722 Knibb employed 10 apprentices who were taken on in the following succession: 1673, Samuel Aldworth; 1675, William Hitchcock; 1679, Thomas Lidbrook; 1681, Mathias Unite; 1682, John Ford; 1686, John Goweth; 1696, John Free; 1698, Thomas Gillett; 1706, George Wentworth; 1710, circa, Humphrey Brickland. For further particulars of these men see the Biographical Dictionary. Unless Hitchcock and Lidbrook failed to complete their full terms, Knibb was employing 3 apprentices between 1681 and 1688 and, in addition, Aldworth as journeyman. Thereafter he did not have more than 2 apprentices at one time. The last recruit, Brickland, after finishing his training, remained as assistant until Knibb died in 1722.

As is summarised below John Knibb sold all sorts of clocks including lantern clocks, 30-hour wall clocks, longcase clocks in the cheaper grades as well as high class movements in the most expensive London-made cases. His bracket clocks, whether time-pieces or elaborate repeaters, were generally of first class workmanship. There seems to be little doubt that most of his best stock was produced in co-operation with his brother, Joseph, at least to the end of the 17th century. One bracket clock is known which is also signed by Joseph. When Joseph retired from London to Hanslope, Bucks, John had an opportunity to acquire some of the stock or unfinished material which was not included in the Suffolk Street sale in 1697. Later in 1712 John inherited Joseph's business at Hanslope and, for some time, continued to manage it. One longcase clock signed *John Knibb Hanslap* is known. Towards the end of his life, when he employed only one apprentice, he seems to have obtained some clocks from Samuel Aldworth, his former journeyman (q.v.). At any rate an arch-dial bracket clock now in the Victoria and Albert Museum, made by Aldworth, has on the dial a superimposed tablet with Knibb's name.

Compared with that of his contemporaries in Oxford John Knibb's output of clocks was copious. He had almost a monopoly of the better class business. Nevertheless very few of his watches are known, possibly because of their lower chances of survival.

A Summary of Types of Clocks by John Knibb

Lantern clocks: with verge escapements, with or without alarm, dolphin or foliate frets, dial centre with leaf scrolling, chapter ring with simplified fleur-de-lys in 3 designs; 8 to 15 inches high. Signed below XII, *John Knibb Oxon* or at base of chapter ring or front fret, *John Knibb Oxon fecit*, (Ashmolean Museum, Dr. H. Baines, R. Weideman, Lady Stott, W. Summers). See frontispiece.

Wall clocks: with verge escapements and lantern type movements, with or without alarm. (a) one-hand, 5 to $5\frac{1}{4}$ inch square dials, centre matte or engraved with tulips or scrolls, spandrels engraved with winged cherub head casts. Signed below chapter ring, *John Knibb Oxon fecit*. Hood cases walnut or ebony veneered, with dome or portico top.

(b) two-hand, 5 to $5\frac{3}{4}$-inch square dials, dial centre matte, spandrels engraved or with cherub head casts. Signed at base of chapter ring, *John Knibb Oxon,* or below it, *John Knibb Oxon fecit*. One case of walnut veneer with a carved pediment (coll. R. Lee; coll. Beeson).

(c) two-hand, 12-inch square dials, centre matte, spandrel pieces cherub head casts, chapter ring as in longcase clocks. Signed below chapter ring, *John Knibb Oxon fecit*. Walnut veneered hood cases (one illustrated *Ant. Horol.*, June 1956).

Bracket clocks: (a) with tic-tac escapement. Dial $6\frac{1}{4}$ inches square, calendar aperture below XII, centre matte, typical spandrels and chapter ring. Signed at base of dial, *John Knibb of Oxford*. Movement with 6 latched pillars and latched dial feet, 8-day striking, internal locking-plate; back plate plain, signed, *John Knibb of Oxford*. Case 12 inches high, ebony veneered, cranked bar handle, dome mount a large winged cherub head, ball finials, door with 2 repoussé mounts, bun feet (coll. Beeson, Plate 21, fig. 42 & Plate 24, fig. 47).

(b) with verge escapements — Timepiece. Pull-repeating, dial $6\frac{1}{4}$ inch square, borders engraved, calendar below XII, signed at base of dial *John Knibb Oxon*. Case walnut veneered, 2 mounts on the door. Plate 22, fig. 43. Hour-striking, dial $6\frac{5}{8}$ inches square, centre matte, engraved calendar aperture below XII, dial borders engraved with leaf scrolling, winding holes ringed. chapter ring $5\frac{7}{8}$ × 1 inches, typical fleur-de-lys and cherub head spandrel-pieces. Ring signed *Jo Knibb Oxoniae*. Movement with 5 latched baluster pillars and latched dial feet, 8-day, outside spoked and numbered locking-wheel, simple verge pendulum cock; back plate engraved with leafy scrolls, signed, *John Knibb Oxoniae fecit*. Case ebony veneered, $12\frac{1}{4}$ inches high, 3 repoussé mounts on dome, 2 mounts on door, sinuate with a winged head (Mrs. G. Hudson).

Others with $6\frac{1}{4}$-$6\frac{3}{4}$-inch dials, calendar aperture below XII or above VI, repeating or not repeating, normal or skeleton chapter ring. One with silver mounted dial and silver hands. Movement with latched pillars, outside locking-plate, plain, decorated or spoked, also rack-striking. Back plates with leaf scrolling or sunflowers or tulips. Signatures, *John Knibb Oxon fecit*, and, *John Knibb Oxon* and *John Knibb Oxoniae*.

Cases veneered ebony or walnut with up to 3 dome mounts, repoussé (early) or cast (later) and up to 3 mounts on the door. Usually a wood fret to the door, rarely also on the inner frame (Plate 22, fig. 44). Various owners and several illustrations in text-books, journals, etc.

Longcase clocks: (a) 30-hour, 11-inch square dial, one hand; 4-post movement with outside locking-plate. Signed, *John Knibb Oxon*.

(b) 8-day, 10-12 inch square dials, calendar aperture below XII or above VI, without or with a seconds ring, winding holes later ringed on a matte surface, chapter ring with $\frac{1}{4}$-hour marks and a narrow minute hand usually numbered in fives, dial borders engraved. Movements usually with latched pillars and dial legs, outside or inside locking-plate, disc or spoked; pendulum with wing nut suspension from a separate cock, the rod hooked on below the crutch fork, or normal type; one second or one and a quarter seconds beat.

Plate 17, fig. 29 shows a longcase clock, 12-inch square dial, bordered, signed at base, *Johannes Knibb Oxoniae Fecit*, seconds ring, date aperture between winding holes. Movement 8-day. Case 8 feet high, burr walnut, hood surmounted by a carved cresting with cherub head and floral garland, ca. 1685 (Brigadier W. E. Clark, Asmolean Museum).

(c) month, $10\frac{1}{4}$ and 12-inch dials. Movements with 2 trains or 3 trains, some with bolt and shutter; pendulum one second. Other features as in 8-day clocks.

Signatures to (b) and (c), *Johannes Knibb Oxoniae fecit*, and, *John Knibb Oxon Fecit*.

Cases: plain oak; walnut veneered, figured or burr, sides panelled with stringing; marquetry on walnut or oyster olive wood, broken arch or oval or hexagonal panels and $\frac{1}{4}$-round corners of bird and flower or vase and flower patterns or medallions of olive and various coloured woods. Hoods usually flat topped or with carved cresting or with ball finials, twist pillars, convex moulding below, lift-up or with hinged door.

Various owners and numerous illustrations in text-books, journals and sale catalogues, (e.g., *Antiq., Horol.,* March 1955, p.vi and June 1959, back cover).

It is not yet possible to assign any chronological significance to the various forms of signature on clocks.

Watches: Verge watch with curiously wrought pillars, ca. 1690, signed, *John Knibb at Oxon*, Guildhall Museum (Britten).

Verge watch, chased silver dial, with two dragons within the chapter ring each supporting a shield with the Arms of the City of London, signed, KNIBB LONDON; movement with tulip pillars, elaborately pierced cock and sides, signed, IOHN : KNIBB : AT : OXFORD, the word Oxford obliterated; plain silver case, SB No. 507, and pair case with silver chain.

A brass watch lent to Anthony à Wood (see above).

Turret clocks: Made the clock for St. John's College in 1690. Repaired Yarnton church clock in 1703.

Maintained Wadham College clock 1673 to 1722.

HANSLOPE: Only one longcase known. 8-day, 11-inch dial, calendar aperture above VI, chapter ring and small seconds ring not silvered, winding holes ringed, ca. 1712. In an earlier case of bird, butterfly and flower marquetry in broken-arch panels on walnut, sides with stringing; hood with twist pillars, flat top. Movement with 4 pinned pillars, inside locking-plate, normal pendulum suspension. Signature on chapter ring, *John Knibb Hanslap* (coll. Beeson).

KNIBB, JOSEPH (1) — OXFORD, LONDON, HANSLOPE (1640-1711)

C. & W., turret clocks. Fifth son of Thomas Knibb, yeoman of Claydon, and Elizabeth his wife, baptised 2 February 1640. He was not apprenticed in Oxford or in London and it is possible that he learnt his trade from his cousin Samuel Knibb (see below). If he joined Samuel about 1655 he could have completed 7 years apprenticeship before Samuel joined the London Clockmaker's Company in 1663. Joseph then decided to start business in Oxford where there were good prospects of the better class trade and also of patronage. He settled first in St. Clement's outside the City's liberties; his name does not appear in any of the Hearth Tax and Poll Tax lists of that period and the only evidence is the statement in a petition by the freemen smiths and watchmakers of the City. In 1665 or 1666 he moved to Holywell Street, outside the City Wall, but then within the City liberties. It was a street in which tradesmen of all sorts had their shops and workshops, most of them as direct leaseholders of Merton College or as sub-lessees. He found accommodation in a house of five tenements known as the Twentyseventh on the south side of Holywell, leased by Merton College to the University and by the University to William Harris, a joiner. (A view of Holywell Street taken from David Loggan's map of 1675 is given in Beeson, 1960).

At this time plague was raging in London; Charles II came to Oxford with his Court, Parliament and Law Courts. The resentment of the City freemen to trading by "foreigners", which, only a year or so previously, had been expressed in a petition to the King, again became active and Knibb was forced to apply for the freedom. On 1 February 1667 the Mayor, William Bayley, proposed his admission but it was respited to the next Council meeting. The smiths and watchmakers then put in a petition opposing Knibb and freedom was refused; the chief objectors were Michael Bird, John and Richard Quelch, and William Young (q.v.).

But Joseph Knibb had an alternative which was to become a privileged tradesman in the employ of the University. He was living as subtenant in a house of which the University was landlord and he occupied a scheduled University shop. He therefore matriculated. The entry in the Matriculations Register for 24 August 1667 reads, *Josephus Knibb an. n. 27 fil. Tho. Knibb de Claydon in par. Cropredy Oxon pl. Hortulanus Coll. Trin.* His choice of Trinity College — only a short distance from the Holywell shop — may have been on the advice of Brome Whorwood of that College, but the designation, hortulanus or gardener, is inexplicable. Actually, admission to the College was not completed, Knibb did not sign the Admission Register and his name is not among the college employés.

Another petition was put in to the City Council on 29 October 1667, this time by the "Clockmakers and Watchmakers of the City", and the Council resolved that Knibb and any others who offend were to be suppressed.

The matter was again debated on 22 January 1668 and further on 11 February 1668 ... *Mr. Mayor acquainted this House that Joseph Nibb Clockmaker who formerly sett upp shopp in the parish of Holywell in the Suburbs of this Citty upon Accompt of being a Gardener to Trinity Colledge did now make his application to this Citty for a freedome waveing the power of the University who formerly endeavoured to Maynteyn him to keepe shopp upon this accompt.* This means that Knibb had offered the Council a compromise solution that he would withdraw his claim to be a privileged tradesman if he were made a freeman of the City by payment of a fine. The plea was accepted and he was admitted for a fine of 20 nobles (£6 13s. 4d.) and a leather bucket.

It so happened that the almost perpetual disputes between the City and the University about their respective privileges were amicably settled (at least temporarily) in the previous month, January 1668, and partly through the mediation of Brome Whorwood, M.P. for the City. The Council seems to have decided it would be better to have Knibb as a freeman under their control than to offend the University again by expelling one of its members. In fact, when one of the Bailiff's serjeants arrested a University privileged person a few weeks later, the Council suspended the serjeant " since they are unwilling to make any breach in the articles of peace lately agreed upon between the City and the University ".

Joseph Knibb was now 28 years old. On 31 July 1668 he took Peter Knibb (q.v.) as apprentice, and on 1 October 1669 another apprentice, Thomas Smith of Bloxham. Business was now prospering and a token was issued, which E. T. Leeds (1923) considered was designed and engraved by Knibb. Obverse, " *Joseph Knibb: Clockmaker in Oxon* " in 4 lines; reverse, A clock dial with one hand and IK; a beaded border round the field on both faces. In 1669 Knibb turned his attention also to turret clocks (see Wadham College, and St. Mary the Virgin).

In January 1670 he was made free of the London Clockmaker's Company. This event occurring so soon after his efforts to join the Oxford freemen needs explanation. It may be that Samuel Knibb died about this time and Joseph had to administer his affairs and perhaps take over his London business. At any rate his presence in London can only have been arranged by leaving his brother John in executive charge at Oxford for 2 or 3 years. In 1672 he took an apprentice in London, Patrick Vans.

Joseph's marriage has not been traced; his wife Elizabeth was born in 1642 and her maiden name was probably West since she refers in her will to a sister-in-law, Mrs. Diana West. Only one child is recorded, Thomas, who died unmarried and was buried at Hanslope on 10 January 1703.

In April 1697 he sold up his London business and retired to Hanslope, Bucks. As he made his will on 23 August 1697 it is possible that serious illness was the main cause of his retirement. At Hanslope he continued to make clocks several of which still exist. He was buried at Hanslope on 14 December 1711, aged 71 and his will was proved on 24 May 1712, by his brother John as superstites. The trusteeship which had reposed in Isaiah Knibb, clerk, son of George Knibb of Farnborough, Warwicks, and in his cousin, Richard Wyse, Yeoman of Foskett, Bucks, did not operate. Joseph Knibb was a man of considerable means. He owned " messuages, lands and tenements " in Hanslope and in Farnborough. He settled an annuity on his wife Elizabeth in full recompense of her dower and gave legacies to his brothers, George and John, to all the children of his three sisters and to the three daughters of John. He also entailed the landed estate so that it passed to John and then to John's eldest son and to yield a further legacy of £200 to each of the other children of John.

After her husband's death Elizabeth Knibb came to live with her brother-in-law, John, in Oxford and, after his death in 1722, with his son, John Knibb junior, until she died at the age of 84 and was buried in St. Cross Church, Holywell (see text-figure, page 119).

No conveyances or other deeds have been discovered to show how Joseph acquired his country properties.

Joseph Knibb's Oxford Clocks

Lantern clock: Centre of dial and spandrels engraved with tulips and other flowers, unusual frets engraved with floral sprays, signed at base of front fret, IOSEPH KNIBB OXON FECIT, alarm dial with arabic numbers. Movement, crown wheel and verge escapement, hammer pivoted on top plate, alarm pulley outside the back plate, rope driven; ca. 1670. Plate 20, figs. 39, 40.

Wall clock: 30-hour, $5\frac{1}{2}$-inch square dial, centre and spandrels finely engraved with flowers and foliage, signed at base, *Joseph Knibb Oxon Fecit*. Plated 2-hand movement, ting-tang $\frac{1}{4}$-striking on 3 bells by vertically pivoted hammers, the verge and crown wheel arbors offset obliquely. Hooded case, walnut veneered on carved bracket. Plate 23, fig. 45 (coll. L. R. Bomford).

Longcase clock: 8-day, $8\frac{1}{4}$-inch square dial, centre with Tudor rose on matte ground, shuttered maintaining power, calendar aperture above VI, signed at base, *Joseph Knibb Oxon Fecit*. Movement with verge escapement and $12\frac{1}{2}$-inch bob pendulum, outside locking-plate, vertically pivoted hammer shaft. Case ebony veneered, hood with portico top and unusual silver gilt mounts, case door and sides panelled, 6ft. 3ins. high, $8\frac{1}{4}$-inches wide trunk; ca. 1670. Plates 17, fig. 28, and 23, fig. 46 (coll. L. R. Bomford).

Turret clocks: Wadham College (Plate 5, figs. 7 and 8). Fitted a pendulum to the clock of St. Mary the Virgin (q.v.).

KNIBB, JOSEPH (2) — OXFORD, LONDON (1695-1722)

Youngest son of John Knibb, clockmaker of Oxford, baptised at St. Cross Church, Holywell, on 20 February 1695. Apprenticed on 15 June 1710 to Martin Jackson, London Clockmakers' Company, for 7 years. There is no record of his having become free or of his return to Oxford. He was alive in 1722 and received only one shilling under his father's will because he benefitted by a legacy of £200 from his uncle Joseph.

KNIBB, PETER — OXFORD, LONDON (1651-1679)

C. & W. Son of George Knibb, yeoman of Farnborough, Warwicks, and Jane, his wife; baptised 30 July 1651. Apprenticed to Joseph Knibb, clockmaker of Oxford, on 31 July 1668 for 7 years. When his master went to London he accompanied him and after 2 years as journeyman became free of the London Clockmakers' Company on 5 November 1677. In 1679 he returned to Farnborough and there married Katherine Shrewsbury on 29 April.

A bracket clock and a watch are known.

KNIBB, SAMUEL — NEWPORT PAGNELL, LONDON (1625-?1670)

C. & W., instrument maker. Third son of John Knibb, yeoman of Claydon and Warborough, his wife, baptised 15 November 1625. Cousin of Joseph and John Knibb (above). He has been confused with Samuel Knibb, baptised 28 December 1637, fourth son of Thomas Knibb of Claydon, and elder brother of Joseph Knibb (1); and also with Samuel Knibb who matriculated at Christ's College, Cambridge in May 1647.

Where he learnt his trade is unknown, or when he started business in Newport Pagnell, Bucks, but it was probably before 1655. There are no entries in the Newport Pagnell registers of his marriage or children. It is inferred that he took his cousin Joseph as apprentice from about 1655 to 1662. He became a member by redemption of the London Clockmakers' Company in 1663 and entered into partnership with Henry Sutton, the instrument maker.

He died about 1670 or a little later and it is suggested that Joseph Knibb was concerned in winding up his affairs or taking on his London business. There is no evidence that Samuel owned the Knibb property in Hanslope, Bucks.

KRATZER, NICHOLAS — OXFORD (1487-1547)

Astronomer. Born in 1487 in Bavaria. Admitted in March 1516 to Corpus Christi College, Oxford, and became a Fellow in 1517; B.A. of Wyttenburg and Cologne; incorporated at Oxford on 9 March 1522, M.A., on 20 March 1522. Astronomer and clockmaker to Henry VIII. Supposed to have designed the astronomical clock at Hampton Court Palace in 1540 (Britten), which is now credited to Nicholas Cursian. Devised column sundials for Corpus Christi College (q.v.) and the church of St. Mary the Virgin, Oxford (q.v.) in 1520. A gilt clock dial made for Cardinal Wolsey and a manuscript drawing of the Corpus Christi polyhedral dial are in the Museum of the History of Science, Oxford.

LAMB, JAMES — BICESTER (1818-1853)

C. & W., glass, china and earthenware dealer. Married at Bicester on 14 March 1818 Ann Neale, children born 1818-1827. In Market Square. Verge watch, enamel dial, front wind, signed, *J. Lamb* BICESTER no. 9953. (Ashmolean Museum).

LAMPREY, BENJAMIN — BANBURY (ca. 1696-1721)

C. & W. Married three times but not in Banbury, first, to Jane about 1696 who died in September 1708, secondly, about 1711 to Elizabeth who died in February 1715 and thirdly, about 1719 to a wife who survived him. Of his 13 children several died young; one son, John, continued the business. He was buried 18 September 1721. Bond, renunciation and inventory of his effects dated 8 June 1722.

Lantern clock signed, *Benj Lamprey Banbury* (late 17th cent.).

LAMPREY, JOHN (1) — BANBURY (1704-1759)

C. & W. Son of Benjamin Lamprey, clockmaker, (above), baptised 7 April 1704. Married Mary about 1725, not in Banbury, children born 1726 to 1742. Attended Borough Vestry Meetings in 1741 to 1748. A freeholder of this name voted in the Poll of 1754.

His will and bond proved 15 August 1759.

Longcase, 30-hour, 10-inch square dial, one hand of crossed loops, central area engraved with weak scrolls, leaves and a Chinese-like house, signed below centre, *John Lamprey* BANBURY; chapter ring with $\frac{1}{4}$-hour divisions and nothing between the hour numerals; spandrel pieces urn in foliage, C. and W. 24. Movement 4-posted with iron plates. Case ebonised oak, flat topped hood, ca. 1750 (coll. Beeson).

Longcase, 8-day, mahogany case, ca. 1740.

Cleaned and mended Claydon church clock in 1746.

LAMPREY, JOHN (2) — BANBURY (1734-1771)

C. & W. Son of John Lamprey (1) above, baptised 25 January 1734. Married Ann Carpenter in Banbury on 11 October 1757.

Repaired the clock of the second Town Hall of Banbury in 1771.

LANE, WRIGHT — OXFORD, LONDON (1676-1689)

C. & W. Son of Nathaniel Lane, cleric late of Dolton, Devonshire. Apprenticed to Michael Bird, watchmaker, Oxford on 8 July 1676 for 7 years. Free of the London Clockmakers' Company in 1687. Freedom of Oxford on 13 September 1689. Greenaway Curtice was made over to him from Michael Bird in Oxford to complete his apprenticeship in 1689. Lane may possibly have been concerned in the take-over of the business of M. Bird, senior.

LANGLEY, THOMAS — OXFORD (1673-1687)

C. & W. Son of Thomas Langley, mercer of Stanford, Berks. Apprenticed to John Quelch, watchmaker of Oxford, on 4 April 1673 (from 24 June last) for 8 years. Freeman of Oxford on 2 December 1687.

LAWRENCE, THOMAS — THAME (1759)

Possibly son of William Lawrence, clockmaker of Thame, to whom he was apprenticed on 21 October 1759 (Registered at P.R.O.).

LAWRENCE, WILLIAM — CUDDESDON, THAME (1744-1759)

C. & W., turret clock maker. Second marriage to Margaret Eales, widow, at Aylesbury on 5 October 1764, then described as clockmaker of Thame. Advertised in Thame in 1744 and 1755 as a maker of large clocks and chimes. Took as apprentice Thomas Lawrence on 21 October 1759.

Longcase, 8-day, square dial, matted central area engraved with birds in flight, large seconds ring, single date aperture, pierced spade hands; chapter ring silvered; spandrel pieces a small cherub head in lose reticulation; signature, *William Lawrance Cudsdon.* Case oak, hood with Chippendale scrolls 3 brass ball and eagle finials, shell marquetry on door.

Longcase, 30-hour square brass dial, signed *Wm. Lawrence Cudsdon*, stained oak case.

Longcase clock, signed *Wm. Lawrence Thame* (no details).

LEVI, BENJAMIN — WITNEY (1770)

Baillie gives the date "from 1770" based on a watch.

LEVI, ISRAEL MORRIS — OXFORD (1853)

C. & W., silversmith, jeweller at 67 St. Giles Street in 1853.

LIDBROOK, THOMAS — OXFORD (1679)

Son of Robert Lidbroocke (also Ladbrooke) yeoman of Great (or Burton) Dassett, Warwicks. Apprenticed to John Knibb, clockmaker of Oxford on 26 July 1679 for 7 years [Robert Lidbrook, then of Knightcote, Warwicks, married Mary Knibb of Claydon in 1657. A Lidbrook was kinsman to the wife of Joseph Knibb].

LOCK, EDWARD — OXFORD (1729-1813)

C. & W., goldsmith, silversmith. Born 1729. Married Hannah Bridge of Bicester on 14 June 1759 at All Saints Church, Oxford. Freedom of Oxford by Act of Council on 2 February 1760. Bailiff in 1766. Elected honorary member of the Taylors' Company in 1792. Mayor of Oxford 3 times in 1776, 1791 and 1806. In business at No. 7 High Street, north side, the shop having a frontage of 20 feet. Leased from the City houses in Castle Street from 1764 to 1806. Took as apprentices Joseph Lock (free 1782) and John Davis (free 1792). Died 14 September 1813 aged 84. His memorial is in All Saints Church, Oxford.

Supplied silverware to colleges between 1772 and 1804.

Lost watches advertised in 1764 and 1777.

LOCK, HENRY — OXFORD (1795-1802)

C. & W. ? Son of Edward Lock, above. Apprenticed to Richard Pearson, watchmaker of Oxford. Freedom of Oxford on 5 July 1802.

MACE, JOHN — TASTON (early 19th cent.)

C. & W., blacksmith. Kibble (1928) mentions "an old style bedstead frame clock" by Blacksmith Mace of Taston, and one signed "Mace Ditchley" once owned by Viscount Dillon.

Longcase, 30-hour, 11-inch square painted dial, calendar. Oak case, shell inlay; hood with scrolls and 3 finials (R. Collier).

MACE, ROBERT HARRIS — TASTON, SPELSBURY

C. & W., blacksmith, shopkeeper in Taston in 1853.

A headstone to Robert Mace of Taston in Spelsbury Churchyard has a slight variation of a popular epitaph to blacksmiths found in other parishes.

> My sledge and hammer lie declined
> My bellows too have lost their wind
> My fire's extinct my coal's decayed
> And in the dust my vice is laid
> My forge is left my iron is gone
> My nails are drove my work is done.

MARRINER, WILLIAM — OXFORD (1774-1782)

C. & W. Son of John Marriner, baker of Oxford, St. Peter in the East. Apprenticed to Thomas Reynolds, whitesmith of Oxford, on 18 April 1774 for 7 years. Freeman of Oxford on 26 April 1782.

MARSHALL, THOMAS R. — OXFORD (1852)

C. & W., goldsmith, optician, engraver at 31 High Street in 1852. His Directory advertisement states, Watches, Clocks, Plate and Jewellery Repaired on the Premises by competent workmen.

MASEY, THOMAS — OXFORD (1550)

Smith. Repaired the 1523 clock in the church of St. Mary the Virgin, Oxford (q.v.) in 1550.

MATHEWS, CHRISTOPHER — OXFORD (1667-?1700)

Painter. In 1667 he was living as a journeyman (wages £8) to Richard Hawkins, painter, at 116 High Street, Oxford. From 1683 to before 1710 he was subtenant in part of 9 and 10 Oriel Street. In 1694 he served as Constable to the City Council.

Painted the wooden dial of the clock of Yarnton Church in 1682.

Painted and gilded Kratzer's dial at St. Mary the Virgin in 1688.

MAY — Birds-in-flight Dials

A popular form of decoration on some 18th century brass dials uses birds on the wing and with or without baskets of fruit. The birds, engraved on a matte ground, have the beak open or holding a leaf, the tail either forked or with expanded feathers, and the wings in various positions. They occur in pairs or up to 8 placed in different parts of the dial within the chapter ring. The basket containing foliage and fruits, apparently apples, may be placed above the date-aperture or in the upper part of the dial. Presumably the symbolism is Peace and Plenty although the dove often resembles a hawk or a bird of paradise.

An early example of 2 birds and a basket occurs on a longcase clock by John May of London and later examples on 8-day and 30-hour clocks by John May of Witney (q.v.). Plate 14, fig. 23. Other makers using these motifs include Nicholas Lambert of London, John Deverell of Winslow, Tobias Gilks of Chipping Norton, William Lawrence of Cuddesdon, Francis Webb of Watlington, Jethro Tull of Newbury, Edward Billington of Market Harborough and Roger Lee of Leicester.

These features were used for some 70 years and fairly widely in the Midlands; the considerable variation in design represents the conceptions of several artists. J. A. Daniell (1952), however, suggests that in Leicester, at least during the first half of the 18th century, one itinerant engraver obtained orders for dial engraving from local clockmakers.

MAY, EDWARD (1) — HENLEY ON THAMES (1680)

Longcase with lantern clock movement, ca. 1680 (Britten).

MAY, EDWARD (2) — HENLEY ON THAMES (1755-1795)

C. & W. Advertised as removed to Market Place, Henley on 30 June 1755. Baillie gives the date 1795.

Longcase, 30-hour, brass one-hand dial, chapter ring signed, *Edwd May* HENLEY.

MAY, EDWARD — OXON (1726)

C. & W. Took Benjamin Thorp as apprentice in 1725. Possibly the same as the following.

MAY, EDWARD — WITNEY (1725)

C. & W. Longcase, 30-hour, one-hand brass dial, ca. 1725 (Britten).

MAY, JOHN — WITNEY (1725-1795)

C. & W., In Witney in 1771. Baillie gives the dates ca. 1725 and 1795, Britten ca. 1700. Possibly two men.

Longcase, 8-day, $10\frac{3}{4}$-inch square dial, central area matte with 2 birds in flight flanking a basket of fruit and scrolls, seconds ring, calendar, chapter ring with $\frac{1}{4}$-hour marks, fleur-de-lys, minute figures in fives, signed John May WITNEY, corner-pieces C. and W., fig. 26; 2 pierced hands. Case mahogany, flat-topped hood (C. L. Whitehead).

Longcase, 8-day, 12-inch square dial, central area matte, seconds ring, calendar, chapter ring silvered, no marks between the hours, large minute figures in fives, signed *John May* WITNEY, corner-pieces C. and W., fig. 22, 2 hands. Movement rack-striking. Case carved oak, flat-topped hood, 6ft. 8ins. high (W. J. Brooke).

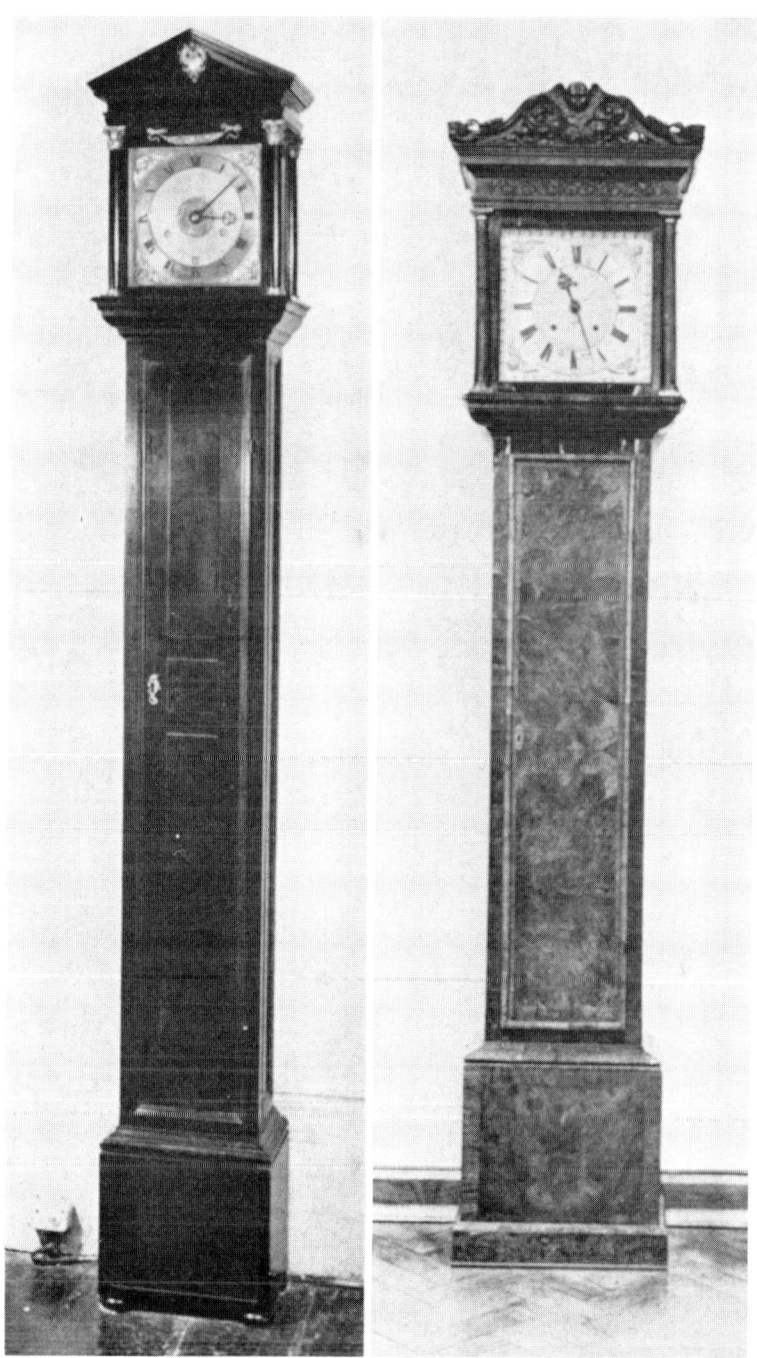

PLATE 17

Fig. 28, JOSEPH KNIBB of Oxford, 8-day verge movement, ebony veneered case, 6ft. 3ins. high. (Photo by courtesy of L. R. Bomford).

Fig. 29, JOHN KNIBB of Oxford, 8-day anchor escapement, burr walnut case, ca. 8ft. high. (Photo by courtesy of the Ashmolean Museum).

Fig. 30   Fig. 31

Fig. 32   PLATE 18   Fig. 33

Figs. 30 and 31, JOHN HARRIS of Oxford, Watch with early balance spring and cock foot, 1680. (Author's photos by courtesy of the Ashmolean Museum).

Figs. 32 and 33, WILLIAM WALLEN of Henley-on-Thames, Verge watch in silver pair case, 1725. (Photos by courtesy of the Ashmolean Museum).

Fig. 34　　　　　　　　　　　　　　　　　　Fig. 35

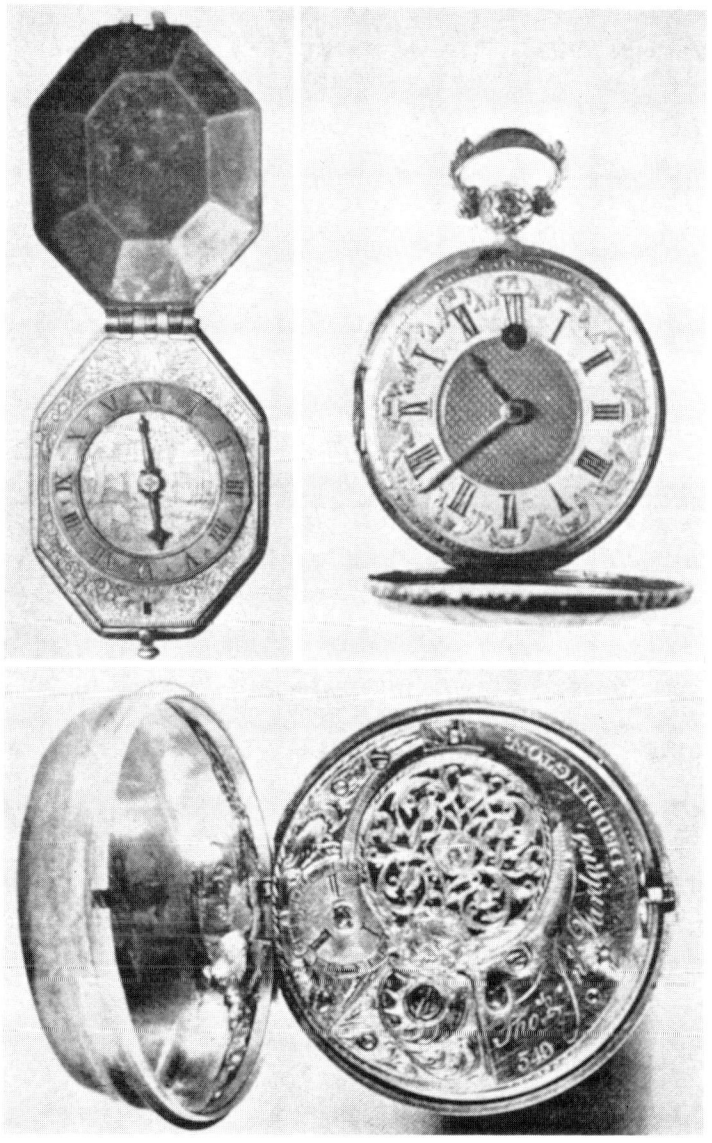

Fig. 36　　　　PLATE 19

Fig. 34, RICHARD QUELCH of Oxford, Silver and gilt case, gilt engraved dial, ca. 1650. (Photo by courtesy of the Ashmolean Museum).

Fig. 35, WILLIAM TASKER of Banbury, Four-colour gold dial and case, 1829. (Author's photo by courtesy of the Ashmolean Museum).

Fig. 36, THOMAS & JOHN FARDON of Deddington, Verge watch, silver pair case, 1801.

Fig. 37 & 38, GEORGE HARRIS of Fritwell (Photos by courtesy of J. M. Surman).

Figs. 39 & 40, JOSEPH KNIBB of Oxford (Photos by courtesy of Oliver Bentley).
PLATE 20

Fig. 41, GEORGE WALKER of Oxford, Double rack-striking movement. (Photo by courtesy of Dr. N. Watson).

Fig. 42, JOHN KNIBB of Oxford, Tic-tac escapement. (Author's photo; block by courtesy of Percy G. Dawson). [see also Pl. 24, fig. 47].

PLATE 21

PLATE 22

Fig. 43. JOHN KNIBB of Oxford, pull-repeater in walnut case. (Photo by courtesy of Percy G. Dawson).

Fig. 44. JOHN KNIBB of Oxford, ebony veneered case. (Photo by courtesy of Ronald A. Lee).

PLATE 23

Fig. 45, JOSEPH KNIBB of Oxford, 30-hour ½-striking movement in walnut veneered case. (Photo by courtesy of L. R. Bomford). Fig. 46, JOSEPH KNIBB of Oxford, Hood and dial of clock shown in Pl. 17, fig 28. (Photo by courtesy of L. R. Bomford).

PLATE 24

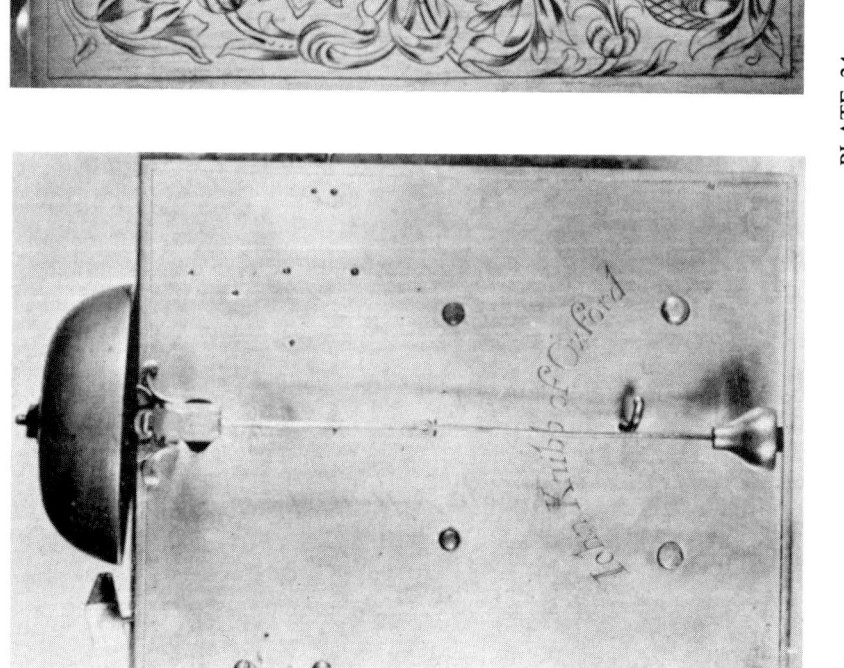

Fig. 47, JOHN KNIBB of Oxford, Back plate of clock shown in Pl. 21, fig. 42.

Fig. 48, SAMUEL ALDWORTH of Oxford, Back plate of 8-day bracket clock (Photo by courtesy of W. Philippson).

Longcase, 30-hour, 10-inch square dial, central area matte with 2 birds in flight and basket of fruit, carved looped hand, chapter ring with $\frac{1}{4}$-hour marks and fleur-de-lys between hours, signed *Ino May Witney;* corner-pieces C. and W., fig. 24. Movement iron 4-posted with brass plates in pine case, ca. 1750 (coll. Beeson, Plate 14, fig. 23).

Longcase, 30-hour, square one-hand dial, centre matte, chapter ring with $\frac{1}{4}$-hour marks and diamond at $\frac{1}{2}$-hour, no marks between the hours, signed, *In. May Witney,* corner-pieces C. and W., fig. 33. Movement with 4 brass posts and plates, ca. 1760 (coll. Dr. P. Latcham).

Longcase, 30-hour, one-hand square dial, centre matte with 2 birds in flight. Tall elaborate oak case.

Longcase, 30-hour, one-hand 10-inch square dial, centre matte, chapter ring signed, *John May* WITNEY. Plain oak case, hood scroll topped, height 6ft. 8ins. (W. Lea).

Act of Parliament clock in the Congregational School, Witney.

N.B.—The dial of a 28-day walnut longcase clock by John May, London (1692-1738) has 2 flying birds and a basket of fruits on a matte ground (London Clockmakers' Company Museum). He came from Witney.

MAY, THOMAS — WITNEY (1790)

C. & W. Longcase, 30-hour, painted dial.

MAZEY, CHARLES — WITNEY, WOODSTOCK (1771-1795)

C. & W. In Witney in 1771. In Woodstock in 1795.

MEAKINS, GEORGE — WITNEY (1767)

C. & W. From London, opened a shop in Corn Street, Witney in October 1767. Not in Directory for 1771.

MEDCALF — WOODSTOCK (1800)

C. & W. In Woodstock ca. 1800.

MERCHANT, THOMAS — BANBURY (1847-1851)

C. & W. In Back Lane, Banbury 1847-1851. Not in 1853 Directory.

[MERN, WILLIAM, OXFORD — BRISTOL]

C. & W. Son of Daniel Mern, potter. Made burgess of Bristol on 29 October 1774 (Nott H. E. and Hudleston, C.R., 1935, *Trans. Bristol & Glouces. Arch. Soc.,* 57, p.187). Wrongly listed by Baillie as an Oxford clockmaker].

MIDWINTER, JOHN — OXFORD (1764-1772)

C. & W. At no. 4 High Street, north side, near Carfax in 1772. John Midwinter, junior, obtained the freedom of Oxford by Act of Council 15 December 1794. In 1764 a John Midwinter married Hannah Pricket of Oxford. Probably father and son of the same name.

Lost watch advertised in 1767.

MOORE, EDWARD (1) — OXFORD (1714-1744)

C. & W. Eldest son of Solomon Moore, gardener of Oxford. Freedom of Oxford on 10 August 1714. Served as Constable, N.E. Ward in 1717. Buried at St. Mary the Virgin church, Oxford on 25 June 1774.

Lantern clock, 6½-inch diameter dial, centre scrolled, chapter ring with fleur-de-lys, scroll-leaf frets; movement 5½-inch square × 6¼-inch high, verge and bob pendulum, rope drive; signed, *Edward Moore* OXON, ca. 1720 (History of Science Mus., Oxford).

Lantern clock signed *Ed. Moore Oxon* with a faked fret dated 1683, converted to anchor escapement.

Wall clock, 8-day, arch dial 4½ × 5½ inches, in lacquered case about 4 feet high, ca. 1735 illustrated in Symonds, *History of Furniture,* and *Country Life Annual,* 1955.

Longcase, 8-day, arch dial 12 × 17 inches, matte centre, seconds ring, calendar aperture bordered by leafy scrolls and a basket of fruit, spandrel pieces C. and W., fig. 20, signed on boss in arch, Ed Moore Oxon. Movement rack-striking, 5 pillars. Case walnut veneered with herringbone stringing, hood arch-topped; ca. 1730 (coll. Beeson).

Repaired St. Martin's church clock 1715 to 1726.

Repaired the clock in Christ Church Cathedral in 1737 and the clock in Tom Tower, Christ Church in 1740.

MOORE, EDWARD (2) — OXFORD (1751-1785)

C. & W. Eldest son of Edward Moore, watchmaker of Oxford. Freedom of Oxford on 8 July 1751. In High Street in 1772 at no. 23 with a frontage of 10 feet and no. 50 with a frontage of 13 feet.

Longcase, 8-day, arch dial, signed on a roundel in the arch, *Edward Moore London,* ca. 1785. In the case is stuck a trade paper, " EDWARD MOORE, Watch and Clock-maker in the Turl, Oxford — Makes mends & cleans Repeating, Horizontal, Seconds & Plain Watches. Also Clocks of all Sorts at the Most reasonable Rates."

A lost watch advertised in 1776.

MOORE AND SON — LONDON

Turret clocks, by John Moore and Son, Clerkenwell, London, in Cropredy church 1831, in Dorchester Abbey, ca. 1868 and in Lincoln College, ca. 1860 and in Jesus College, 1831.

MORRIS, JOHN — CHIPPING NORTON (1770)

C. & W., ca. 1770.

Longcase, 8-day, 13-inch arch dial, seconds ring, calendar steel hands, corner pieces, mask in close reticulation, dolphins in the arch, signed on roundel in arch, John Morris Chipping Norton; case oak, flat top to hood.

MUSSELWHITE, WILLIAM — BICESTER (1787-1791)

C. & W. At Market End in 1791. Took as apprentice Thomas Field on 5 February 1787 (Registered at P.R.O.).

NETHERCOTT, GEORGE — FYFIELD, ? also WANTAGE, Berks (ca. 1770)

Longcase clock, 18th century, (Baillie); ca. 1770 (Palmer, 1935).

Act of Parliament clock, 2 foot, circular yellow dial, plain black and gilt case with Chinoiseries, signed, George Nethercott Wantage.

Wall clock, 7½-inch square brass dial, alarm, engraved spandrels, 4-post movement with tic tac escapement by George Nethercott, Wantage.

NETHERCOTT — CHIPPING NORTON
Longcase, mid 18th century, (Baillie).

NETHERCOTT, JOHN — LONG COMPTON (ca. 1750)
John Kibble, 1928, states " Jno. Nethercote's name is on various clocks. He was a wandering star and various places are given as his residence ".

Longcase, 30-hour, 10-inch square dial, Quaker type, central area with 2 concentric rings and contiguous zigzags, one hand 4-inches long, corner pieces cherub head winged with sunflowers, C. and W. 7, signed *Jno Nethercott Longcompton;* 4-post movement, iron posts, brass plates; oak case, flat-topped hood; ca. 1750. (coll. Beeson).

A " Nethercot ", no locality, repaired Tredington church clock in 1707.

NETHERCOTT, JOHN — DRY SANDFORD (?)
Longcase, 30-hour, $9\frac{3}{4}$-inch brass dial, centre with foliage and a crowned bird carrying a leaf, 2 later gilt hands, ca. 1800, chapter ring without $\frac{1}{4}$-hour marks, minutes by punched dots, spandrel pieces the 4 seasons. Movement made from a lantern clock originally with a central fluke-bob pendulum, now with anchor and long pendulum. In the striking train the 3rd wheel is a bevel wheel engaging a worm on a turned vertical arbor carrying a heavy brass fly (this requires constant lubrication to work). Painted pine wood case with a circular pane in the door (coll. Beeson).

NETHERCOTT, JOHN — STANDLAKE (ca. 1750)
Longcase, 8-day, brass arch-dial, 12 × 16 inches, strike-silent in arch; mahogany case.

Longcase, 8-day, brass arch-dial, $11\frac{1}{2}$ × 16 inches, sun in splendour inlaid with red colour on boss in arch, dolphin spandrel-pieces; central area of dial matte, signed on a plaque below XII, *Ino Nethercott Standlake,* chapter ring without $\frac{1}{4}$-hour divisions and nothing between the hours, two steel crossed-loops hands, spandrel pieces C. and W. 26. Case, oak, of unusual design, hood with a 4-tiered top, door not hinged but held in place by one spring catch so removeable; very narrow waist with an arch-top door, short base; overall about 8ft. high; ca. 1750.

NETHERCOTT, WILLIAM — LONG COMPTON (ca. 1750)
Longcase, 30-hour, one-hand 10-inch square dial, central area with a symmetrical foliage scrolls and a bird in flight, chapter ring with $\frac{1}{4}$-hour marks, diamond between the hours, corner pieces cherubs supporting crown and cross; plated movement; case oak, flat topped hood, ca. 1750.

OAKELY, JAMES — OXFORD (1731-1749)
C. & W. also Oakley, probably son of John Oakly below. Married Mary Brennan, widow, of Oxford in 1731. Looked after the turret clock in Wadham College from about 1733 to 1744, and the clock in St. Mary the Virgin from 1742 to 1748. Signed the church accounts 1742-1745; was Churchwarden 1746 and 1747. Buried at St. Mary the Virgin on 20 January 1749.

Verge watch movement, enamel dial, black tulip hands; back plate signed, *James Oakely Oxon,* square baluster pillars, ca. 1740 (coll. Beeson).

OAKLY, JOHN — OXFORD (?1685-1705)
C. & W. also Oakley. Possibly the John Oakley apprenticed to William Dent, London, on 6 April 1685 for 7 years; he did not join the London Clockmakers' Company.

Mended the clock and chimes at St. Martin's church in 1704-1705.

Lantern clock, dial centre engraved with leafy scrolls and signed above, John Oakly Oxon, chapter ring $6\frac{3}{8}$ inch diameter, dolphin fret in front, scrolling at sides; movement originally verge escapement, converted to anchor, the crutch inside the back plate. Doodling on the back of the dial includes " 20 October 1704 " (coll. Beeson).

ORTELLI, A. — OXFORD (?1790-1846)

C. & W. Also barometers some signed Ortelli & Co. In High Street, All Saints' parish 1843-1846, not in 1852.

Longcase, 8-day, painted arch-dial, signed Ortelli Oxford,, mahogany inlaid oak case with arched pediment.

Verge watch movement, matte silvered dial and raised gold numerals; signed, A. Ortelli Oxford No. 15468, ca. 1790 (Ashmolean Museum).

Verge watch movement, enamel dial, plain balance cock, signed, A. & D. ORTELLI Oxford 2426, ca. 1825 (coll. Beeson).

PADBURY, MATTHIAS — BURFORD (1751-1785)

C. & W. A Quaker, son of Matthias Padbury, merchant of Sibford Gower (died 1790) and Hannah (died 1801); born 3 August 1751; married in 1774 Susanna daughter of Anthony Minchin, butcher of Burford at the Meeting House, Milton under Wychwood; married again in 1785 Susanna daughter of Edward Baker gentleman of Burford at Burford Meeting House.

Longcase, 30-hour, square brass dial with engraved corners; 4-posted movement; case oak with arched pediment.

PAGE, EDWARD — OXFORD (1684)

Son of John Page, yeoman of Goring, Berks., apprenticed to John Quelch, watchmaker of Oxford on 24 July 1684 for 7 years.

PAGET, WILLIAM — BURFORD (an. 1771)

Lost watch advertised 17 June 1771.

PAINE, JOHN — BANBURY, HOOK NORTON (1824-1842)

C. & W., A Quaker, son of Robert and Ann Paine of Milton, Oxon; married on 15 October 1824 Dinah Lamb at Sibford Gower, children born 1826-1834. In Hook Norton in 1842.

Longcase, 8-day, painted arch dial, inlaid and veneered oak case, signed, John Paine Banbury.

Longcase, 30-hour, 12-inch square painted dial, seconds ring, chain drive; oak case; signed, J. Paine Banbury.

Longcase, 8-day, painted arch dial, seconds and calendar rings, perforated gilt hands; oak case, inlaid mahogany, hood with scrolls and brass rosettes, waist with recessed fluted pilasters, signed Jno Paine Hook Norton.

Longcase, 8-day, 12-inch square painted dial, convex centre, corners gilt and blue flowers, seconds and calendar circles, hands gilt; movement plated with extra plate for dial fixing. Case oak with mahogany veneers, shell inlay in door and panel below, circle in basal panel. Hood scrolled with 3 ball and spike finials, waist sides with inset fluted capped pilasters.

Longcases, several, 30-hour, 12-inch square painted dial, seconds ring and/or calendar, gilt hands; dial by Walker & Hughes, signed John Paine Hook Norton; case oak with mahogany banding and shell inlay, hood scrolled. (L. Palmer, L. W. Mathews).

PANTIN, NICHOLAS — LONDON (1651-1663)

C. & W. Apprenticed in 1651 to Humfrey Downing, London, on 24 March 1663, Lady Falkland, wife of the Lord Lieutenant of Oxfordshire, tried unsuccessfully to have him made a freeman of Oxford.

PARSONS, JOSEPH — CHIPPING NORTON (1853)

C. & W. At Tite End, Chipping Norton in 1853.

PAYNE, WILLIAM PETTY — BANBURY (1818-1897)

C. & W., jeweller, silversmith. Son of Thomas Payne, clockmaker of Abingdon, born 1818. In High Street, Banbury 1842 to 1853. Also in business in Wallingford, Berks, from 1859 to 1897. Died in 1897. His son George Septimus started the business of Payne and Son in Oxford in 1880.

PEACOCK, WILLIAM — BANBURY (1788-1823)

C. & W. Married Elizabeth Pinfold at Banbury on 26 February 1788. In Bridge Street from 1798 to 1823. Served as Overseer of the Poor 1795 and 1796.

Longcase, 8-day, painted dial; mahogany case.

Act of Parliament clock (Plate 16, fig. 27) circular white dial diam. 24ins. black case lacquered with Chinese scene. Unusual 2-train movement striking on wire gong, ca. 1800 (coll. Beeson).

Repaired the Bodicote church clock 1798 to 1802, and the Claydon church clock 1801 to 1809.

PEARSON, HAWTIN — OXFORD (1825-1853)

C. & W. Third son of Richard Pearson, watchmaker of Oxford (below). Freeman of Oxford on 10 August 1825. At 50 High Street to 1853. Son born 1828 baptised at St. Peter in the East.

PEARSON, HENRY — OXFORD (1827-1862)

C. & W. Second son of Richard Pearson, watchmaker of Oxford (below). On 29 October 1827, then of St. Peters in the East, married Phoebe Egglestone at Chipping Norton. Freeman of Oxford on 2 August 1830. In High Street in 1853-1862.

Watchpaper: Winged Father Time and pillar with an urn clock; in an oval, " PEARSON Watch & Clock Maker OXFORD "; chequered border; repair date 1840.

PEARSON, RICHARD — OXFORD (1778-1823)

C. & W. Apprenticed to Robert Tawney, watchmaker of Oxford. Freeman of Oxford on 16 December 1785. Took as apprentice Henry Lock about 1794. Business at the Corner of King Street and High Street until 1823 or later. In 1796 he advertised an exhibition of his Mechanical Microcosm. His four sons all became freemen of Oxford, Richard the eldest in 1812, Henry in 1830, Hawtin in 1825 and John in 1825 also. Hawtin and Henry were apprenticed as watchmakers.

Verge watch, silver pair case, enamel dial painted with an agricultural scene; movement signed, " Rich Pearson OXFORD No. 21042 "; back wind. (Ashmolean Museum).

Installed the clock in Garsington church (q.v.) by John Thwaites in 1796.

PHILLIPS, CHARLES — BANBURY (1832)
C. & W., jeweller, silversmith. In Church Lane in 1832.

PHILLIPS, MICHAEL — BANBURY, (1853)
C. & W. In Monument Street in 1853.
Wall clock, circular dial, case mahogany inlaid with brass, ca. 1850.

PINFOLD, THOMAS — BANBURY, MIDDLETON CHENEY (1751-1789)
C. & W., turret clocks. Probably born in Middleton Cheney, Northants. A list of men of that village liable to serve in the Militia, dated 20 November 1762, includes his name and that of Thomas Pinfold, blacksmith. On 5 November 1764, then a clockmaker of Lower Middleton Cheney, he married Mary Turner of Banbury by licence. In 1768 he was Overseer of the Poor in Banbury and attended Vestry meetings up to 1783. On 8 December 1770, as a widower, he married Elizabeth Bloxham of Banbury. Took as apprentice Matthew Wise (q.v.), registered at the P.R.O. on 22 April 1776. He died in 1789.

Longcase, 8-day, dial 13 inches in diameter, seconds ring, date aperture, signed, Thomas Pinfold Middleton Cheney; case plain oak, 8 feet high, hood flat topped (N. M. Belcher).

Longcase, 8-day, dial only 11-inch square brass, calendar lunette, centre signed, *Thos. Pinfold* BANBURY.

Longcase, 30-hour, 8-inch square dial, central area matte with leaf sprays, signed, *Thos Pinfold* BANBURY; chapter ring silvered with large minute figures, spandrel pieces female bust and scrolls, steel hands; movement 4-posted. Case of elm throughout painted black and gold, hood flat topped, ca. 1765.

Made the church clock at Chipping Warden, which was converted to dead beat and reconverted to recoil escapement in 1952.

Repaired the Claydon Church clock 1751 to 1758, and the Kings Sutton church clock in 1759-1793, and the Bodicote church clock in 1768-1789. Cast bell brasses for Claydon church in 1768.

Watchpaper: In centre, " Watches, *Clock & Jacks Made Mended & SOLD BY Thos. Pinfold* BANBURY "; borders with Sun faster, Sun slower, and lobed calendar.

POWELL, JAMES — WITNEY (1852)
C. & W. Repairer in Corn Street in 1852.

POWLEN, THOMAS — THAME (1488-1500)
Smith. Repaired the clock in St. Mary's Church, Thame (q.v.) on occasions between 1488 and 1500.

PRUJEAN, JOHN — OXFORD (1676-1689)
Mathematical Instrument Maker. Lived in New College Lane. Published an explanation of Gunter's Quadrant in 1676. Made a paper astrolabe signed, Joha Prujean Fecit Oxon.

Painted the mural sundial of the church of St. Peter in the East, Oxford in 1689. Made the drawings for painting Kratzer's column dial outside the church of St. Mary the Virgin in 1688.

QUELCH, JOHN — OXFORD (1652-1695)

C. & W. Son of Richard Quelch (1) watchmaker of Oxford. Apprenticed to his brother Richard Quelch (2) on 29 November 1652 for 9 years from 20 June last. Freedom of Oxford on 7 September 1663. Lived in All Saints' parish near the church apparently in the house previously occupied by Richard Quelch (2) (Poll Tax 1667, Subsidy Tax 1667). Children born between 1668 and 1677 were baptised at All Saints' Church. On 30 September 1686 he was chosen one of the 24 Common Councillors and at the same time was given a Bailiff's place as Mayor's Child, holding the post of Bailiff until 1694. According to Baillie he died in 1699 but his name is not in the Window Tax list of 1696.

Apprentices: Thomas Langly in 1673, Henry Godfrey in 1676, Edward Page in 1684 and Joseph Rustin in 1686.

Only one watch known.

QUELCH, JOSEPH — OXFORD (1684)

C. & W. Occupied a tenement on the east side of Broadgates Hall in 1684 described as watchmaker (Salter, 1915).

QUELCH, MARTIN — OXFORD (1650-1653)

Probably the son of Richard Quelch (1) watchmaker of Oxford, apprenticed to him on 16 February 1650 and transferred to Richard Quelch (2) on 18 February 1653 as from 29 September last.

QUELCH, RICHARD (1) — OXFORD (1608-1652)

C. & W. Son of Richard Quelch of Wallingford, Berks. Apprenticed to Triumph de St. Paule, clockmaker of Oxford, on 9 August 1608 for 7 years. Freedom of Oxford on 2 September 1616. Martin Quelch was apprenticed to him on 16 February 1650 and was transferred to Richard Quelch (2) as from 29 September 1652.

In February 1653 he was described as lately dead.

Three of his sons John, Richard and Martin became clockmakers.

Watch (Plate 19, fig. 34), silver and gilt metal octagonal case, the gilt dial engraved with a landscape within a floral border, one hand, silver chapter ring, sides of case engraved with flowers and winged heads between gilt brass mouldings; verge movement with very large crownwheel, Egyptian pillars, ca. 1650 (Ashmolean Museum).

QUELCH, RICHARD (2) — OXFORD (1652-1667)

C. & W. Son of Richard Quelch (1) above and apprenticed to him. Freedom of Oxford on 26 November 1652. Lived in All Saints' parish near the church in a house paying tax on 2 hearths (1665). His name is replaced by John Quelch in the Poll and Subsidy Tax lists of 1667.

Apprentices: The residue of Martin Quelch's apprenticeship in 1653, John Quelch in 1652. Thomas Creed in 1657, James Browne in 1663 and William Fowler in 1667.

RANDALL, HENRY PEARCE — WITNEY (1852)

C. & W. In Market Place.

RANKLYN, THOMAS — OXFORD (1604-1658)

Locksmith, turret clocks; also Ranckle, Francklin. Apprenticed to John Winckle, smith of Oxford. Freedom of Oxford on 17 August 1604 paying

4s. 6d. and 2s. 6d. Lived in the parish of St. Mary Magdalen, paying subsidy tax in 1648. Apprentices: Richard Carter, ca. 1605, Richard Bradford ca. 1611, and John Shewell ca. 1650.

Worked for the University and the City as blacksmith, locksmith and on armour from 1604 to 1658. During the Civil War was employed on the drawbridges and gates of the fortifications of the City. He died in 1658.

Repaired and maintained the clock and chimes of St. Martin's church from 1611 to 1622 and also on the bells. Again in charge of St. Martin's clock from 1641 to 1658. The City Council voted a payment to him in April 1653 " for setting up ye chimes at Carfax ".

Repaired the church clock at Yarnton in 1648 and 1657.

RAYE, JOHN — OXFORD (1617-1648)

Smith, turret clocks; also Rayer, Reyer. The Oxford Council Acts for 23 August 1620 state, John Raye the smith of Magdalen parish to be a freeman paying 4s. 6d., 2s. 6d. and 40s. But a later entry for 5 July 1631 states, " Richard Rayer apprentice to John Rayer, and by him assigned over to serve the father of the said Rayer who was not free, is to pay 40s. and the officers' fees ". The Chamberlains' Accounts for Michaelmas 1630 state, " Raye the smith his freedom £2 ". Paid subsidy tax in 1648 for property in St. Mary Magdalen parish near Broad Street where, according to baptismal records, he had been since 1628.

Repaired the clock and chimes at St. Martin's church several times between 1626 and 1630. From 1640 Goodman Raye was on an annual contract for maintenance.

Employed at St. Mary Magdalen church from 1617 to 1640.

Made a new clock and quarter clock at the church of St. Mary the Virgin for the University in 1640 for £22. See the Accounts of these churches in Part One, Turret Clocks.

Employed at St. John's College 1631-1640.

REYNER, THOMAS — OXON (1740)

A bracket clock ca. 1740 (Britten).

REYNOLDS, HENRY — THAME (1753)

Son of Robert Reynolds, cook of Thame, apprenticed to John Mitchell, London Clockmakers' Company, on 8 October 1753, fee £5 5s. 0d.

REYNOLDS, JOHN — OXFORD, HAGBURN, Bucks. (1702-1733)

Blacksmith, turret clocks. Applied for the freedom of Oxford in August 1702 but it was respited; apparently it was granted at a later date (see Thomas Reynolds, below).

The Overseers' Records for Bampton for January 1733 record a payment of a contract with John Reynolds who made a clock and chimes for the church (q.v.).

REYNOLDS, THOMAS — OXFORD (1745-1799)

C. & W. Blacksmith, whitesmith, turret clocks. On 6 May 1745 he informed the City Council of Oxford that he was bound apprentice to his father, John Reynolds, with whom he served out his apprenticeship, but because his father then lived outside the City jurisdiction and had not given bond to pay "scot and lot" with the other citizens, his freedom has been refused. The house after consideration agreed that he be admitted free.

He was in business in Holywell Street in 1745 and occupied or rented various houses in that street during the following 50 years. Half of no. 8 Holywell Street was in his occupation in 1753 and this lease was renewed by Merton College in November 1768 for 40 years (rent 5s. per annum plus a bushel of good sweet wheat). In 1758 nos. 12 and 50 Holywell Street were also in his name. In 1771 married Hannah Stevens of Oxford. At the survey of 1772 he rented nos. 12 and 59, the former on the north side, the latter on the south.

Apprentices: In indentures he is sometimes described as whitesmith and sometimes as clockmaker, viz. John Hawting in 1745 (whitesmith); 1750, William Almond; 1755, William Gardner (clockmaker), this apprentice absconded in 1758; 1765, Richard Brooks (whitesmith); 1770, Richard Aulkin (whitesmith); 1774, William Marriner (whitesmith); 1779, William Carter (whitesmith and clockmaker); ca. 1785, Thomas Earle who became his partner and successor.

He died in 1799.

Long case, 8-day, arch dial 12 × 17 inches, central area scrolled with recessed seconds dial, chapter ring without marks between the hours, spandrel pieces coarse scrolling; boss in arch signed, Thomas Reynolds OXFORD. Case mahogany inlaid with light woods, hood with scrolled top, short door, and raised panel below it, base oak with mahogany borders, ca. 1760.

Watchpaper: Centre inscribed, *All sorts of* Clocks & Watches *made mended & sold* by THO. REYNOLDS *in Holywell* OXFORD; lobed calendar border.

Repaired the clock in Tom Tower, Christ Church in 1745 and the Christ Church Cathedral clock in 1746.

Repairs and retainer on the clock of St. Mary Magdalen from 1746 to 1799.

Repairs and maintenance of the clock, jacks and chimes of St. Martin's church from 1750 to 1799.

Repaired the Wadham College clock in 1755 and affixed an indicator dial with his name on it (Beeson 1957).

Made a turret clock and chimes for Swalcliffe Church, ca. 1756.

RIGGINS, EDWARD — OXFORD (1725-1730)

Carpenter. Made a dial board for the clock of St. Mary Magdalen in 1725, the bill for which was disputed (see Turret Clocks).

RIXON, JOHN — OXFORD (1660-1676)

Painter. Apprenticed to Richard Hawkins, painter of Oxford. Freedom of Oxford on 12 March 1660. Constable in 1662.

Painted the dial of St. Martin's church clock in 1676.

ROLEWRIGHT, THOMAS — OXFORD (1584-1587)

Blacksmith. Apprenticed to Richard Cakebread, blacksmith of Oxford. Freedom of Oxford on 17 July 1584, paying 4s. 6d. Served as Constable to the Council in 1586 to 1588.

Repaired St. Martin's church clock in 1586 and Christ Church Cathedral clock in 1587.

ROSE, T. — OXFORD (1772)

C. & W. Lived in St. Clement's parish in a house with a frontage of 16 feet, at the east end, north side. Died in 1772 and was succeeded by John Kennington.

Watchpaper: In centre, *T. Rose* Watch-Maker *St. Clement's* OXON, in scrolling and an outer border of three rings.

ROWELL, GEORGE — OXFORD (1770-1834)

C. & W. Born in 1770. Lived in Broad Street from ca. 1800 to 1834; the house was rated in the name of Mrs. Mary Rowell until 1845. George Augustus Rowell married Maria Barrett in 1829. Succeeded by his son Richard (below). Died at the age of 64 and was buried on 20 February 1834 at St. Peter in the East.

Longcase, 8-day, 12-inch arch top white painted dial, spandrels with coloured flowers, calendar and seconds ring; case oak with mahogany stringing, hood with Chippendale scrolls.

Wall clock, 30-hour, brass dial 10-inches high with alarm disc; verge escapement, weight driven, ca. 1795 (coll. Dr. B. Maclean). Skeleton timepiece ca. 1825 (Exeter College).

Watch movement, enamel dial, gold hand, seconds dial, steel hand, duplex escapement, bimetallic compensated balance, diamond endstone, full plate and cap signed, ROWELL OXFORD No. 1256 (coll. Dr. R. V. Mercer).

Verge watch, silver pair case, H. M. London 1790, signed, *Geo. Rowell Oxford no. 507*, (coll. W. H. Bentley).

Verge watch, enamel dial, movement signed, *Rowell* OXFORD no. 211, back wind (Ashmolean Museum).

Watch movement, enamel dial signed, *Gregson A PARIS*; jewelled lever, engraved cock, signed *Rowell OXFORD No. 4653*, ca. 1785 (coll. Beeson).

Maintained the St. John's College clock from 1806 to 1834, and the clock of St. Mary Magdalen from 1810 to 1839 and that of Wadham College from 1811-1834.

Installed the clock by Moore and Son, London in Jesus College in 1831.

ROWELL, RICHARD ROUSE — OXFORD (1834-1865)

C. & W., silversmith, jeweller. Son of George Rowell (above). Freedom of Oxford on 15 December 1834. Continued the business of George Rowell at 36 Broad Street, of which he was ratepayer from 1846.

Watch movement, silver dial, lever with club roller, signed, Rowell and Son, Oxford, No. 1237, ca. 1835 (coll. R. Miles).

Annually maintained from 1835 the clocks of Balliol, Jesus, St. John's, Magdalen, Wadham and Worcester Colleges; and of Christ Church, Tom Tower; and of the churches of St. Martin and St. Mary Magdalen.

Installed the clock by Evans, Handsworth in Great Haseley church in 1865.

RUSTIN, JOSEPH — OXFORD (1686)

Son of Henry Rustin, yeoman of Laurence Hinksey, Berks, apprenticed to John Quelch, watchmaker of Oxford, on 30 July 1686 for 7 years.

DE SAINCT PAUL, JOHAN — OXFORD (1576-1596)

C. & W., turret clocks: also John the Frenchman. In May 1576 married Elizabeth Smyth at All Saints' Church, Oxford; she died in December 1589.

In November 1590 he married Alice Jones at All Saints' Church. On 6 June 1596 he was buried at All Saints' church. Succeeded by his son Triumph (below).

Worked on the St. John's College clock from 1578 to 1596.

Repaired the Christ Church Cathedral clock in 1586 and did further work on it until 1595.

DE ST. PAUL, TRIUMPH — OXFORD (1601-1651)

C. & W., turret clocks. Son of Johan de Sainct Paul, clockmaker of Oxford. Freeman of Oxford on 18 February 1601 paying 10s., 4s. 6d. and 2s. 6d. Married Ursula Ewen before 1610, who died of plague in July 1643. Their children were baptised at All Saints' Church up to 1618. Took Richard Quelch (q.v.) as apprentice on 9 August 1608; and on 20 April 1639 his son Richard de St. Paul for 7 years. Was appointed Constable to the City on 3 October 1614. During the Civil War Royal orders were issued for the collection and deposit of arms in the Guildhall, Oxford. Triumph and his son Lewes contributed a holbert, a coliner and bandeleers. His house was at the site of 140 High Street, St. Martin's parish; it was inherited by his wife, Ursula, from her father Stephen Ewen. He paid 9d. subsidy tax in 1648. His name appears in Suitors to the Court of Hustings in 1645 and 1651.

Repaired the Magdalen College clock of 1505 in 1602.

Maintained the St. John's College clock from 1608 to 1630.

SAUNDERS, CHARLES — BANBURY (1794-1823)

C. & W. Married Kitty Ward at Banbury on 31 July 1794. Notified as bankrupt on 8 January 1805. In Bridge Street in 1823. G. Herbert states that " Mr. Thomas Saunders (sic), a watchmaker and a very clever man (in Parsons Lane early in the 19th century) made a machine for tagging laces such as stay- and boot-laces. He emigrated to America, his brother and nephew having preceeded him ".

His business was taken over by J. H. Durran (q.v.).

SAUNDERS, RICHARD — OXFORD (1674)

Son of Ralph Saunders, son of Thomas Saunders late Chirurgion, decd., was apprenticed to Michael Bird, watchmaker of Oxford on 14 December 1674 for 7 years.

SAVAGE, T. — OXFORD (?1750)

C. & W. Longcase, brass dial, 30-hour movement; oak case.

SAYER, MATT — OXON

Britten gives Matthew Sayer, Exon 1763, and Matthew Sage, Oxon, watch ca. 1760, and Matt Sayer, Oxon, watch ca. 1757. The first is an Exeter maker; the other names have not been verified.

SHEWELL, JOHN — OXFORD (1632-1689)

Locksmith, turret clocks, also Showells. Born at Whitfield, Worcs., in 1632. Apprenticed to Thomas Ranklyn, smith of Oxford. Freedom of Oxford on 14 September 1657. Lived, with wife and 3 children in the parish of St. Peter the Bayly, Queen Street near St. Ebbes Lane (Taxes of 1665 and 1667). Worked on the Sheldonian Theatre.

Repaired and maintained the clock and chimes of St. Martin's church from 1665 to 1676.

SIMMS, BENJAMIN — WITNEY (early 19th cent.)

C. & W. Possibly the same as Benjamin Simms of Chipping Norton. Longcase, 8-day, painted arch dial; mahogany case, hood with scrolls.

Longcase, 8-day, painted and gilt dial; carved case (H. Cox).

SIMMS, CHARLES P. — CHIPPING NORTON (1820-1910)

C. & W. A Quaker. Son of Samuel Simms, clockmaker of Chipping Norton (below), born 1820. Carried on the family business in High Street after 1853. His son Daniel Rutter (1864-1954) succeeded him at the same address.

Imported American shelf clock with his name.

SIMMS, FREDERICK — CHIPPING NORTON (1816-1894)

C. & W. In Middle Row.

Longcase with his printed label pasted on the back.

SIMMS, JOHN (1) — CHIPPING NORTON (1772-1779)

C. & W. In High Street. His son, John (2) carried on the business.

Verge watch, gilt copper cases, the outer with tortoiseshell, gold arrow hands; movement signed Ino Simms Chipping Norton, no. 2369, late 18th century.

SIMMS, JOHN (2) — CHIPPING NORTON (1757-1823)

C. & W. Son of John (1) above. In High Street.

Longcase, 30-hour, 11-inch square painted dial, calendar, 2 hands. Signed, JNO SIMMS CHIPPING NORTON; oak case, hood flat-topped with cube border, free standing pillars, late 18th century.

Verge watch, silver case, London H.M., 1792, movement signed, *Ino Simms Chipping Norton,* no. 2226.

SIMMS, SAMUEL — CHIPPING NORTON (1790-1869)

C. & W., jeweller. Son of John (2) above and continued the business in High Street.

Wall clock, circular white dial, matching moon hands, English fusee movement.

Longcase, 30-hour, 12-inch square painted dial, calendar, gilt hands. Case oak, hood with Chippendale scrolls, mahogany veneering, sides of waist with fluted pilasters, ca. 1830.

Verge watch movement, signed, Saml Simms Chipping Norton, No. 1947, ca. 1830 (coll. R. Miles).

SIMMS, WILLIAM — CHIPPING NORTON (1785-1844)

C. & W. Not in the High Street business.

Longcase, 10-inch square dial, painted with gilt spandrels, signed, *Wm. Simms Chippingnorton,* from a 2-hand, 30-hour clock (coll. Beeson).

SINFIELD, T. — OXFORD (1777)

C. & W. Based on a lost watch an. 1777 (Baillie). Not in the 1772 survey.

SLY, ROBERT — OXFORD (1823-1853)

C. & W. In Magdalene Place in 1823; in New Inn Hall Street in 1853.

SMITH, ABRAHAM — OXFORD (1771)

Son of Abraham Smith, Oxford, apprenticed to Robert Tawney of St. John Clerkenwell in the county of Middlesex on 31 May 1771, for 7 years (registered in Hanasters Book, Oxford).

SMITH, ROBERT — THAME (1442-1455)

Smith, turret clock; also Roberd Smyghth.

Repaired the clock in St. Mary's church, Thame (q.v.) from 1442 on several occasions; also did locksmith's work. Another smith, Thomas Smith, was employed for heavy ironwork.

SMITH, THOMAS — OXFORD (1669)

Son of John Smith, yeoman of Bloxham, apprenticed to Joseph Knibb, clockmaker of Oxford and on 1 October 1669 for 7 years. Probably accompanied Knibb to London as there is no record of his being turned over to another master.

SMITH, VINCENT — POUNDON, Bucks. (1711)

Made the turret clock for the church of Ambrosden, St. Mary in 1711.

SOWTER, JOHN — LONDON, OXFORD (1810-1853)

C. & W., silversmith, jeweller. According to an advertisement of 17 October 1818, came from London after 8 years with Mr. Clements, and opened a shop at 38 High Street, Oxford; there in 1853.

Bracket clock, lancet-topped figured mahogany case, sides with cornucopiac and ring, and gilt metal grills, brass ogee feet, circular silvered dial signed, SOWTER *OXFORD;* heavy 8-day strike-silent movement (coll. Beeson).

Bracket clock, broken arch top, mahogany, 17 inches high, side handles and brass grilles, bun feet; circular white dial; 8-day movement, strike/silent, (Mrs. N. Waterhouse).

Wall clock, 8-day timepiece, large circular white dial; case mahogany veneered, door in base.

Wall clock, plain brass arch dial, 7 inches high, signed in arch *SOWTER OXFORD;* movement enclosed lantern type, verge escapement, alarm, two hands, rope drive (coll. Beeson).

Repaired Oriel College clock from 1825 onwards.

STEELE, JOSEPH — OXFORD (1831-1846)

C. & W., Jeweller. Freedom of Oxford by Act of Council on 22 July 1831. Lived in High Street. Married Frances, children born 1833-1846. Succeeded Samuel Denton (q.v.), ca. 1828-31.

Bracket clock, 8-day, silvered arch dial, signed, *Steele Oxford;* case rosewood, recessed top handle, ca. 1830.

Watchpaper: " I STEELE (*late Denton*) WATCH & CLOCK Maker *High Street* OXFORD Mourning Rings Jewellery &c. "; inner border with " Musical & Foreign Watches & Clocks carefully repaired "; outer border dentate.

STEVENS, JOHN — GREAT MILTON (1784-1819)

C. & W. Married Catherine Saunders both of Great Milton on 27 February 1784.

Repaired or cleaned the church clock at Stadhampton 1800 to 1819.

In 1810 recast the treble bell of Great Milton church.

In 1819 repaired the church clock at Stanton St. John.

STEVENS, ROBERT — GREAT MILTON

Longcase clock at Wantage, Berks. Baillie gives " 1781-95, part of Gravesend ", which seems doubtful.

STEWARD, JOHN — HENLY

Not further identified. Lantern clock in engraved case with pierced foliate frets, 14 inches high, one primitive hand, signed in capitals on upper part of dial centre; movement a longcase type in plates, anchor escapement.

STOCKFORD, JOSEPH — THAME (1770-1774)

Bellhanger, whitesmith. Signed Churchwardens' Accounts in 1773 and 1774. Made the church clock at Ewelme. Advertisement in June 1770.

STOCKFORD, THOMAS — GREAT HASELEY (1764)

C. & W. Succeeded Thomas Holloway, clockmaker of Great Haseley on 3 March 1764.

STOCKFORD, THOMAS — THAME (1778-1785)

C. & W. Probably the same as the above. Advertisement dated 14 March 1778. Signed Churchwardens' Accounts in 1781 and 1785.

STONE, JOHN — THAME (1764)

C. & W., goldsmith, jeweller. The date 1764 is based on a watch. Watchpaper: *"John Stone Watch and Clockmaker At Thame Oxon Goldsmiths & Jewellers Work done in the neatest Manner "*; border of rococo scrolls and foliage from which watches are suspended.

STONE, JOHN — HENLEY-ON-THAMES (1795)

C. & W. A watch ca. 1795.

STONE, JOHN — AYLESBURY (1764-1789)

C. & W. Possibly the same as John Stone of Thame (above). His trade card states that he was " from London " and in business at The Dial in Aylesbury. There are also marriage records of a John Stone described as a clock and watchmaker, viz. (a) on 12 October 1764 (then aged about 24) to Elizabeth Deane of Thame, at Aylesbury, (b) in 1766 to Martha Eustace, both of Aylesbury and (c) on 12 December 1789 to Susanna Line, both of Aylesbury.

An enumeration of the wholesalers who supplied him with material is included in the Historical Review, p.23. His collection of bill headings and watchpapers is in the Guildhall Library, London. His business was amalgamated with that of T. W. Field in 1802.

Watchpaper: *" All Sorts of Clocks & Watches Made Mended & Sold By John Stone AYLESBURY Sells Gold Rings & all sorts of Silver Goods "*; lobed calendar border.

STONE, RICHARD — THAME (1761-1783)

C. & W., Turret clocks. Son of Edward Stone, sadler of Thame, apprenticed to Charles Howse, London Clockmakers' Company on 8 September

1761 for 7 years, fee £8. W. D. Palmer mentions "a handless clock by Richard Stone, Thame, 1783," i.e. a turret clock, and in the Oxford Times for 10 December 1926, writes, "Richard Stone, Thame about 1771 . . . Quite recently I purchased from a marine store in Tring a church clock made by Richard Stone, Thame, 1783 which I believe is now in going order fixed to a barn at Dunstew, Deddington".

A watch, H.M. 1771.

Made the clock now in the church of St. Nicholas, Marston (q.v.).

STONE, THOMAS — THAME (1801-1809)

C. & W. Named in the Churchwardens' Accounts 1801-1809.

Verge watch movement, enamel dial, front wind, signed, THOS STONE Thame No. 504, (Ashmolean Museum).

STRANGE, THOMAS — BANBURY (1823-1866)

C. & W., turret clocks. In Market Place 1823 to 1853; in a local exhibition in 1861. An advertisement in the County Directory for 1853 states that the business was "established upwards of eighty years", that is in 1773, which is based on the fact that he took over the business of C. W. Drury (q.v.) between 1823 and 1837. He sold "gold and silver watches of the best qualities both English and Foreign. English and French clocks, Jerome's American clocks warranted perfect from 15s. upwards. Turret, House and Ornamental Clocks wound up and kept in repair by the year".

Bracket clock, 8-day, chiming, silvered arch dial. Case much carved with applied mouldings. Another 8-day, circular painted dial in mahogany case inlaid with brass, spade hands (M. M. Gillett).

Longcase, 8-day, circular white dial, calendar and seconds. Oak case with mahogany veneer, hood with Chippendale scrolls. Another in mahogany case.

Wall clock, 8-day, circular white dial. Case lacquered red and gold.

Longcase, 30-hour, painted arch dial, calendar. Case oak with mahogany veneer, hood with Chippendale scrolls.

Longcase, 30-hour, 12-inch square painted dial, calendar. Plain oak case (C. Field).

Turret clocks: Repaired the Adderbury church clock and added two 2-hand dials in 1839, and maintained it annually until 1866.

Improved the clock in Bodicote church (q.v.) about 1843.

Made the clock in Sibford Gower church (q.v.) in 1841

STREET, RICHARD — CHARLBURY (1795)

C. & W. Married Mary Cooper at Banbury on 5 December 1795

TASKER, WILLIAM — BANBURY (1813-1853)

C &. W., jeweller, silversmith. In High Street to 1853. Witnessed a marriage in 1813. Married Anne Moore both of Banbury on 4 November 1817.

Longcase, 30-hour, square painted dial; case oak veneered and inlaid.

Verge watch, case carved and engine turned 18-carat gold, dial four-colour gold with festooned numerals, front wind. London H.M. 1829; movement with plain balance cock, signed, *W. Tasker* BANBURY 5894". (Plate 19, fig. 35, coll. Beeson).

TAWNEY, ROBERT — OXFORD, LONDON (1755-1776)

C. & W. Son of Robert Tawney, carpenter of Oxford, apprenticed to John Herbert, watchmaker of Oxford on 10 June 1755 for 7 years. His father, and from 1769, his widowed mother leased a house in Holywell Street from Merton College. He may have worked in Oxford for some time but in May 1771 was then in business in St. John Clerkenwell, Middlesex. Abraham Smith was apprenticed to him on 31 May 1771 and Richard Pearson a few years later, both men from Oxford.

A lost watch advertised in 1776.

TAYLOR, WILLIAM — OXFORD (1795-1854)

Turret clock and chime maker, bellfounder. Elder son of Robert Taylor (1759-1830) bellfounder of Oxford; born 1795 at St. Neots, died unmarried at Oxford in 1854. Continued the business of Taylor and Sons, Oxford, with his brother John, under the style of W. & J. Taylor, Oxford, in St. Ebbe's Street and in Blackfriars Road. A rack-striking 8-day church clock with a 5ft. 6ins. copper dial could be obtained from this firm in the 1830s for £120. The clockmaking ceased with his death and the factory was closed in 1854 and the bell-founding business was transferred to the foundry of John Taylor (1797-1858) which he had established in Loughborough, Leicestershire about 1839. This became the present firm of John Taylor & Co., Loughborough, in family hands for two hundred years.

Rack-striking turret clock for All Saints, Churchill in 1826.

Rack-striking clock for St. Peter and St. Paul, Deddington in 1833.

Work for Christ Church, Oxford ca. 1840.

Turret clock and dial for the second church of St. Martin, Oxford in 1848.

Pinwheel escapement, motion work and 2 skeleton dials for the 1741 clock of St. Mary the Virgin, Oxford in 1851.

The two bells for the quarter-jacks on Carfax Tower, Oxford, were made by J. Taylor & Co., Loughborough in 1898.

TERRY, JOHN — BANBURY (?1700-1736)

C. & W. Married Patience — ? in 1700. Buried 2 May 1736 described as a clockmaker.

THORNDELL, RICHARD — BAMPTON (1762-1768)

C. & W., ironmonger. Advertisement of 17 September 1768. Married Martha Ridge of Aston in 1762.

THORP, BENJAMIN — OXFORD (1726)

Son of Thomas Thorp, decd., apprenticed to Edward May, clockmaker " OXON " in 1726.

THWAITES, AYNSWORTH and JOHN. THWAITES & REED, LONDON

Turret clock makers. Clocks made by Aynsworth Thwaites (free London Clockmaker's Company in 1751) and by John Thwaites (free L.C.C. in 1802) or their successors for buildings in Oxfordshire include those for Balliol College, 1838, Garsington Church, 1796, Merton College, 1813 and Worcester College 1856 (see notes in the section on Turret Clocks). Several Oxfordshire clocks by other makers have been repaired or replaced in more recent times by Thwaites & Reed.

A biography of John Thwaites by G. T. E. Buggins and F. B. M. Devereux appeared in *Antiquarian Horology* for June 1956.

TOMLINSON, JOB — BICESTER, THAME (1819-1865)

C. & W., turret clock, gunsmith. Married Mary Anne before 1826. Children baptised at Bicester in 1826 and 1840, latterly described as watchmaker in Thame. In High Street Thame to about 1865.

Longcase, 30-hour, 11-inch square dial, silvered and engraved central area, and a harlequin motif, 2 hands, silvered chapter ring, without marks between the hours, gilt spandrel pieces an eagle in scrolling, signed below hands, TOMLINSON THAME. 4-posted movement, case oak, ca. 1800.

Longcase, 30-hour, square painted dial, oak case, signed TOMLINSON THAME.

Worked for St. Mary's church Thame, 1819-1838. In 1828 put new copper figures on the clock dial, and renewed the escapement. A Job Tomlinson undertook annual maintenance of the clock to 1865.

Watchpaper: "*J. Tomlinson* Watch & Clock Maker *Silversmith &c* THAME", on landscape with trees and 3-masted ship at sea, a woman with the right arm resting on a large anchor; above the signature a clock; in outer border, " MUSICAL, REPEATING, HORIZONTAL, DUPLEX. PATENT LEVER & PLAIN WATCHES REPAIRED ".

TOMLINSON, JOHN — BICESTER (1843)

C. & W. No biographical history. Query the same as John Tomlinson below.

Longcases 30-hour, 12-inch square painted dials, usually arabic hour numerals, signed, *John Tomlinson Bicester;* cases plain oak or veneered and inlaid with mahogany (Miss E. M. Hitchins).

Repaired the Fritwell Church clock in 1843.

TOMLINSON, JOHN — THAME (1772-1819)

C. & W., whitesmith. Worked for St. Mary's church Thame 1772-1819.
Repaired Thame church clock 1772 to 1782.

TONGE, GEORGE — OXFORD (1730-1803)

C. & W., silversmith. Son of Henry Tonge; baptised at All Saints' Church, Oxford on 3 April 1730. Apprenticed to John Wilkins, goldsmith of Oxford. Freedom of Oxford on 12 November 1759. In High Street (Kemp Hall, the site of the present Closed Market) in a house with a frontage of 27 feet adjoining the yard of the King's Head, from 1760 to 1803.

Apprentices: Richard Hickman (free 1776), William Jones (free 1772) and Thomas Turner (free 1784). Bailiff in 1770, Mayor in 1781 and 1795; eligible for Alderman in 1802 but died before election.

Bracket clock, brass arch dial, matte centre, chapter ring not silvered, pendulum lunette, spandrel pieces scrolling, signed on boss in arch, *Geo Tonge OXON;* movement, timepiece, verge escapement, pull quarter-repeater on 3 bells, back-plate engraved with leaf scrolls and 2 birds; and a basket of fruit; ebonised bell top case 14 inches high, gilt metal side grilles, ca. 1770 (coll. Beeson).

Supplied silver to Oxford Colleges and plate to St. Mary's church, 1764 to 1774.

TURNBULL, CHARLES — OXFORD (1573-1581)

Admitted to Corpus Christi College, Oxford in 1573. Born in Lincolnshire. Wrote a Treatise on the use of the Celestial Globe. Designed the column sundial at Corpus Christi in 1581 (see Sundials).

TYLER, RICHARD — ADDERBURY (1733-1800)

C. & W. A Quaker, born 1733, died 11 June 1800 aged 67. Described in the Friends' Register as a clockmaker of Adderbury, and may have been an apprentice or assistant to Richard Gilkes (q.v.).

VAUGHAN, RICHARD — WATLINGTON (1732)

Son of William Vaughan, yeoman decd., of Watlington, apprenticed to Francis Webb, clockmaker of Watlington, on 17 May 1732 (registered at P.R.O.).

VERNON, JOSEPH — WATLINGTON (1852)

C. & W. Joseph Vernon and Son in Couching Street in 1852.

VESEY, ROBERT — OXFORD (1682-1694)

C. & W. Son of Robert Veasey, tailor of Oxford; apprenticed to John Harris, watchmaker of Oxford on 9 March 1682 for 7 years. Freedom of Oxford on 11 September 1691. Children baptised at All Saints' Church from December 1691 to 1694. Lived in High Street, possibly no. 10 or 11.

VULLIAMY, BENJAMIN L. — LONDON (1775-1820)

C. & W. Turret clocks. A short biography of Benjamin Lewis Vulliamy by S. Benson Beevers appeared in *Antiquarian Horology* for March 1954. Turret clocks were supplied by him to Oriel College and other places in Oxfordshire (see section on Turret Clocks).

UNITE, MATTHIAS — OXFORD (1681-1688)

C. & W. Son of Mathias Unite, minister of Fenny Compton, Warwicks.; apprenticed to John Knibb, clockmaker of Oxford, on 18 October 1681 for 7 years.

Lantern clock, 13 inches high, $6\frac{1}{4}$-inch diameter chapter ring, signed, *Matthias Unite,* without locality, probably an apprentice master-piece (coll. W. R. Gillett).

WAGSTAFF, THOMAS — BANBURY, LONDON (1724-1802)

C. & W. A Quaker. Younger son of Thomas and Sarah Wagstaff of Banbury, born 1724; married at Milton under Wychwood Meeting House. At Carey Street and 33 Gracechurch Street, London, 1756 to 1793. Member of the Merchant Taylors' Company but not the London Clockmakers' Company. When visiting London members of the Society of Friends used to lodge at Wagstaff's house; American visitors frequently returned to the States with one of his clocks. Author of several publications including *Piety Promoted or Brief Memorials of the Quakers,* parts 8 and 9, also an account of the life and gospel labours of William Beckett of Wainfleet, Lancs. (1776), and another book dealing with 66 records of deceased Friends (1796). When over 70 years old he retired to Stockwell in Surrey and continued his literary work. Finally he went to Chipping Norton, where he had relatives, and died there in 1802 (Williamson G. C., 1921). Williamson illustrates a silhouette portrait of Wagstaff and one of his bills dated 29th 11th 1777 concerning an engraved gold watch in a tortoiseshell pair case, no. 7945, supplied to George Fox for £28 7s. 0d., and credit £8 8s. 0d. for a Wagstaff watch no. 7519 returned.

Watches in several museums including the British Museum and the Metropolitan Museum, New York. Musical bracket clocks. Longcase clocks in the H.P. Strause collection and others in the United States. Mahogany longcase illustrated in Hayden, A., 1917, and *The Connoisseur,* December 1928.

WALFORD, HENRY — BANBURY (1853-1861)

C. & W., silversmith, jeweller. Henry Walford and Son in High Street Banbury in 1853, and in a local exhibition in 1861. Baillie gives the date of a Henry Walford of Banbury and Oxford as ca. 1775 but this cannot be confirmed.

WALFORD, JOHN GEORGE — BANBURY (1814-1853)

C. & W., silversmith, jeweller. In High Street Banbury 1814-1853. Baillie gives 1790 as the earliest date but Walford & Son, Banbury, the modern successors state the business was established in 1814. The trade directory for 1853 gives Henry Walford, John Walford and Walford & Son all watchmakers and jewellers in High Street.

Longcases, 8-day and 30-hour, with 11- and 12-inch square painted dials usually signed, J. G. Walford Banbury, in veneered and inlaid oak cases many of which are said to have been made in Banbury (see Edward Holloway); the dials were supplied by Walker and Hughes.

Wall and dial clocks.

Watchpaper: A woman with a large anchor leaning against a wall on which is a draped urn, background with a sailing ship at sea; on the wall, " WALFORD Watch & Clock *MAKER High Street* Banbury "; inside border, " Silver Plate Gold Rings Jewellery Mourning Rings &c made to Order ✶ Ready Money for Old Gold & Silver ✶ "; outer border of denticles and leaf sprays.

WALKER, GEORGE — OXFORD (1689)

C. & W. Possibly the son of George Walker, maltster of Oxford (1668-1700). Described as a watchmaker when elected Constable of the S.E. Ward on 30 September 1689.

Lantern clock, 15 inches high, chapter ring $6\frac{1}{4}$ inches in diameter, $1\frac{1}{4}$ inches wide, central area scrolled, signed above, *George Walker Oxon,* simple looped hand, foliate frets not chased. Movement with an unusual type of rack-striking, the snail and one curved rack in front and a second straight rack in the striking train; anchor escapement. Plate 21, fig. 41 (coll. Dr. N. Watson).

WALKER, THOMAS — OXFORD (1665-1715)

Blacksmith, locksmith. In St. Ebbe's district from 1665 leasing from the City Council a cottage and land in a lane leading from St. Ebbe's to St. Aldate's church, which lease was renewed to his widow in 1726. Received loans from the City in 1670, 1678 and 1685.

Appointed Constable of the S.E. Ward in 1683. An unnamed apprentice lived with him in 1667 and another, George Haddocke, was made free in 1685. Died between 1715 and 1726.

Repaired the chimes of St. Martin's church clock and did other work in the church from 1683 to 1697.

WALKER & HUGHES — ?BIRMINGHAM

Painted clock dials were supplied by this firm in the early 19th century to north Oxfordshire clockmakers. The name is stamped in capitals on the back of the dial plate, and the month wheel is stamped W. & H.

WALLEN, J. — HENLEY-ON-THAMES (1790)
C. & W. A watch dated by Britten ca. 1790.

WALLEN, WILLIAM — HENLEY-ON-THAMES, READING (1725-1756)
C. & W. Britten gives the date 1725 for Henley based on a watch. Baillie gives 1756 for Reading based on an advertisement for his shop.

Verge watch, small silver pair case, enamel dial, black beetle hands; movement signed, Wm Wallen Henley No. 5141, back wind, ca. 1725 (Plate 18, figs. 32, 33) (Ashmolean Museum).

WALTER, EDMUND — BURFORD, WITNEY, FARINGDON (1853)
C. & W., silversmith, engraver, insurance agent. In Market Place, Witney, and High Street, Burford and Market Place, Faringdon in 1853.

Verge watch, silver case, movement signed, E. Walter Witney no. 445-00.

Watchpaper: above, an eagle holding a ribbon labelled, TEMPUS FUGIT, "E. WALTER Watch & Clockmaker Silversmith Jeweller & Engraver WITNEY (below in oval) Old Gold & Silver bought or exchanged. A General Assortment of Fishing Tackle &c. Table and Fine Sheffield Cutlery", in outer border.

WANGLER, LUKE — OXFORD (1852)
C. & W., silversmith. At 10 Castle Street.

Dial wall clock, circular mahogany frame, "Black Forest" movement.

WANGLER, M. — BANBURY (1842)
C. & W. In Bridge Street in 1842 and possibly earlier.

WARNER, RICHARD — BURFORD (1790)
C. & W. Probably connected with the clockmaking families of Warner in Chipping Campden, Glos. Took as apprentice John Faulkner on 16 November 1790.

WATTS, CHARLES — OXFORD (1823-1829)
C. & W., goldsmith, silversmith. In High Street 1823-1829. Freedom of Oxford on 2 August 1824. Married Elizabeth, children born 1825-1829.

WEBB, CHARLES — HORNTON (1850-1853)
C. & W. In 1850 repaired the derelict turret clock, from Cropredy church and installed it in Horley church (q.v.). He was described by the Curate of Horley, who himself gilded the dial, as "a self instructed watchmaker".

WEBB, FRANCIS — WATLINGTON (1710-1732)
C. & W. Married Mary Dobinson of Watlington in 1729. Took as apprentice Richard Vaughan in May 1732.

Bracket clock, 8-day, striking, ebonised bell top case, arch dial, ca. 1730, illustrated in *The Connoisseur,* May 1925.

Wall clock, 30-hour, 7-inch square dial, central area matte, alarm disc with Tudor rose and 6 projections, chapter ring silvered, fleur-de-lys between hours, one pierced steel hand; spandrel pieces head in scrolling, signed, *Fra Webb Watlington;* 4-posted movement with short pendulum, rope drive in stained wood box-like case, ca. 1715 (History of Science Museum).

Longcase, 30-hour 10-inch square dial, central area matte, chapter ring silvered with fleur-de-lys, signed, *Fra Webb Watlington,* one pierced steel hand, spandrel pieces pair of cherubs holding crown (cross removed, C. and W., 8); 4-posted iron frame, anchor escapement, rope drive, ca. 1710. Later case (coll. Beeson).

Longcase, 30-hour, 9-inch square dial, centre matte with 2 birds and a V-scroll, chapter ring with fleur-de-lys, one looped steel hand, spandrel pieces. C. and W. 18, signed, *Frans Webb* WATLINGTON; 4-posted movement, deal case (coll. Professor P. Espinasse).

WEBB, THOMAS — HOOK NORTON (1795-1819)

C. & W. Longcases, 30-hour, square painted dials, 4-posted movements late 18th century.

Repaired Wigginton church clock in 1819.

WEBB, ? — BAMPTON (1776)

C. & W. Advertisement of 4 September 1776.

WELLS, JOHN — BANBURY (1774)

C. & W. Married Mary Page on 28 January 1774, both of Banbury.

WELLS, JOHN — SIBFORD GOWER, SHIPSTON ON STOUR (1785-1809)

C. & W. A Quaker, son of Thomas and Elizabeth Wells of Byfield, Northants; married Mary French at Sibford Gower Meeting House on 11 August 1785. In business in Shipston on Stour. Died in 1809. His business was carried on by his two sons Thomas and John as John Wells & Co.

Longcase, 30-hour, 12-inch square painted dials, 2 hands with pierced diamond head, signed, *Jno Wells Shipston;* case mahogany, flat-topped hood, rectangular door, ca. 1795.

Verge watch, silver pair case, HM 1789, case mark 1R, enamel dial with minutes at quarters, beetle hands, signed, *Jno Wells Shipston 1397* (Beeson 1958, fig. 11).

Watchpaper printed in red: In centre, "All Sorts of CLOCKS *and Watches carefully repaired* By JOHN WELLS *at Shipston Upon* STOWER"; inner border, Sun Slower — Sun Faster, lobed calendar outer border. Repair dates 1804-1809.

WELLS, JOHN — CHIPPING NORTON, SHIPSTON ON STOUR (1787-1847)

C. & W. A Quaker, son of John Wells of Shipston on Stour (above); born 1787. After their father's death in 1809 the business was carried on by the brothers Thomas (1786-1855) and John (1787-1847). John had a business in Chipping Norton also ca. 1823.

Longcases, square or arched painted dials, in veneered and inlaid oak cases, signed, *John Wells, Shipston* or *Wells, Shipston* (Mrs. A. K. Gardner, Mrs. L. V. Jones).

Longcase, 30-hour, 10-inch square painted dial, signed, *John Wells Chipping Norton,* ca. 1820.

WELLS, THOMAS — BANBURY (1832-1853)

C. & W., musical instruments, bell-hanger. In Parsons Street in 1832-1853.

Longcases, 30-hour, 12-inch painted dials, veneered and inlaid oak cases.

WENTWORTH, GEORGE — OXFORD (1692-1747)

C. & W., goldsmith. Son of Thomas Wentworth, watchmaker of the City of Sarum, Wilts., born 1692. Apprenticed to John Knibb, clockmaker of Oxford on 2 April 1706 for 7 years. Freedom of Oxford on 15 May 1713. Married Dinah Mose of Oxford in 1713. Served for a year in 1715 as Constable of the N.E. Ward. Elected Councillor in 1719, fee £3 10s. 0d. Chosen Mayor's Chamberlain in 1727, fee £2, and put in charge of the waterworks, i.e., cleansing streams and letting out fishing. In May 1738 he purchased from the City 2 derelict cottages in St. Michael's parish for £15 15s. 0d. plus the amount due to him from the City on account of his expenditure on waterworks. In September 1738 he was appointed Senior Bailiff, fees £5 and £13 6s. 8d., and from 1740 to 1743 was also Fairmaster.

Before 1731 he leased a house belonging to the City which was south of Broad Street built against the City Wall " in the Town Ditch ", abutting on what is now Turl Street opposite Trinity College. In July 1737, after a dispute with his neighbour, he was allowed to erect a chimney on the Wall. The only apprentice registered was Doiley Wise, indentured on 6 August 1720.

He died in 1746 or 1747 and the lease of his house was renewed in 1748 to his widow, Dinah.

Early Act of Parliament clock, octagonal 32-inch wide dial, black and gilt, the door of the trunk lacquered, ? Bacchus astride an oval cask; 8-day movement in rectangular plates, anchor escapement, block weight; signed on the basal mouldings, GEORGE WENTWORTH OXON, ca. 1745.

Longcase, 8-day, arch dial 11 × 16 inches, centre matte, seconds ring and calendar aperture, chapter ring with $\frac{1}{4}$-hour divisions and no marks between the hour numerals, spandrel pieces female head in scrolling, C. and W. 21, dolphins in arch, signed on boss, *George Wentworth* OXON; movement rack-striking in 5-pillar plates. Case oak 8ft high, hood arched top with 3 wooden ball and spike finials, free-standing plain pillars, waist door broken arch top, ca. 1735 (coll. Beeson).

Repaired the Christ Church Cathedral clock in 1730.

WHITERN, WILLIAM — HENLEY-ON-THAMES (1806)

Apprenticed to James Coster, clockmaker of Henley on 19 August 1806.

WHITE, JAMES — OXFORD (1804-1812)

C. & W. Apprenticed to Thomas Earle, clockmaker of Oxford. Freedom of Oxford on 7 October 1812.

WILKINS, GEORGE — LONDON, OXFORD (1768-1798)

C. & W. Was workman to John Holmes, clockmaker of The Strand, London, for 18 years, whence he came to Oxford and opened a shop in High Street (*Oxford Journal,* 4 February 1786). In June 1798 he advertised removal from opposite the Swan Court to opposite the Angel Inn, High Street.

WILLIAMS, JOSEPH — ADDERBURY (1762-1835)

C. & W. A Quaker, son of William, a cordwainer, and Hannah Williams of Adderbury East, born 6 September 1762. Married Hannah (died 25 March 1812 and was buried at Adderbury).

Children born 1793 and 1794. In business from 1788. His house was in Adderbury East as customary tenant of the Manor of Eadburby. Died 25

October 1835, aged about 73 years and was buried at the Adderbury Meeting House. His son William (below) carried on the business.

Longcase, 30-hour, 11½-inch square painted dial, arabic hour numerals, calendar lunette, 2 gilt crescent hands, a balloon above centre, signed, *Jos Williams Adderbury*. Case oak inlaid with mahogany and light wood stringing, hood with Chippendale scrolls and central finial, free-standing pillars, short waist with raised panel below door, sides with fluted recessed pilasters, ca. 1800 (coll. Beeson).

The first balloon ascents in England were made from the Physic Garden, Oxford, on 4 October 1781 and 12 November 1784 by James Sadler of High Street, Oxford and a medal was struck describing him as the first English Aeronaut. A balloon was thereafter used as a decoration of clock dials and as a motif in watch cocks.

Maintained the Adderbury church clock from 1788 to 1827.

WILLIAMS, WILLIAM — ADDERBURY (1793-1862)

C. & W., jeweller. A Quaker, son of Joseph Williams (above), born 8 April 1793. Worked with his father until the latter's death in 1835 and continued this business in Adderbury East until 1862. He was a customary tenant of the Manor of Eadburby and mortgagee of cottages in Church Lane, Adderbury East from 1832 to 1847. He was buried at Adderbury Meeting House on 11 February 1862, aged 69.

Longcase clock, painted dial (J. G. Bennett).

Maintained the Adderbury church clock from 1828 to 1839 and thereafter probably as an agent of Thomas Strange (q.v.).

Watchpaper: green paper, WILLIAMS Clock & Watch Maker ADDERBURY. Jewellery and Silver Articles, Umbrellas, Locks and Guns neatly repaired; simple border.

WISE, DOILEY — BODICOTE (1720-1788)

Son of William Wise, husbandman of Bodicote. Apprenticed to George Wentworth, watchmaker and goldsmith of Oxford, on 6 August 1720 for 7 years. Returned to Bodicote; in 1733 married Mary North (died 1766) and was occupied as a metal worker, plumber and glazier. Died at Banbury and was buried in the chancel of Bodicote church on 13 February 1788. His will proved July-October 1788, describes him as a yeoman.

WISE, JOHN — LONDON (1675-1723)

The clock in Bodicote church (q.v.) is inscribed " Tho Bradford & Rich Wise then Churchwardens John Wise Londini 1700 ". This John Wise is apparently he who was apprenticed in 1675 and was a member of the London Clockmakers' Company from 1683 to 1723.

See also Richard Wise, another clockmaker related to the Wise families of Bodicote, under Sundials, Bodicote Weeping Cross.

WISE, MATTHEW — BANBURY, DAVENTRY (1759-1786)

C. & W. Son of Matthew and Rachael Wise of Bodicote, baptised 2 August 1759. Apprenticed to Thomas Pinfold, clockmaker of Banbury, registered 22 May 1776. Married Ann Pinfold of Middleton Cheney, Northants, at Banbury by licence on 6 May 1781. Later in business at Daventry, Northants from ca. 1786.

WISE, THOMAS — CHARLBURY (early 19th cent.)

C. & W. In Charlbury early in the 19th century, also Sam Wise his son (Kibble J., 1927). W. D. Palmer unaccountably assigns the dates 1620 and 1650 to these names.

Longcase, 30-hour, square painted dial, one hand; plated movement; deal case (coll. Dr. P. Latcham).

WOOD, RICHARD — OXFORD (1673-1700)

Stone cutter. Freedom of Oxford on 27 November 1676. Mayor in 1694. Erected a sundial at the South Bridge in 1673. Died ca. 1700.

WREN, CHRISTOPHER — OXFORD (1632-1723)

The horological interests of Christopher Wren were shown during the years of his closer connection with Oxford, before his knighthood in 1673. He took his B.A. on 18 March 1651 at Wadham College; here he associated with that group of scientific scholars, whose Philosophical Meetings, held in the tower room over the gateway of Wadham, preceeded the institution of the Royal Society. He was a Fellow of All Souls from 1653 to 1657, and, after holding the Professorship of Astronomy at Gresham College, London, became Savilian Professor of Astronomy at Oxford in 1661. From then onwards he was frequently in Oxford on architectural projects, viz. the Sheldonian Theatre 1664-9, All Souls 1664, Trinity 1668, St. John's 1670, Queens 1671-2 (Colvin, 1954).

In 1661 Charles II commanded him to demonstrate the latest discoveries made with the telescope and the microscope and to make a model of the moon. One of his earliest inventions was a working model of the periodic motions of the earth, sun and moon. His early experiments with pendulums are enumerated by Thomas Sprat (1667). In 1663 he devised a pendulum weather clock which was probably the earliest attempt to correlate horology and meteorology. It was actually constructed and was seen by M. de Monconys at his visit to Wren in June 1663 and was among his inventions exhibited at the first assemblies of the New Philosophers at Wadham College. Later on Robert Hooke added 4 or 5 extra meteorological motions to it so that eventually it needed a weight of " $\frac{3}{4}$ of a 100 lib " to run it for a week (Grew, 1681). Wren's design is described and illustrated by H. A. Lloyd.

Wren is credited with having designed for the grounds of Hanwell Castle, near Banbury, a water-driven clock, showing the time on the principle of Mario Bettino, which was later described by Robert Plot (1677). Perhaps his most significant contribution to Oxfordshire clockmaking is the Wadham College turret clock made with an anchor escapement in 1670 by Joseph Knibb (Plate 5, figs. 7, 8); this is still referred to as Wren's clock (Beeson 1957). And it is very probable that his advice resulted in the fitting of an anchor escapement to the clock of St. Mary the Virgin, Oxford, also by Joseph Knibb in 1670 (Beeson, 1961).

Appreciation of Wren's early abilities is expressed with unusual graciousness by Robert Hooke (*Micrographia,* 1665), " there scarce ever met in one man, in so great a perfection, such a Mechanical Hand and so Philosophical a Mind ".

YOUNG, SAMUEL — BONBURY (sic)

Bracket clock, bell-top ebonised case, arch-dial signed, Samuel Young, Bonbury (sic), in the possession of Mr. R. C. Hart, Chiddingfold Manor House, Surrey, illustrated in *Antique Collector,* February 1960. The name and place have not been identified.

YOUNG, WILLIAM — OXFORD (1656-1695)

Locksmith, blacksmith, turret clock and chime maker. In Catte Street, St. Mary's Parish before 1656. Paid tax on 2 hearths in 1665, and Poll Tax for wife, Jane, and 3 children in 1667. Chosen by the City Council to receive a loan to bind an apprentice in 1679. In March 1686 he was ordered to pull down an encroachment he had built on the street in front of his house. He was working as late as 1695 but the Window Tax of 1696 records the house in Catte Street as occupied by his son Thomas.

Worked for the University of Oxford from 1668 at the Divinity Schools, Old Ashmolean Museum, Physic Garden, Bodleian Library, etc.

Mended and kept the clock and chimes at New College, 1661-1682.

Repaired the clock in Tom Tower, Christ Church in 1686.

Worked at the church of St. Mary the Virgin from 1657 to 1684, set up and repaired the chimes several times; and was clock-keeper from 1661 to 1673.

Maintained the clock and chimes in St. Martin's church from 1676 to 1695, and made a pendulum and anchor escapement in 1686.

Made the chimes for Burford church in 1670.

Cox, J. C., 1913, records, William Young, locksmith of Oxford, covenanted in 1673 with the churchwardens of St. Laurence, Reading, for the sum of 20s. in hand and a further sum of £29 to supply and set up " a firme good substantial and tuneable sett of Chymes to two tunes "— the tunes of the CXLVIII and CXII Psalms, or any other two tunes " — best approved by the Wardens — to strike upon all the eight bells in the tower of equall and good notes ". He further covenanted to make " a good and substantiall Quarterne clock to strike on the aforesaid bells in an orderly manner " and also to put the clock then standing in the tower in thorough repair. The work was finished in 1680.

For details of his work on Oxfordshire clocks see the section on Turret Clocks.

TOPOGRAPHICAL LIST

Clock and Watch Makers, Smiths, Apprentices, Painters, etc., arranged by Places.

ADDERBURY
Fardon, Thomas
Gilkes, Richard
Tyler, Richard
Williams, Joseph
Williams, William

AYLESBURY, Bucks
Stone, John

BAMPTON
Thorndell, Richard
Webb —

BANBURY
Buller, Robert
Burditt, J. W.
Carpenter, William
Drury, Charles W.
Drury, James
Drury, William
Durran, Eustace
Durran, James H.
Dutton, T.
Flowers, John
Hide, Charles
Holloway, Edward
Howse, Charles
Kalabergo, John
Kenning, William
Lamprey, Benjamin
Lamprey, John
Merchant, Thomas
Paine, John
Payne, William P.
Philipps, Charles
Philipps, Michael
Pinfold, Thomas
Roberts, William
Saunders, Charles
Strange, Thomas
Tasker, William
Terry, John
Wagstaff, Thomas
Walford, George
Walford, Henry
Walford, John G.
Wangler, M.
Wells, John
Wells, Thomas
Wise, Matthew
Young, Samuel

BICESTER
Ball, William
Camozzi, Charles
Field, Thomas
Hemins, Edward
Hemins, Joseph
Lamb, James
Musselwhite, William
Tomlinson, Job
Tomlinson, John
Tomlinson, Thomas

BLOXHAM
Arsborn, Thomas

BODICOTE
Wise, Doiley
Wise, John
Wise, Richard

BURFORD
Este, William
Faulkner, John
Gilkes, John
Griffin, John
Harris, John
Head, James
Padbury, Matthias
Paget, William
Walter, Edmund
Warner, Richard

CHARLBURY
Fairbrother, James
Gilkes, Thomas
Harrison, William
Kendall, Larkum
Street, Richard
Wise, Thomas

CHIPPING NORTON
Atkins, Joshua
Atkins, William
Bassett, Joseph
Coles, Richard
Gilkes, Thomas
Gilks, Tobias
Morris, John
Nethercott —
Simms, C. P.
Simms, John
Simms, Samuel
Simms, William
Wells, John

CUDDESDON
Lawrence, William

DEDDINGTON
Edwards, Thomas
Fardon, John
Fardon, Thomas
Gibbs, Joshua
Harris, Thomas

DRAYCOTT
Warner, John

FINMERE
Clarke, James

FINSTOCK
Fairbrother, James

FIFIELD
Nethercott, George

FRITWELL
Harris, George
Harris, Nicholas
Jennings, Thomas
Jennings, William

HASELEY, GREAT
Holloway, Thomas
Stockford, Thomas

HENLEY-ON-THAMES
Archer, John
Atte Lee, John
Barton, John
Beck, I.
Coster, Charles
Coster, James
Gainsborough, Humphrey
Grayson, John
Grayson, William
Harington, Thomas
Hyde, Thomas
Jackson, John
May, Edward
Steward, John
Stone, John
Wallen, J.
Wallen, William
Whitern, William
White, Dan

HOOK NORTON
Bedford, William
Paine, John
Webb, Thomas

HORLEY
Blundell, John

HORNTON
Webb, Charles

LONDON
Moore, John and Son
Thwaites, Aynsworth
Thwaites, John
Vulliamy, B. L.

LONG COMPTON, Warwicks
Nethercott, John
Nethercott, William

MARLOW, GREAT Bucks
Coster, James

MILTON, GREAT
Stevens, John
Stevens, Robert

MILTON UNDER WYCHWOOD
Green, William

155

OXFORD
Adams, Thomas
Aldworth, Samuel
Almond, William
Aulkin, Richard
Belcher, John
Bignell, Thomas
Bird, Michael
Bird, Nathaniel
Bird, Wright
Brickland, Humphrey
Broadwater, Hugh
Brooks, Richard
Browne, James
Bull, Lionel
Bull, Thomas
Cakebread, Richard
Carter, Henry
Carter, William
Cartwright, William
Chapman —
Corbet, Hugh
Creed, Thomas
Curtice, Greenaway
Denton, Robert
Denton, Samuel
Denton, William
Drury —
Earle, Thomas
Este, William
Ford, John
Fowler, William
Free, John
Free, Penelope
Gardner, William
Gillett, Thomas
Godfrey, Henry
Gomes, William
Goweth, John
Green, George S.
Green, John
Gurden, Joseph
Haines, Robert
Harris, John
Harvey, Robert
Hawkins, Richard
Hawting, John
Hawting, William
Heath, John
Herbert, John
Heyten, Richard
Hickman, Edward
Hitchcock, William
Hobdell, Henry B.
Hodges, Anthony
Hunt, Thomas
Jeeves, Anthony
Jeffs, John
Jones, John
Kennington, John
Knibb, John
Knibb, Joseph
Knibb, Peter
Knibb, Samuel
Kratzer, Nicholas
Lane, Wright
Langley, Thomas
Levi, Israel M.
Lidbrook, Thomas
Lock, Edward
Lock, Henry
Moore, Edward
Marriner, William
Marshall, Thomas
Masey, Thomas
Midwinter —
Mathews, Christopher
Oakley, James
Oakley, John
Ortelli, A.
Page, Edward
Pearson, Hawtin
Pearson, Henry
Pearson, Richard
Quelch, John
Quelch, Joseph
Quelch, Martin
Quelch, Richard
Ranklyn, Thomas
Raye, John
Rayner, Thomas
Reynolds, Thomas
Reynolds, John
Riggins, Edward
Rixon, John
Rolewright, Thomas
Rose, T.
Rowell, George
Rowell, Richard R.
Rustin, Joseph
De St. Paul, Johan
De St. Paul, Triumph
Saunders, Richard
Savage, T.
Sayer, Matt
Shewell, John
Sinfield, T.
Sly, Robert
Smith, Abraham
Smith, Thomas
Sowter, John
Steele, Joseph
Tawney, Robert
Taylor, William
Thorp, Benjamin
Tonge, George
Unite, Mathias
Vesey, Robert
Walker, Thomas
Walker, George
Wangler, Luke
Watts, Charles
Wentworth, George
White, James
Wilkins, George
Wood, Richard
Wren, Christopher
Young, William

POUNDON, Bucks
Smith, Vincent

SANDFORD
Nethercott, John

SHIPSTON ON STOUR, Warwicks
Gilkes, John
Gilkes, Thomas
Wells, John
Wells, Thomas

SIBFORD FERRIS & SIBFORD GOWER
Enoch, Ezra

Gibbs, Richard
Gilkes, Thomas
Wells, John

SOULDERN
Gibbs, Joshua

SPELSBURY
Mace, Robert

STADHAMPTON
Jordan, James
Jordan, Thomas

STANDLAKE
Nethercott, John

TASTON
Mace, John
Mace, Robert

THAME
Bowles, Edward
Buckland, William
Cottiswold, Thomas
Harris, William
Heth —
Humphris, William
Lawrence, Thomas
Lawrence, William
Powlen, Thomas
Reynolds, Henry
Smith, Robert
Stockford, Joseph
Stockford, Thomas
Stone, John
Stone, Richard
Stone, Thomas
Tomlinson, Job
Tomlinson, John

WALLINGFORD, Berks
Gunn, William

WARWICK
Paris, Thomas

WATLINGTON
Hewitt, Owen
Hine, John
Tomlinson, John
Vaughan, Richard
Vernon, Joseph
Webb, Francis

WANTAGE, Berks
Nethercott, George

WITNEY
Ansley, Richard
Barrit, George
Fairbrother, James
Green, Samuel
Harris, James
Harris, Joshua
Jeffrey, James
Keyte, Richard
Levi, Benjamin
May, Edward
May, John
May, Thomas
Mazey, Charles
Meakins, George
Powell, James
Randall, Henry P.
Simms, Benjamin
Walter, Edmund

BIBLIOGRAPHY

BAILLIE, G. H., 1947, *Watchmakers and Clockmakers of the World*, 2nd ed.
1951, *Clocks and Watches, An Historical Bibliography*.
BEESLEY, A., 1842, *The History of Banbury*.
BEESON, C. F. C., 1957, 'A History of the Wadham College Clock', *Antiquarian Horology*, June, pp. 47-50.
1958, 'Quaker Clockmakers of North Oxfordshire', *Antique Collector*, October, pp. 185-190.
1960, 'John Knibb, Citizen of Oxford', *Antiquarian Horology*, December, pp. 132-135.
1961, 'Knibb and The Anchor Escapement', *Antiquarian Horology*, June, pp. 208-209.
BEDINI, S. A., 1959, 'Rolling Balls Clocks in England', *Horological Journal*, pp. 778-784.
BLOMFIELD, J. C., 1884, *History of Bicester*.
1887, *History of Finmere*.
BOASE, C. W., 1894, 'Registrum Collegii Exoniensis', *Oxf. Hist. Soc.*, xxvii.
BRIERS, P. M., 1960, 'Henley Borough Records, 1395-1543', *Oxford. Rec. Soc.*, XLI.
BRITTEN, F. J., 1932, *Old Clocks & Watches and Their Makers*, 6th ed.
BROOKS, C. C., 1929, *History of Steeple Aston and Middle Aston*.
BROWN, R. A., 1960, 'King Edward III's Clocks', *Antiquarian Horology*, December, pp. 124-126.
BUCKLEY, F. & G. B., 1929, 'Clock and Watchmakers of the 18th century in Gloucestershire and Bristol', *Trans. Bristol and Glos. Arch. Soc.*, LI.
BUGGINS, G. T. E. & DEVEREUX, F. B. M., 1956, 'John Thwaites', *Antiquarian Horology*, June, pp. 161-163.
BURN, J. S., 1861, *History of Henley*.
CESCINSKY, H. & WEBSTER, M., 1914, *English Domestic Clocks*.
COLVIN, H. M., 1954, *A Biographical Dictionary of English Architects, 1660-1840*.
COX, J. C., 1913, *Churchwardens' Accounts*.
DANIELL, J. A., 1952, 'The Making of Clocks and Watches in Leicestershire and Rutland', ex *Trans. Leics. Arch. Soc.*, XXVII.
DICKINS, M., 1928, *A History of Hook Norton*.
DROVER, C. B., SABINE, P. A., TYLER, C. & COOLE, P. G., 1960, 'Sand-glass Sand', *Antiquarian Horology*, June, pp. 62-72.
ELAND, G., 1931, *The Purefoy Letters 1735-1753*, Vol. I.
ELLIS, W. P., 1901-1915, 'The Churchwardens' Accounts of the Parish of St. Mary Thame', *Berks, Bucks and Oxon Archaeol. Journ.*, VII-XX.
FEEK, P. G., 1955, 'Further Note on Worcestershire Church Clocks', *Trans. Worcs. Archaeol. Soc.*
FLETCHER, C. J. H., 1896, *History of the Church and Parish of St. Martin, Oxford*.
FOWLER, T., 1893, 'The History of Corpus Christi College', *Oxf. Hist. Soc.* XXV.
FOX, C. A. O., 1947, *An Anthology of Clocks and Watches*.
FREEBORN, M. E., n.d., *Twixt Cherwell and Glyme*.
GIBSON, J. S. W., 1959, 'Index to Banbury Wills, 1542-1858', *Oxford. Rec. Soc.*, XL. *Banbury Hist. Soc.*, 1.
1960, 'Marriage Register of Banbury, 1558-1724', *Banbury Hist. Soc.*, 2.
GREEN, D., 1951, *Blenheim Palace*.
GREW, N., 1681, *Museum Regalis Societatis, Catalogus*, pp. 357-8.
GUNTHER, R. T., 1923, 'Early Science in Oxford', vol. II, *Oxf. Hist. Soc.*, LXXVII.
HAYDEN, A., 1917, *Chats on Old Clocks*.
HEARNE, T., 1884-1918, 'Remarks and Collections of Thomas Hearne', ed. C. E. Doble and others, *Oxf. Hist. Soc.*, 11 vols.
HERBERT, G., 1958, *Shoemaker's Window*.
HILARY, E., 1952, 'Thomas Dallam's Great Musical Clock', *Horological Journal*, February, pp. 118-119.
HISCOCK, W. G., 1946, *A Christchurch Miscellany*.
HOBSON, M. G., 1939, 'Oxford Council Acts 1666-1701, *Oxf. Hist. Soc.*, N.S. II.
HOBSON, M. G. & SALTER, H. E., 1932, 'Oxford Council Acts, 1626-1666', *Oxf. Hist. Soc.*, XCV.
HUGHES, D., 1960, 'Bells and Bellfounding', *Antiquarian Horology*, September, pp. 96-101.
JACKSON, T. G., 1897, *The Church of St. Mary the Virgin, Oxford*.

KIBBLE, J., 1928, *Historical Notes on Wychwood Forest*.
LEEDS, E. T., 1923, 'Oxford Tradesmen's Tokens', in Surveys and Tokens, *Oxf. Hist. Soc.*, LXXV for 1920.
LLOYD, H. A., 1957, 'Lady Diana Field:ng's Watch Repairs', *Antiquarian Horology*, June, pp. 53, 57.
1958, *Some Outstanding Clocks Over Seven Hundred Years, 1250-1950*.
MONK, W. J., 1891, *History of Burford*.
MOORE, C., 1896, *Brief Notes relating to Municipal Oxford*
OWST, G. R., 1952, *The Destructorium Viciorum of Alexander Carpenter*.
PESHALL, Sir J., 1773, *City of Oxford*.
PLOT, R., 1677, *The Natural History of Oxfordshire*.
POTTS, W., 1958, *A History of Banbury*.
PUGIN, A. G., 1777, *Specimens of Gothic Architecture*.
RASHDALL, H. & RAIT, R. S., 1901, *New College* (College Histories).
ROBINSON, T. R., 1957, 'Verge and Pendulum Tower Clock', *Horological Journal*, June, pp. 358-359.
SALTER, H. E., 1915, 'A Cartulary of the Hospital of St. John the Baptist', vol. II, *Oxf. Hist. Soc.*, vol. LXVIII.
1919, 'Mediaeval Archives of the University of Oxford', *Oxf. Hist. Soc.*, LXXIII.
1920, 'Surveys and Tokens', *Oxf. Hist. Soc.*, LXXV.
1925, 'Oxford City Properties', *Oxf. Hist. Soc.*, LXXXIII.
1927, 'Oxford Council Acts 1583-1626', *Oxf. Hist. Soc.*, LXXXVII.
1933, 'The Churchwardens' Accounts of St. Michael's Church Oxford', *Oxf. Archaeol. Soc., Trans. No. 78*.
SHARPE, F., 1946-1951, 'The Church Bells of Oxfordshire', *Oxford. Rec. Soc.* XXVIII, XXX, XXXII, XXXIV.
SKELTON, J., 1843, *Oxonia Antiqua Restaurata*.
SPRAT, T., 1667, *History of the Royal Society*.
STAPLETON, B., 1893, 'Three Oxfordshire Parishes', *Oxf. Hist. Soc.*, XXIV.
TURNER, W. H., 1880, *Records of the City of Oxford 1509-1583*.
TYSSEN, A. D., 1863, 'The Old Churchwardens' Account-Books of St. Peter-in-the-East, Oxford', *Oxf. Archaeol. Soc.* I, pp. 286-302.
WEINSTOCK, M. M. B., 1940, 'Hearth Tax Returns Oxfordshire 1665', *Oxford. Rec. Soc.*, XXI.
WILLIAMS, W., 1733, *Oxonia Depicta*.
WILLIAMSON, G. C., 1921, *Behind my Library Door*.
WOOD, A., 1891-1900, 'The Life and Times of Anthony Wood', ed. A. Clark, *Oxf. Hist. Soc.*, 5 vols.
ZINNER, E., 1954, 'Aus der Frühzeit der Räderuhr', *Deutches Museum, Abh. und Berichte*, 22, Heft. 3.

ORIGINAL MS DOCUMENTS

Churchwardens' Accounts:
Adderbury, St. Mary; Bodicote, St. John; Burford, St. John; Chinnor, St. Andrew; Claydon, St. James; Fringford, St. Michael; Great Haseley, St. Peter; Horley, St. Etheldreda; Kidlington, St. Mary; South Newington, St. Peter; Oxford, St. Clement; Oxford, St. Cross; Oxford, St. Mart:n; Oxford, St. Mary Magdalene; Oxford, St. Mary the Virgin; Stanton, St. John; Swalcliffe, SS Peter & Paul; Thame, St. Mary; Wardington, St. Mary Magdalene; Yarnton, St. Bartholomew; King's Sutton (Northants).

College and University Accounts:
New College; St. John's; Wadham. Vice-Chancellor of Oxford.

Account Books of Thwaites and Reed.

Council Minute Books:
Banbury Borough Account Book and Vestry Book; Chipping Norton Corporation Oxford City Council; Henley Vestry Books.

Lease Books and Leases:
Merton College; Oxford City Council; Oxford University.

Parish Registers:
Adderbury, St. Mary; Banbury, St. Mary; Bicester, St. Edburg; Bodicote, St. John; Chipping Norton, St. Mary; Claydon, St. James; Fritwell, St. Olave; Mollington, All Saints; Oxford, All Saints; Oxford, St. Mary Magdalene; Oxford, St. Mary the Virgin; Society of Friends, Banbury Region; Farnborough (Warwicks); Fenny Compton (Warwicks); Newport Pagnell (Bucks).

Oxford Archdeaconry Papers, Marriage Bonds and Intentions

Wills:
District Probate Registry, Birmingham; Somerset House, London

LIST OF ILLUSTRATIONS

PLATES

Frontispiece. JOHN KNIBB of Oxford. Lantern clock, 8¾ inches high. Photo by courtesy of W. Summers.

Plate
1. Fig. 1. HANWELL, St. Peter. Clock made by George Harris of Fritwell in 1671. Author's photo.
2. Fig. 2. Sixteenth century Jacks formerly on St. Martin's church, OXFORD, now in the Town Hall. Author's photo.
   Fig. 3. Inset. Copies of the original Jacks and the quarter bells below the modern clock dial on Carfax Tower. Photo by courtesy of *The Oxford Times and Oxford Mail*.
3. Fig. 4. OXFORD, St. Mary the Virgin. Clock made by Thomas Paris of Warwick in 1741. Author's photo.
4. Fig. 5. BICESTER. Clock made by Edward Hemins of Bicester, ca. 1735. Author's photo.
   Fig. 6. CHURCH HANBOROUGH, SS. Peter and Paul, 17th century clock. Photo by courtesy of A. W. Cox and J. Brice.
5. Figs. 7 and 8. OXFORD, Wadham College. Clock made by Joseph Knibb of Oxford in 1670. Author's photos; blocks by courtesy of *Antiquarian Horology*.
6. Fig. 9. GREAT MILTON, St. Mary. Clock made by Nicholas Harris of Fritwell in 1699. Author's photo.
7. Fig. 10. OXFORD, Magdalen College. Early 18th century carillon barrel. Author's photo.
   Fig. 11. OXFORD, Trinity College. Clock made by John Hawting of Oxford in 1787, showing the anchor escapement. Photo by courtesy of L. S. Northcote.
8. Fig. 12. THAME, St. Mary. Platform and pendulum casing of the church clock. Inset the anchor escape wheel. Photos by courtesy of L. S. Northcote.
   Fig. 13. WATLINGTON, Town Tall. Late 17th century clock. Photo by courtesy of G. F. Shiffner and block by the Watlington Parish Council.
9. Fig. 14. JOHN HAWTING of Oxford. Act of Parliament clock in Radcliffe Infirmary, Oxford, ca. 1775. Author's photo.
10. Fig. 15. RICHARD GILKES of Adderbury. Movement of a 30-hour one-hand wall clock, ca. 1740. Author's photo.
11. Fig. 16. THOMAS GILKES (1) of Charlbury. Dial, 11-inch square of a 30-hour clock. Author's photo.
    Fig. 17. THOMAS HARRIS of Deddington. Dial, 10-inch square of a 30-hour clock. Author's photo.
12. Fig. 18. RICHARD GILKES of Adderbury. Dial, 9-inch square of a 30-hour wall clock. Author's photo.
    Fig. 19. RICHARD GILKES of Adderbury. Dial, 10¼-inch square of a 30-hour clock. Author's photo.
13. Fig. 20. RICHARD GILKES of Adderbury. Dial, 11-inch square of a 30-hour clock. Photo by courtesy of Mrs. C. A. Allitt.
    Fig. 21. RICHARD GILKES of Adderbury. Dial, 11-inch square of a rack-striking plated movement. Author's photo.
14. Fig. 22. FRANCIS WEBB of Watlington. Dial, 10-in square of 30-hour clock. Author's photo.
    Fig. 23. JOHN MAY of Witney. Dial, 10-inch square of a 30-hour clock. Author's photo.
15. Fig. 24. THOMAS GILKES (2) of Charlbury. Dial, 10-inch square of a 30-hour clock. Author's photo.

Plate		
	Fig. 25.	JOHN FARDON (2) of Deddington. Dial, 10-inch square of a 30-hour clock. Author's photo.
16.	Fig. 26.	THOMAS FARDON of Deddington. Act of Parliament clock in mahogany case. Author's photo.
	Fig. 27.	WILLIAM PEACOCK of Banbury. Act of Parliament clock in lacquered case; hour-striking movement. Author's photo.
17.	Fig. 28.	JOSEPH KNIBB of Oxford. Longcase clock, 8-day, verge movement, height 6-ft. 3ins. Photo by courtesy of L. R. Bomford.
	Fig. 29.	JOHN KNIBB of Oxford. Longcase clock, 8-day, anchor escapement, height ca. 8-ft. Photo by courtesy of the Ashmolean Museum.
18.	Figs. 30 and 31.	JOHN HARRIS of Oxford. Watch, silver champlevé dial and case; back plate with early balance spring and irregular cock-foot; ca. 1680. Author's photos by courtesy of the Ashmolean Museum.
	Figs. 32 and 33.	WILLIAM WALLEN of Henley on Thames. Watch, silver paircase, enamel dial; back plate; ca. 1725. Photos by courtesy of the Ashmolean Museum.
19.	Fig. 34.	RICHARD QUELCH (1) of Oxford. Watch, silver and gilt case, gilt engraved dial; ca. 1650. Photo by courtesy of the Ashmolean Museum.
	Fig. 35.	WILLIAM TASKER of Banbury. Watch, four-colour gold case and dial; HM 1829. Author's photo by courtesy of the Ashmolean Museum.
	Fig. 36.	THOMAS & JOHN FARDON of Deddington. Watch, silver pair case; HM 1801. Author's photo.
20.	Figs. 37 and 38.	GEORGE HARRIS of Fritwell. Lantern clock, front and side views. Photos by courtesy of J. M. Surman.
	Figs. 39 and 40.	JOSEPH KNIBB of Oxford. Lantern clock, front and side views. Photos by courtesy of Oliver Bentley.
21.	Fig. 41.	GEORGE WALKER of Oxford. Lantern clock, movement with rack-striking. Photo by courtesy of Dr. N. Watson.
	Fig. 42.	JOHN KNIBB of Oxford. Bracket clock (see also Plate 24, Fig. 47). Author's photo; block by courtesy of P. G. Dawson.
22.	Fig. 43.	JOHN KNIBB of Oxford. Bracket clock, timepiece. Photo by courtesy of P. G. Dawson.
	Fig. 44.	JOHN KNIBB of Oxford. Bracket clock with key. Photo by courtesy of R. A. Lee.
23.	Fig. 45.	JOSEPH KNIBB of Oxford. Wall clock. Photo by courtesy of L. K. Bomford.
	Fig. 46.	JOSEPH KNIBB of Oxford. Longcase, details of hood and dial of Plate 17, Fig. 28. Photo by courtesy of L. K. Bomford.
24.	Fig. 47.	JOHN KNIBB of Oxford. Bracket clock, back plate of Plate 21, Fig. 42. Author's photo.
	Fig. 48.	SAMUEL ALDWORTH of Oxford. Bracket clock, back plate. Photo by courtesy of W. Philippson.

TEXT FIGURES

Diagram showing the annual numbers of master clockmakers and apprentices, Oxford, 1650-1725	page 21
Scroll-ended finials of turret clocks	page 25
Hand of lantern clock by George Harris of Fritwell	page 84
Memorial stone to the Knibbs, St. Cross, Oxford	page 119

PART THREE
ADDENDA

TURRET CLOCKS

[see also pp. 25 - 75]

ADDERBURY — St. Mary (p. 26)

Recently discovered Churchwardens' Accounts reveal that there was a striking clock in the church in 1611 and that it had already given some service. In 1617 it was sent away for repairs which cost 21s., cloth, nails and leather (cost 2s. 5d.) were also needed to repair the "fanne". (The use of cloth and nails to make the vanes of the fly apparently on a wooden frame was not an unusual practice in the 16th century.) In 1621 further repairs, 18s., were done by Emanuel Croker, not a local man, and the fan again received attention. In 1626 the "clockmaker" was paid £2 5s. 2d. plus 7d. for an oak board. An item in 1631 records *seting up the Diall and Stufe for it, 16s. 4d.*, but it is not clear if this was a sundial or a clock dial. From then until 1645, when the Accounts end, there was very little expenditure on the clock and it seems to have functioned until late in the century as a new clock was ordered in 1684 (p. 26).

An hour glass with an iron stand was bought for 7s. 4d. in 1629; the glass was replaced in 1633 for 9d. and again in 1641 for 9d.

BARFORD ST. MICHAEL — St. Michael (p. 30)

It is now known that this clock was not originally in Bloxham Church; see under Wigginton.

BECKLEY — St. Mary

A 19th-century 8-day clock with a Lepaute pinwheel escapement and an 8-foot pendulum swinging in a wooden casing like that shown in Plate 8, fig. 12. The maker's name, BAILEY ALBION WORKS SALFORD MANCHESTER, is cast with the frame.

The church has an 18th-century hour glass in a wrought iron wall-bracket.

BENSON — St. Helen

A new ring of 8 bells was installed in 1781 and a new clock in 1795. It is a tingtang, ¼-striking, 30-hour clock with three trains of brass wheels in a massive iron frame with cast cylindrical ball-ended pillars joined by 3" and 2" wide bars, a recoil anchor escapement beating 2 seconds and maintaining power. A smaller frame with a chime barrel is attached at one end. The hour indicator dial is inscribed "Made by John Thwaites Clerkenwell London 1795 for James Higgs Wallingford", and the minute indicator dial is inscribed "1827 Repaired & Made as New BY THWAITES & REED Clerkenwell London 1795" with the names of 4 Churchwardens. (L. S. Northcote). There is an Act of Parliament clock by James Higgs in the Lamb Inn, Wallingford.

The specification of the original clock in Thwaites and Reed's Daybook for March 1795 is :

"Wallingford for the Parish Church of Benson near Wallingford Mr. James Higgs.

To a new large 30 Hour Quarter Church Clock to show one outside Diall Hours & Minutes & to strike the Hours on a Bell of about 16 cwt Quarters on 6 lefs Bells in Proportion the Watch Great Wheel 11 In Diameter the Striking Great Wheel 1F 3ins Diameter Quarter Great Wheel 1f 6in Diameter the Rest in Proportion & all of Brafs with Quarter Pin Barrell to have 10 Different Changes with Spiral & long Leavers to Pump it the Pinions & Pallets all Hardened & Hammer Work & everything Compleat Except Bell Diall Plate Package Carriage & Fixing. £84.

To a new Copper Diall Plate of 5 Foot Diameter with an Anticke Moulding round the Edge Painted Black with Gilt Figures & Moulding & with Proper Bolts & Brafses to fasten it up, £15.15. ".

5 Packing cases, carriage and 17 days fixing, £56 7s. 0d.

BUCKNELL — St. Peter (p. 33)

A late 17th or early 18th century clock in a frame 32 × 21 × 16 inches, the finials outwardly curved and ending in sub-pyramidal knobs. Trains side-by-side, the striking train with the pivot bars tenon-wedged, iron wheels with 4 straight lap-welded arms, external count wheel, peg and lever control. The going train great wheel is original but the 6-armed second and escape wheels are of brass and the anchor has removeable pallets; the pivot bars are fixed with nuts. The clock was converted to dead beat escapement with a wood-shafted pendulum beating 2 seconds and maintaining power added. There was a single counterbalanced 18th century hand (now nailed to one of the church doors) until about 1894 when a 2-hand dial with appropriate motion work was substituted.

BURFORD — St. John the Baptist (p. 33)

A late 18th-century clock movement is now displayed in the Bell Chapel. The iron bar frame, 33 × 16 × 24 inches, with brass finials, has 2 side-by-side trains of brass wheels and solid pinions in pivot bars fixed by nuts. The A-shaped anchor has fabricated pallets but the pendulum is missing. The striking train has a brass count wheel, hoop-wheel as well as peg and lever control and a centre hour wheel with clutch plate and 2 hand-grips on the arbor; the fly is 4-vaned. A white enamel circular dial is used for time-setting. The clock is probably as late as 1790 (see presentment orders mentioned on p. 33).

CHARLBURY — St. Mary (p. 33)

The 18th-century clock was replaced in 1885 by a large flat bed, $\frac{1}{4}$-striking clock with a dead beat, 2-second pendulum by Henry Sainsbury, Walthamstow. The shaft is an iron tube carrying a heavy cylindrical iron bob, 11 × 9 inches, on which is a wooden platform holding various metal weights the sole means of rating (L. S. Northcote).

CHARLBURY — Lee Place

The clock in a lantern on the stable building with dials dated 1725 was a very small striking clock in a wrought iron frame of unusual construction in that the vertical bars of the going train were of brass. It is likely that these were added at a later date than that of its original construction (J. E. H. Smith).

It was replaced in 1960 by an electrical model by John Smith & Sons, Derby.

CHISELHAMPTON — St. Katherine

A small frame of iron bars rivetted at the corners, the pivot bars fixed by nuts. Two side-by-side trains of brass wheels, a recoil escapement with a low curved anchor spanning 6 or 7 teeth, the pendulum beating 36 to the minute, with a flat iron rod and heavy lenticular bob. The brass indicator dial with 60 minute divisions is inscribed " Repaired by R. Taylor & Sons Oxford 1827 ".

The square dial on the west face of the wooden clock turret is painted blue with gold hands, numerals and spandrels and the date 1762. The north and south dials are painted similarly without numerals or date. (L. S. Northcote). Before 1827 there was a single circular dial.

The clock is contemporary with the church which was built in 1763.

DITCHLEY PARK — The Great House

A very finely made clock in the cupola on the east wing with a bell lantern above; the similar cupola on the west wing with 4 dummy dials has no clock. The iron bar frame, $32 \times 24 \times 16$ inches, has corner-standards of square block ends joined by a turned column, with solid brass urn finials screwed on. All bars are fixed by cast decorated nuts on screw pins neatly finished by turned knob tips. The side-by-side trains have wooden barrels grooved for ropes and brass ratchets. All wheels are of brass, the main wheels with 4 straight arms and the 2nd and 3rd wheels with 4 narrow flask-shaped or lanceolate arms typical of Hemins' work (see pp. 28, 29 and 65). The very small dead beat escape wheel has a brass anchor with adjustable steel pallets pivoted in an elaborate iron sub-frame decorated with brass caps similar to those on the tips of the standard finials. The pendulum suspension has a very short spring and a pair of rating screws for the peg of the crutch on the wooden pendulum rod. In the striking train the brass count wheel is alongside the main wheel, the 2nd wheel has 8 lifting pins, the 3rd wheel one stop pin, peg and lever control; fly with 2 adjustable vanes. Leading off by a central contrate wheel (flask-shaped crossing-out) and vertical rod to 4 dials each with a single counter poised gilt hand. Plate 25, Figs. 49, 50.

The silvered setting dial with a filed tailed hand is signed *Edward Hemins Bisiter Fecit*. As the cupolas on the two wings of the House are part of the design by James Gibbs in 1722 the clock is presumably contemporary. It is one of the earliest turret clocks with the Graham escapement invented in 1715.

DITCHLEY PARK — The Model Farm

In a cupola on the stable building is an 8-day rack-striking clock by John Moore & Son, Clerkenwell, dated 1860. The frame has cast iron corner-standards of the cylindrical column and ball type. Pendulum with a wooden rod, wing-nut suspension, a pair of beat-adjusting screws for the crutch and a rating nut below the large disc bob. Four dials.

DUCKLINGTON — St. Bartholomew

An iron bar frame, $29 \times 21 \times 14$ inches, its cross-bars mortice-welded to corner-standards ending in a weak out-turned finial with a conical boss; bearing bars nutted. All wheels of iron with 4 straight lap-welded arms and leaf pinions. The brass recoil escape wheel has a shallow arched anchor the arbor of which is laterally moveable on loosing a turnscrew. The one-second pendulum is in two parts, the upper sited in the cock is a short rod dependent from the spring and entering the crutch fork, and the lower a rod threaded above carrying a winged rating nut, attached to the upper piece and also passing through the fork. A small brass dial of 15 divisions has a pointer fixed at the end of the arbor of the 2nd wheel. The striking great wheel with 8 lifting-pins has an outside count wheel and peg and lever control.

The 6 top and bottom and middle cross-bars of the frame have each a median slot serving no obvious purpose, but apparently part of the original mid-18th century design.

ENSTONE — St. Kenelm

An 18th century clock existed which was repaired in 1810-11 by Thomas Fardon of Deddington for £19 2s. 0d. This was replaced in 1857 by one made by Frederick Dent, London, successor to E. J. Dent, at a cost of £105 met by a legacy of Mrs. Esther Oakley. Dent's receipt states " A Turret Clock fit to strike upon a bell of 14cwt. with a pinwheel escapement, going barrel and compensation pendulum, to strike the hours only, and with 1 cast iron skeleton dial of 3 ft. 6in. diamr and dial with hands complete and including workman's time fixing the clock ".

EWELME — St. Mary (p. 37)

The church clock was discarded about 1950 and went to a scrap iron dealer and eventually was acquired by Mr. H. N. Edwards of Rawlins Farm, Ramsdell near Basingstoke. In July 1962 it was installed in an outbuilding with a wooden cupola showing two modern dials. The nutted iron bar frame, $45 \times 26 \times 28$ inches, has two side-by-side trains of brass wheels, a recoil anchor escapement with an adjustable pendulum beating approximately 1 second. Striking control is by peg and lever without a hoop wheel and a count wheel that has projecting teeth instead of notches; the locking-piece lifts from the rim and locks on the tip of a tooth; the count wheel is turned by an externally toothed applied wheel. The setting dial is inscribed " The Gift of John Jacob of this place. Mr. Chas Eyre Mr. Ino Lane Mr. Jas Howse Mr. Ino Warner Trustees AND MADE BY STOCKFORD OF THAME 1770 ".

EYNSHAM — St. Leonard

In 1640 the Churchwardens' Accounts record *For the Clock exchaynged, £6. 10. 0. For himselfe & his man's charges, 2s. 6d.* This indicates an earlier clock taken in part payment for a new one; it is possible that the earlier one came from Eynsham Abbey, dissolved in 1539 and rapidly destroyed in subsequent years. A boarded clock-house was made in 1650 at a cost of 20s. 5d. The maker of the clock was probably Thomas Ranklyn (p. 135).

About the middle of the 18th century a new clock was acquired which has functioned for more than 200 years. This is in a frame, $36 \times 18 \times 27$ inches, of flat iron bars without finials bolted together, housing trains of wrought iron side-by-side with a brass escape wheel and a pendulum of about 46 inches. Striking control is by peg and levers and a count wheel outside the frame. A two-hand dial. It was probably supplied by John May of Witney (p. 128) as he had an annual contract of 3s. for keeping the clock repair until 1798. A new dial board was put up in 1799 at a cost of £2 7s. 8d.

From 1808 to 1832 the clock was maintained by Joshua Harris of Witney (p. 109) at annual fees of 7s. then 8s. He was followed by Richard Treadwell at 10s. p.a. until 1863 (end of Accounts). Recently John Smith and Sons, Derby have supplied a new clock and the old one is displayed in the church.

FINMERE — St. Michael

The clock was cleaned in 1738 by a Mr. Bradford and by him on 8 other occasions at 5s. a time until 1775. This man was either Thomas or James Bradford, clockmakers of Buckingham. The clock winder was paid 6s. a year from 1761 to 1795 and at 10s. a year to 1827. In 1776 a Mr. Packer was engaged at 5s. for what is described as "Doing ye Clock" and this arrangement continued until 1825 with the wage raised to 10s. a year in 1810. William Packer was a clockmaker in Buckingham at the end of the 18th century and John Packer had his workshop in Tingewick, Bucks, the next village to Finmere, in the 19th century. He made the clock in Newton Purcell church (q.v.).

In 1827 William Stanton clockmaker of Buckingham, was responsible cleaning and repairing the clock until 1844, when William Bayliss of Finmere (p. 92) replaced him and remained in charge of the clock for the next 40 years with Gabriel Friday as winder. During the period of 130 years from 1731 clock lines and wire were bought on 16 occasions and "2 new wheels" in 1811; as the latter cost only 5s. 6d. they were probably pulleys.

The dial was repainted in 1783 and 1808 by Gunn and 1831 by Tappin. It is a wooden hexagon painted white with slender black hands.

IDBURY — St. Nicholas

The remains of an early wood-framed clock were found by Mr. L. S. Northcote and the author in the tower below the bell frame. Originally the oak frame of 3 vertical compartments (Type 5 of *Antiquarian Horology*, Vol. IV, Dec. 1963) measured about $4 \times 3\frac{1}{4}$ feet. The ends of the 2 outer compartments have been sawn off and have disappeared together with important parts of the movement.

In the left hand compartment the going train barrel was wound by 4 wooden hand-grips and a horseshoe spring click, its wheel is of iron with 4 lap-welded spokes, 102 teeth and one let-off pin. It is planted at the top of the train; below it is one arbor with lantern pinion (D in adjoining text-figure) and no wheel; the pivot bolt (K) for a 3rd lowermost arbor remains in the post of the middle compartment. The original escapement was most probably a vertical crown wheel and underslung foliot, as was usual in wood-framed clocks of this type. The 18th-century escapement which has replaced it is of a unique design and is mounted on a $13\frac{1}{2}$ inch post rising from the basal member of the frame in the left hand compartment. This post is clearly a substitute or an addition as it is fixed in place by nailed iron plates and is not morticed as are the other vertical posts.

A description of the surviving parts of this escapement and reconstruction of the escape wheel and a discussion of its action are given below by Mr. L. S. Northcote.

The remaining parts of the clock comprise an externally notched count wheel turned by a leaf pinion in the middle section. The striking train barrel and wheel with 8 lifting pins are like those of the going train but are at the bottom of the train. The hoop wheel, lantern pinion and arbor are complete but only one pivot bolt for the fly, etc. remains. In an outrigged pair of iron brackets rising from the two middle posts are pivoted the lifting, locking and stop detents.

The iron pivot bolts, inserted through the posts, have simple brass bushed loops and are tenon-wedged at the opposite ends. One pivot of the striking main wheel has a removeable wedged block instead of a loop, allowing the wheel to be detached. Weights are of stone and there was no dial.

THE ESCAPEMENT by L. S. Northcote

All that remains of the escapement is an iron arbor (A) 15¾ inches long between pivots and 7/16 inch in diameter, mounted in an iron bracket (B) on a short post (C) at right angles to the train, through which it passes between the great and second wheel arbors. (See adjoining text-figure). Fixed to it just in front (i.e., on the winding side) of the point where it crosses over the second wheel arbor (D), is a curiously shaped anchor (E) with two flat "flukes", ⅛ inch thick and ⅜ inch apart, parallel to each other but not quite at right angles to the arbor either on plan or elevation. At rest this anchor hangs down just clear of the second wheel arbor.

The flukes are rivetted and brazed to opposite sides of a short square shaft taken through the arbor and rivetted in a depression on its upper surface. Their upper edges overlap by ¾ inch but, as their opposing edges are cut away at an angle of 45°, there is a small gap at their lowest points. These inclined edges are ground to an angle of 45°, both bevels facing to the front. The near fluke is set slightly higher than the other and the lower edges of both are cut in curves which correspond roughly to arcs of circles centred on the arbor. Looked at from the side the arbor is horizontal and the anchor is set at an angle of 2½° from the vertical, raking towards the second wheel arbor. On plan it is turned 7° anticlockwise from the position at right angles to the arbor.

It must be remembered that although dimensions and angles are given as accurately as possible, they cannot be regarded as precise owing to the rust eroded surfaces and the considerable wear. The bevelled edges of the flukes for instance are not flat but very rounded.

Near the far end of the arbor, is rivetted a piece of iron (F) ¼ inch square, not quite in line with the anchor, the angle between them being about 5°. This piece is bent twice at right angles, first backwards and then vertically downward. Overall it reaches down about six inches, the lower part being flattened and drilled with five holes parallel to the arbor. Counting from the top the first hole is ¼ inch diameter plain, the second 5/16 inch countersunk and the rest ¼ inch countersunk. All the countersinking is on the back. The pivot near this "tail" is ¼ inch diameter and in good condition, nicely turned with chamfered shoulder and tip. It is clearly a replacement having been inserted in the end of the arbor and secured by punching through the arbor when hot. The other pivot is very worn, tapering from ¼ inch at the root to 3/16 inch with flat shoulder and tip.

Debaufre Escapement. Although puzzling at first it was realized eventually that this is a type of Debaufre escapement (originally produced in 1704), as improved by Sully in 1721 for use with a single escape wheel and balance, and further modified by an unknown maker in this instance for use with a pendulum.

As originally designed it was a dead beat frictional rest escapement, but the 7° twist of the anchor has changed this example into an ungainly recoil. The flukes are of course the pallets and as they are parallel the following rather odd sequence of events occurs. The entry pallet (G) receives (in theory at least) first a very slight impulse, which it cancels almost immediately by forcing the tooth to recoil until, as the latter falls off the bevel, the pallet receives full impulse. The exit pallet (H) on the other hand first gives a very slight recoil; then as the swing reverses receives an equally slight impulse until, as the tooth falls off the bevel it receives full impulse. In fact, the angle being so small and the friction so great, it is doubtful if any impulse is given before the tooth reaches the bevel. It may be that the maker, or a later repairer, wanted to reduce the amplitude of the swing and found this 7° twist sufficient for his purpose without going to the trouble of re-setting the pallets at opposing angles to give a straightforward recoil at each swing. Punch marks round the square shaft suggest that in fact it has been shifted from its original setting in the arbor. But more probably it was moved to ease the passage of the teeth through the pallets, thereby reducing the amplitude, without considering the slight recoil effect. The replacement of the rear pivot was no doubt due to the continual grinding end pressure which in time would have allowed a fatal mis-locking as the pallets moved forward over the wheel.

On measuring the remains of the train as carefully as possible it was found that the 2½° slant of the anchor, together with the slightly deeper exit pallet, brings both pallets on the circumference of a circle centred on the pivot bolt (K) of the missing escape wheel arbor. Only a socket for the opposite pivot hole for this arbor is on the short vertical post (C). The anchor swings over the escape wheel at a point on the circumference 5½° beyond the centre of this circle.

Pendulum. At first it was thought that the twice bent "tail" was screwed to a wooden pendulum rod of some length and this may have been the case initially. But on considering the problems presented by the form of the anchor and the train as a whole, it seems more reasonable to suppose that a very short rod was used

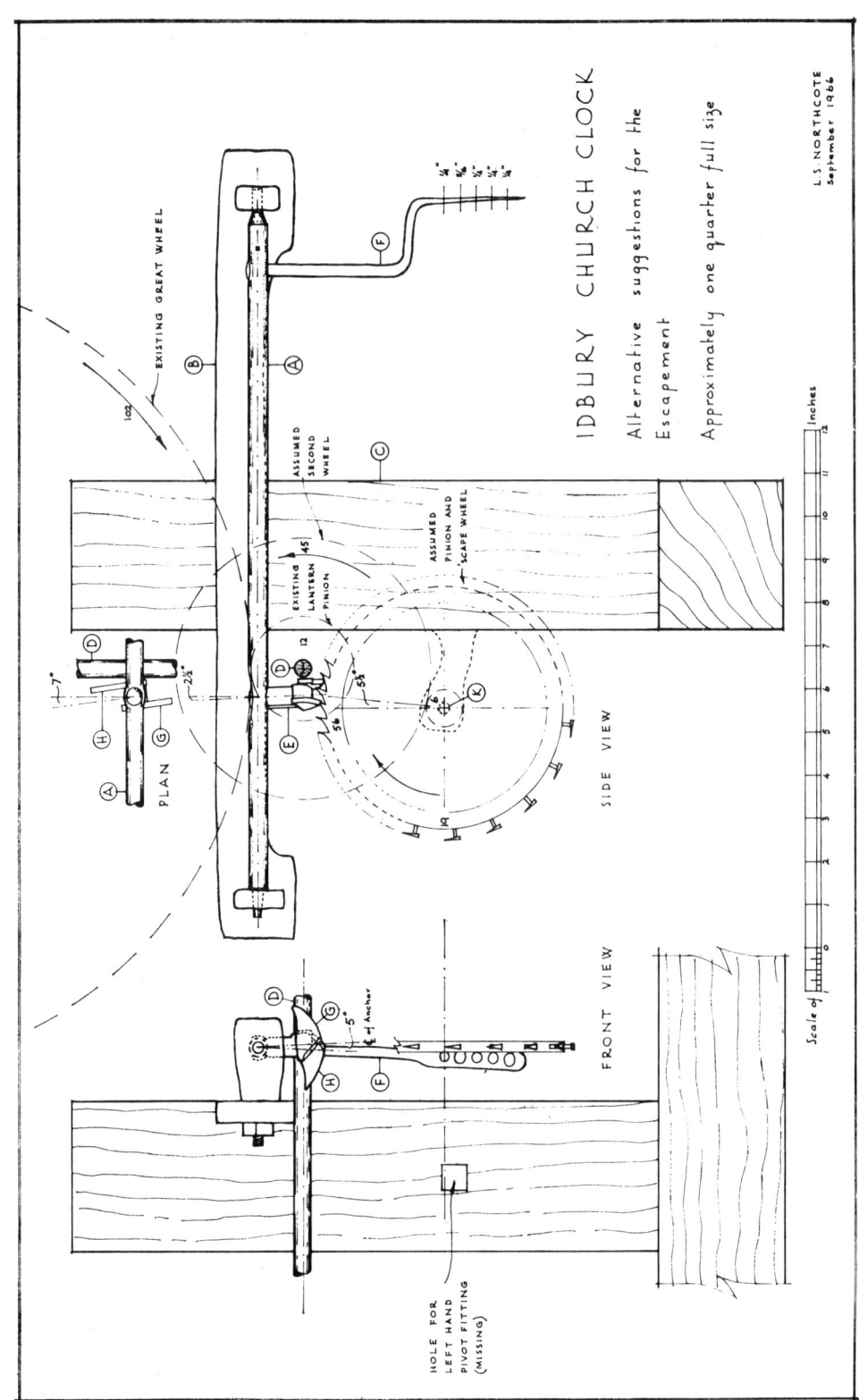

and that the number of holes was merely for rough adjustment of length. It is noticeable that the holes are equidistant and, since only two would be needed for fixing, several alterations in length are possible using the same holes in the rod.

Escape Wheel. The teeth of the escape wheel had to be robust and hard enough to withstand the wear across the front surface as the pallets slid past and the wear on each side as the teeth dropped off the opposing bevels. So it is imagined that they had slightly rounded fronts and tips with flat backs and were not less than $\frac{1}{8}$ inch wide at the tips. Even this width of tooth in passing through the pallets would have caused a full arc of some 11°, which points to the use of a fairly short pendulum. Sully used teeth similar to those on an ordinary recoil type of escape wheel and it seems probable that his example would have been followed.

To prevent excessive drop the wide spacing of the pallets requires either a tooth which nearly fills it or a much thinner tooth so that two may be contained in it. Either method is possible but, if the second was used, why was the width not made to suit one thin tooth in the first place? The answer may be that more than one attempt, by the same man or his successor, has been made to produce a more reliable escapement.

Some years after Sully produced his Debaufre type escapement using a single escape wheel with thin teeth, Julien le Roy, about 1735, made what he considered to be a further improvement. On the single escape wheel were teeth hooked at the tips, like inverted L's, pointing towards the pallets. No doubt there was less friction as only the tip of the hook rested on the locking face of the pallet. It is not clear, but probable, that the impulse was divided between the sides of the hook and the opposing bevels of the pallets, the upright portion of the teeth not coming into contact with the pallets at all. A tooth of this shape needs much more room between the pallets than the thin tipped type and it has the disadvantage of requiring higher train ratios if the pendulum length is to be kept within bounds, since so few teeth can be accommodated on the 6 inch diameter wheel.

In this instance with a drop of slightly under 1/32 inch only 19 teeth are possible. The ratio of the great wheel to the lantern pinion is 102:12 or 8·5:1 and the ratio of the missing second wheel to the missing escape pinion cannot have been much higher as the centres are only $3\frac{1}{2}$ inches apart. The great wheel, carrying the strike letting off pin, revolves once an hour and the time for one vibration with such a train is 1·32 seconds, the pendulum being 5 feet 8 inches long.

With such a pendulum length the arc must have been small; only to be obtained with this escapement by using very narrow teeth working on the extreme edges of the pallets. Mislocking and lack of impulse would have followed the smallest degree of wear and, although the design of the pallets indicates that this arrangement was intended, it could not have worked for long.

The inference is that the escape wheel was changed for one with a higher number of teeth after the Sully pattern. No doubt it was in hard brass and with a drop of about 1/32 inch it is possible for two teeth to be accommodated between the pallets; one having dropped off the entry pallet and the other locked on the exit pallet. On the reverse swing the forward tooth drops off the exit pallet, the next is a little more than half way between the pallets and the following tooth is locked on the entry pallet. The teeth may be about 1/64 inch thick and up to $\frac{1}{8}$ inch wide at the tips with clearance behind them, after dropping off the entry and exit pallets, of 1/64 and 1/32 inch respectively. Between 54 and 58 teeth may be used depending on the depth, but as there would be unsafe locking and lack of impulse with 54 and danger of binding on the central block with 58, the optimum number seems to be 56. It has the additional advantage of being easier to set out.

With this number of teeth on the escape wheel the whole situation is much better. The ratio of second wheel to pinion can be lowered to 7·5:1 or possibly less, and the pendulum is short, about 10 inches with a half second beat.

The Train. The following tentative suggestions for the whole train are therefore put forward:

Great wheel	102	Existing — iron
Lantern pinion	12	
Second wheel	45	Probably brass
Solid escape pinion	6	Iron
Escape wheel	56	Almost certainly brass

The time for one vibration is $\dfrac{3600 \times 12 \times 6}{56 \times 2 \times 102 \times 45} = 0\cdot504$ second.

The pendulum length is $\left(375\cdot4 \div \dfrac{60}{0\cdot504}\right)^2 = 9\cdot94$ inches.

Note that a pendulum of this effective length would have swung clear of the top of the stand or table on which the clock was fixed.

A final observation, which seems to add weight to the above argument, is that two of the holes in the "tail", the second and fifth from the top, are very worn. They are elongated in the direction of the swing, indicating that the screws or rivets worked loose. The degree of strain to cause this wear is very understandable if a fairly heavy bob were swinging quickly in a wide arc but hardly to be imagined with a long slow pendulum moving in a narrow one.

It is hoped that these notes and conjectures will be of interest, especially to those with a practical knowledge of escapements, and that one day an authoritative report will be written on this interesting relic of eighteenth century ingenuity.

Provenance of the Clock. At Idbury, a very small village on the Gloucestershire-Oxfordshire border, the clock is well outside the Midlands region in which wood-framed clocks have so far been located (*Antiquarian Horology, loc. cit.*), and is unlikely to have been made in the near neighbourhood. There is, moreover, no instance of a similar escapement being made by any Oxfordshire clockmaker. The Lords of the Manor of Idbury in the 16th and 17th century were the family of Logan or Loggin, the last of whom to hold the Manor was Elizabeth Logan who married a Papist, Charles Fortescue of Husbands Bosworth, Leicestershire. In the early 18th-century, according to Mrs. E. Goshawk (*Idbury History*, 1961) " there is a general suggestion of poverty and neglect given in papers dealing with Idbury at this time ". Bishop Secker's Visitation of 1738 (*Oxf. Rec. Soc.*, XXXVIII, 1957) states that there was no resident parson then and Mr. Fortescue, the absentee landlord, visited only twice a year to receive rents, at which times a Papist priest officiated in a chapel in the Manor House. Nevertheless, a new treble bell for the church was supplied by Abel Rudhall of Gloucester in 1749. Someone must have paid for it, presumably a Fortescue, whose family had the patronage and advowson until 1863.

In view of the village's connection with Husbands Bosworth it is possible that the clock originated in that area and was transferred to Idbury by one of the Fortescue-Turville family after being discarded from its former church. This must have occurred in the early 18th-century after the new escapement had been fitted. The clock very soon fell into disuse and its existence was unknown to the present inhabitants.

KIDDINGTON HALL — (p. 44)

The clock in the cupola on the stable building has a small iron bar frame $19 \times 14\frac{1}{2} \times 12$ inches, and operates 4 two-hand circular slate dials. The dead beat escapement with fabricated pallets, set laterally to the wheel, beats about 39 per minute; the crutch has adjusting screws to the wooden pendulum rod which carries a very heavy bob. Instead of notches the external silvered count wheel has 12 spaced pegs. The hand-setting dial marked for minutes is also silvered. Maintaining power. Plate 26, Fig. 52.

There is no signature or indication that it was made by B. J. Vulliamy.

MIDDLETON STONEY — All Saints (p. 46)

A late 18th-century clock in a frame $22 \times 16 \times 22$ inches with cast iron ball-ended cylindrical pillars, the 2 trains with brass wheels, a recoil anchor with fabricated pallets, maintaining power and a pendulum beating 52 to the minute. Striking control uses a centre wheel with a brass snail and an arched steel rack, with the levers mounted on the middle pivot bar. The barrel is as in the going train but has an additional inner brass wheel turned by an extra pinion on the winding arbor. Hand-setting is by means of a brass wheel with clutch, a chapter ring marked I-XII with hour and minute pointers and is inscribed *Jullion Brentford*.

The firm of Jullion was operating as early as 1760. The contemporary 2-hand dial is diamond-shaped.

MIXBURY — All Saints (p. 47)

The frame measures 25 × 14 × 22 inches. The great wheel of the going train has an applied chapter ring numbered 0·30·60 repeated 3 times, and 3 let-off pegs and turns once in 3 hours with the later brass second and escape wheels.

NEWTON PURCELL — St. Michael (p. 49)

The frame is of nutted iron bars, 27 × 12 × 18 inches, with a pendulum beating 45 to the minute. The setting dial is signed PACKER TINGEWICK. It operates two circular dials on the west and south of the small bell turret. The clock probably dates from the beginning of the 19th-century; it is not accessible for close examination and is wound once a week.

NORTHMOOR — St. Denys

Richard Lydall, who died 20th April, 1721, left money " to buy and sett up a good Church Clock to strike on the said Great Bell . . . and that money shall be set apart and applyd for keeping . . . the clock in good repair and for making such yearly allowance to some propper person for taking care of the said Bells and Clock ".

But it was not until 1785 that a clock was installed. It has an iron bar frame 32 × 24 × 14 inches with cast cylindrical corner-standards ending in urn finials; the cross-bars are morticed and fixed by nuts and the bearing bars are also fixed by nuts. All wheels are brass with straight 6-arm crossing-out, the great wheels with a brass ratchet wheel, its detent with a semi-circular spring. The dead beat escape wheel has a $\wedge$-shaped anchor with adjustable pallets. The pendulum has a very long flat iron rod in a triangular wooden casing similar to that on Plate 8, Fig. 12 which extends 16 feet below the clock room floor with 4 feet of pendulum rod above. The striking train has an iron count wheel to which is attached a brass wheel, and control is by stop pin, levers and a peg on the fly-wheel arbor.

The old setting dial is inscribed " *Hawting* OXFORD 1785 " and the later hand setting dial is inscribed in capitals " The dead beat escapement pendulum and minute hand fited by W. White Fyfield 1863 ". There is a 2-hand circular skeleton dial.

For John Hawting see p. 111. W. White was a clockmaker and bellhanger, who installed a 3-train, $\frac{1}{4}$-striking turret clock at Besselsleigh Manor also in 1863, and was a member of the White family of bellhangers at Besselsleigh and Appleton, Berkshire.

OXFORD — Magdalen College (p. 51)

Evidence of an earlier clock than that put in the bell tower in 1505 has been found in a payment to a carpenter, Thomas Walshe, who repaired the wooden housing of a clock in St. John's Chapel in 1496. The chapel was built between 1474 and 1480.

Walshe also worked with Robert Carrow, the master carpenter, and James Lynche installing the bell tower clock in 1505-6.

The carpenters Robert Boys and Geoffry Carpenter made a frame for this clock in 1511, work which took them 12 days (wages 7s. 11d.); the wood had been previously sawn up by one Morten. In 1512 Boys put the weight on the clock removing it from the upper chamber of the tower to a lower one. In 1518 repairs to the timber framing were done by Roger Wryght (*Oxoniensia*, XVII, 1954, pp. 112-184).

OXFORD — New College (p. 52)

The earliest evidence of the clock in the bell tower at New College occurs in the payment to John Philipps, carpenter, in 1505 for making a " setting " for the clock (*Oxoniensia, loc. cit.*).

ROLLRIGHT, GREAT — St. Andrew

This clock is by Joyce of Whitchurch, dated 1899, an 8-day, flat-bed, tingtang ¼-striking with an Amant pinwheel escapement and a one second pendulum.

ROTHERFIELD PEPPARD — All Saints

Another clock by Joyce of Whitchurch with a pinwheel escapement.

SHIPTON UNDER WYCHWOOD — St. Mary (p. 67)

There is no trace of the early clock. That at present in use is a flat-bed, 2 train, rack-striking clock with a dead beat escapement, 1-second pendulum and maintaining power. A small hand-setting dial is engraved *Thomson-Profaze 25 New Bond Street LONDON*. It is on a floor above the bells and the weight lines go up almost to the apex of the steeple. Thomson and Profaze succeeded Adam Thomson in 1860. The circular slate dial is painted blue with gilt late 19th-century hands.

SOUTH LEIGH — St. James the Great

An iron bar frame $30 \times 23 \times 18$ inches, with the cross-bars morticed and welded to the corner-standards which have traces of buttress mouldings and no finials; the bearing bars are fixed with nuts. All wheels in the side-by-side trains iron with 4 straight welded arms. The going great wheel has 2 lifting-pins. The brass recoil escape wheel and the shallow anchor are pivoted in a lateral L-shaped bracket on the bearing bar resembling the method at Horley (p. 42) and Hornton (p. 43) and Mollington (p. 47). The one-second pendulum has a cylindrical bob. The striking train with peg and lever control has an unusual mounting for the fan vanes which are held between stout iron wires instead of fixed to a rod. The dial on the south wall of the tower is a painted diamond-shaped board bearing the inscription " Ye know not what hour your Lord doth come ", made and gilded by a Vicar at the beginning of the present century.

There is no history of the clock to account for the planting of the going train which may be an alteration to a mid-18th-century design. It is said to have been imported from Prestbury, Gloucestershire.

WARBOROUGH — St. Laurence

The clock was supplied by Gillett and Bland, Croydon, in 1871, the gift of Sidney Beisley. It is a flat-bed type with a dead beat escapement at 40 to the minute, and count wheel hour-striking with one dial on the west side of the tower. The setting dial is silvered and engraved " Christie

Wallingford ", (L. S. Northcote). In the same year the third bell was recast and the ring of 6 bells rehung.

WIGGINGTON — St. Giles (p. 73)

The recently discovered Churchwardens' Accounts for 1716 to 1893 throw more light on the history of the clocks in Wiggington, Barford St. Michael (p. 30) and South Newington (p. 47). In his brochure, "The Church of Saint Giles" (1965), F. D. Price states: "Local tradition has it that this clock 'came from Bloxham'; but this tale almost certainly arises from confusion between the name of the adjoining village and the name of a man. The Wiggington churchwardens' accounts reveal that from 1717 until the 1740's sums were frequently paid to Samuel Bloxham for clock maintenance; and in 1733 - 4 the clock was evidently rebuilt, since the considerable sum of £5 3s. 0d. was then paid to Samuel Bloxham and to a Chipping Norton clockmaker named Tobias Gilks. It was no doubt then that the clock 'came (home) from Bloxham'. It may be surmised that a similar explanation can be given at Barford St. Michael where likewise it has been said that the clock came from Bloxham ".

It may be deduced that in 1743 conversion to anchor escapement occurred and this late date suggests that the previous escapement was a crown wheel and verge, not a foliot. From 1788 Thomas Webb of Hook Norton (p. 149) undertook repairs and was given an annual contract of £1 for maintenance from 1812 to 1834. This contract was taken over by John Paine of Hook Norton (p. 132) from 1835 to 1855. Thereafter until 1875 repairs were done as needed by Thomas Osborne (Arsborn) of Bloxham (p. 86). Daily winding was done, usually by the parish clerk, for 10s. a year in the mid-18th-century rising to £1 by the mid-19th-century.

WITNEY — St. Mary

A stray Churchwardens' record for 1588 states, *whereof they have layde oute for repayring the Churche the bells the Clocke the glass wyndowes,* etc. There is no further history of this clock or of its successor until 1783 when Thomas Webb supplied a spring for the pendulum, 5s. This clock had "chimes" which were kept in repair by John Baker on an annual contract of £1 1s. 0d. from 1787 to 1800; he was paid extra in 1789 £5 5s. 2d. and during the rest of the period £3 18s. 8d. He was succeeded in charge of the chimes by Thomas Moss, who held the contract until 1830. Moss became sexton in 1817 with the duty of winding the clock as well as the chimes.

Maintenance of the clock itself was charged separately. In 1792 Charles Maxcy (p. 129) was paid £3 1s. 9d. for repairs and in the following year a new dial plate was put up at the cost of £17 17s. 0d. In 1796 the Witney clockmaker, William Harris (p. 110) was engaged on an annual contract of £1 1s. 0d. to keep the clock in repair. This he did until 1823 when his widow, Dinah Harris, continued the work; in that year repairs to the chimes cost £2 7s. 6d.

In 1827 a new clock costing £40 was acquired through Joshua Harris, William's son (p. 109). The business of Dinah Harris maintained this clock until 1839, when it was taken over by James Harris (p. 109) on the annual fee of one guinea. In 1842 the dial was taken down and replaced at a cost of £1 8s. 0d., and £15 13s. 0d. was paid to Harris for "clock work", which probably refers to motion work for 2 hands. He looked after the clock until 1859, the fee being raised to £2 10s. 0d. in 1853.

Up to the 1870's E. Walter and Son, Witney clockmakers (p. 148) undertook any repairs needed. The clock was replaced in 1875 by a new one with a carillon playing 14 tunes (now reduced to 7) supplied by Gillett and Bland, Steam Clock Factory, London. The church has had a ring of 8 bells since 1765.

WITNEY — The Blanket Hall

In 1711 a charter was granted by Queen Anne to form a Weavers Gild in Witney which held its first meeting on 10th August, 1711, in the Staple Hall Inn. In August, 1720, the Gild resolved to build the Blanket Hall in the main street; this was completed and occupied by December, 1721. Included in the building expenses is the item, *1722, Paid for the Clock, £21 3s. 4d.* The clock bell is dated 1721.

This small clock differs from others of this period in Oxfordshire, in having corner-standards of cast iron, cylindrical with cubical blocks; the finials and 2 pairs of pendent bearing bars have lathe-turned decorations instead of wrought iron work; the count wheel has spaced lateral hour pins instead of peripheral notches; the pendulum rod has a rectangular gap to clear the hour wheel arbor extension. It cannot have been made by Richard Keyte, as is generally supposed.

The local clockmakers John and Thomas May (p. 128) do not seem to have been employed to maintain the clock although they were paid by the Gild for repairing roasting jacks (1757) and engraving knives (1772). One Thomas Haynes was in regular charge of the clock. The hand-setting dial has been made from the chapter ring and two hands of an arch dial bracket clock and its circular name-plate signed RICH KEYTE WITNEY (p. 117). This must date from the time, probably ca. 1770, when a pair of bevel wheels and leading off rod to the one-hand external dial were added.

In 1776 the escapement was altered to semi-dead beat by John Hawting of Oxford (p. 111), *May ye 18 1776, for swing wheel and pallets, new stopping up the holes, mending and Cleaning the Great Clock, £3 3s. 0d.*

In the 1790's Thomas Robinson, Charles Maxey (p. 129) and Thomas Wells were caretakers of the Blanket Hall Clock. It was assessed as a public clock for taxation under William Pitt's Act of Parliament of 1797, which act, however, was repealed in April 1798. The tax receipt voucher states, *1 clock, 1 watch, 5s. 7½d. . . . Clock and watch took off . . .5s. 7½d.*

No accounts are available after 1813 to explain the erection of the present circular dial which seems to be of mid-19th-century make but retains a single hand.

WOODEATON — The Holy Road

A long disused, rusted clock. Frame of flat iron bars 30 × 15½ × 16½ inches, the corner standards with traces of buttress mouldings and finials bent outwards and upwards from a stepped corbel and ending in a roughly cubical boss. The outer cross-bars and bearing-bars are fixed by square nuts, the middle cross-pieces are welded. Trains side-by-side, all wheels brass with 4 arms; recoil escape wheel and the anchor a flat arc without a neck. The one-second pendulum is suspended by a short spring in an elevated chop, its rod expanded in a loop to clear the winding square. Peg and lever controls of the striking are pivoted in the middle of the frame, the count wheel external. No time-setting facility.

Leading-off rod rises vertically to the bell-loft above and connects by pinion and crown wheel to the single hand on a diamond shaped dial.

In its old fashioned features this clock resembles that in Islip church (p. 44) but as the bars are nutted and not wedged it may be several years younger, ca. 1700.

The ring of 5 bells is by Henry Bagley, 1680.

YARNTON—St. Bartholomew (p. 74)

The corner-standards of the iron bar frame 32×29×17 inches have buttress mouldings above and below ending in outwardly curved finials with round bun-shaped tips. They are joined on 4 sides by tenon-wedged cross-bars. The two side-by-side trains are now pivoted in nutted bearing bars. Both iron great wheels have 4 spokes with a U-shaped bifurcation of the outer half lap-welded to the annulus. Similar crossing-out is used in the clocks at Cumnor, Berkshire and at Farthinghoe, Northamptonshire. The brass recoil escape wheel has an arched anchor of wide span with a long stem and a one second pendulum probably made in 1682. The maintaining power is a later addition. In the striking train the second wheel differs from that of the going train, being a one piece forging with broad arms. Here there is a lantern pinion; all the others are leaf pinions. The count wheel is external and the pieces of the peg and lever control are pivoted in nutted brackets (Plate 27, Fig. 53).

From the history of the clock (p. 74) made in 1641 it is clear that it has undergone several changes and repairs, and is not now in its original position. It no longer operates a dial and all motion work has disappeared. Originally there was a crown wheel and foliot escapement for which some round and square holes in one top cross-bar probably represent the site of suspension brackets. Normally clocks with a foliot have end-to-end trains but there is no evidence in the frame that the trains have been rearranged.

BIOGRAPHICAL DICTIONARY

[see also pp. 84 - 153]

ALDWORTH, SAMUEL — OXFORD, LONDON, CHILDREY (p. 85)

In 1703 Samuel Aldworth, City of London, clockmaker, obtained a licence to marry Elizabeth Knibb of Collingtree, near Northampton. Presumably she was the daughter of John Knibb (p. 117) and was then 23 years old.

The longcase clock mentioned on p. 85 is fully described and illustrated in *Antiquarian Horology*, Vol. IV, pp. 48, 49.

The bracket clock of which the back plate is shown in Plate 24, fig. 48, is illustrated in R. A. Lee, *The Knibb Family*, Plates 184, 185, after restoration of the verge escapement. The arch dial bracket clock mentioned on p. 85 is illustrated *tit.cit.*, Plate 186.

ARSBORN, THOMAS—BLOXHAM (p. 86), see OSBORNE, THOMAS

BALL, JOHN—BICESTER (ca. 1700 - 1736)

C. & W. Son of William and Joan Ball who were residing in Launton in October 1701. An expulsion order dated 20 October 1736 reveals that John Ball, clockmaker of Market End, Bicester, deserted his wife Elizabeth and two daughters aged 4 and 2 years leaving them a charge on the parish. They were removed and conveyed to the charge of the Churchwardens and Overseers of the parish of Launton (informaton from David Watts).

BALL, WILLIAM (1)—BICESTER (p. 86)

Other 30-hour longcase clocks with anchor escapement: $9\frac{1}{8}$ inch square dial, one hand, chapter ring signed *Wm Ball Bister*, spandrels C. and W. 23; movement 4-post with brass plates, rope drive, oak case (E. F. Bunt). 10-inch square dial, 2 hands, signed on a plaque Willm Ball BISTER, matt centre with 3 large birds in flight, chapter ring without $\frac{1}{4}$ hour marks spandrels C. and W. 8; movement 4-post with brass plates, chain drive, oak case.

Wall alarm clock, arch dial $7\frac{1}{2}$ inches high, one hand, scroll and flower engraving and, feather border signed on an oval in the arch *Wm Ball Bicester*; movement 4-post, verge, loop and spikes, rope drive (coll. Beeson).

8-day longcase movement with brass 12-inch arch dial, roundel in arch inscribed Tempus fugit, raised seconds circle below XII, small date aperture, chapter ring with arcaded border between minutes, spandrels C. and W. 26, signed Willm Ball Bister (L. S. Northcote). The only other mid-18th century clock with arcading round the numerals is by Robert Denton, (q.v.).

BALL, WILLIAM (2)—BICESTER (p. 87)

30-hour longcase, 10-inch square dial, 2 hands, centre engraved with leaf scrolls signed above WILLM BALL BICESTER, spandrels C. and W. 37; movement 4-post. Another with the dial centre engraved and signed

PLATE 25

DITCHLEY PARK, The Great House Clock, Figs. 49 and 50, Front and back views of pendulum suspension and sub-frame for anchor.

PLATE 26

Fig. 51, KIDLINGTON CHURCH, Clock, rack-striking.

Fig. 52, KIDDINGTON HALL Stables Building Clock, count wheel.

PLATE 27
Fig. 53, YARNTON CHURCH Clock.

PLATE 28
Fig. 54, EDWARD HEMINS of Bicester, ¼-chime repeater on 8 bells in walnut case.

PLATE 29
Fig. 55. JOHN SOWTER of Oxford, Bracket clock in mahogany case.

PLATE 30
JOHN KNIBB of Oxford, Fig. 56, above, gilt cast engine-turned case, early 18th cent.
Fig. 57, below, silver case, tortoiseshell outer case inlaid with silver, ca. 1680.

PLATE 31

JOHN KNIBB of Oxford. Fig. 58, left, silver pair case, champlevé dial. Fig. 59, centre, back plate of Fig. 58. Fig. 60, right, silver pair case, back plate, ca. 1700.

PLATE 32

Fig. 61, JOHN KNIBB at Hanslope, ca. 1712, in earlier marquetry case.

Fig. 62, JOHN HAWTING, of Oxford, black and gold japanned case, ca. 1770.

PLATE 33

Fig. 63. JOHN MACE of Ditchley, 8-day clock in oak case, ca. 1775. (Photo by courtesy of L. S. Northcote).

Fig. 64. RICHARD GILKES of Adderbury, 30-hour clock with loop and spikes hung on backboard of oak case, ca. 1760. (Photo by courtesy of Miss M. Cherry).

PLATE 34
Fig. 65, THOMAS FARDON, of Deddington, mahogany case, ca. 1790.

above Willm Ball BICESTER; both with anchor escapements, late 18th century.

The 8-day arch dial longcase mentioned on p. 87 is now in coll. Beeson.

8-day longcase, arch dial 12 × 16 inches with strike/silent, dial centre engraved with leafy scrolls and signed *Willm Ball* BICESTER, spandrels of the arch and corners late 18th century rococo, chapter ring without ¼ or ½ hour marks, pierced steel hands, engraved seconds ring; movement rack-striking in large plates, pendulum flat brass rod. Case oak, flat top hood, elm backboard (Mrs. M. Campbell).

BIRD, MICHAEL—OXFORD (p. 88)

Verge watch, silver champlevé dial with matt gilt metal centre, the hours in square lozenges with dancetty arrangement surrounded by a calendar disc with days of the month, signed *Michael Bird Oxon Fecit*. Plain silver case, outer case leather with silver piqué, ca. 1680.

BROWN, RICHARD—CHALGROVE (1831)

C. & W. with others vouched for the characters of persons accused of machine breaking at Stadhampton in 1831. (*Oxf. Rec. Soc.* XLV, 1966).

BUSBY, RICHARD—WITNEY (1823)

C. & W. In Witney in 1823, not in 1851.

CLEMENTS, JOHN—OXFORD (p. 92)

Verge watch, silver hunting case, HM 1808, case maker IR, gold hands, front wind; movement signed *In° Clements* Oxford, No. 209, flat steel balance, engraved cock and foot extension with pointing hand, round pillars, dust cap (coll. Beeson).

DAVIS, ABRAHAM—OXFORD (1851 - 1858)

C. & W., jeweller, silversmith, at 54 High Street in 1850, at 140 High Street in 1858. English lever watch, silver pair case H.M. 1858, signed A. Davis 140 High St OXFORD No. 15680, plain bridge cock, dust cap, back wind (coll. Beeson).

DENTON, ROBERT—OXFORD (p. 93)

8-day longcase, arch dial with *Tempus Fugit* on boss, dolphin spandrels, centre of dial matt, seconds ring with numerals only, square date aperture, chapter ring arcaded round the hour numerals, no ¼ or ½ hour marks, signed *Robert Denton Oxon Fecit* on each side of VI, spandrels C. and W. 26. Oak case, hood top broken arch with crenolate border and 3 ball and spike finials, plain brass capped pilasters, case door arch topped, ca. 1750 (Mrs. M. I. Brooks).

DENTON, SAMUEL (1)—OXFORD (p. 94)

8-day longcase, 12-inch square brass dial, seconds and calendar rings simply engraved, signed Samuel Denton Oxford. Movement rack-striking, ca. 1780.

Wall alarm clock, arch dial 7 × 5 inches, roundel in arch signed Sam Denton Oxford, simple fret cresting at sides, one hand, applied chapter ring, alarm dial, spandrels C. and W. 17; verge movement, loop and spikes, ca. 1760 (Mrs. Gill).

Bracket clock, arch dial, signed *Sam Denton* OXFORD on silvered disc in arch, matt centre, gilt rococo spandrels; verge movement timepiece and pull alarm, back plate not engraved. Case mahogany bell top with handle, 18 inches high, sides with elaborate wood frets, ca. 1770 (coll. Beeson).

DISTON, THOMAS—CHIPPING NORTON (ante 1724)
C. & W. Buried Chipping Norton 24 October 1724.

DRURY, CHARLES WILLIAM—BANBURY (p. 94)
Repaired Brailes church clock, Warwickshire in 1823 for £8 17 0. In May 1821 was witness to a marriage in Banbury.

DRURY, WILLIAM—BANBURY (p. 95)
Bracket clock, silvered arch dial engraved all over, strike/silent in arch, centre signed *Wm Drury* BANBURY; verge movement with suspended pendulum, rack-striking, pull repeater, back plate not engraved. Bell top mahogany case with handle, 4 pineapple finials, 19 inches high, ca. 1790 (coll. Beeson).

Bracket clock, painted arch dial with moon work in arch, corners gilt scrolling, hour ring signed W. Drury, BANBURY; movement with strike/silent lever at side, heavy bob anchor pendulum, back plate engraved with scrolls and signed *Drury Banbury*. Bell top mahogany case with handle, sides with wood frets, brass ogee feet, 21 × 12 inches, ca. 1795 (E. C. Fortescue).

Longcase 30-hour, painted arch dial 14 × 20 inches, arabic hour numerals, seconds ring, date ring, corners with women representing the four seasons, matching moon hands, dummy winding squares, painted moon work in the arch with a fixed lunar month scale, dial signed Drury Banbury and stamped Walker & Hughes on the moon disc, false plate stamped Walker & Hughes Birmingham and also Wilson. Case oak with mahogany veneering and banding and Sheraton inlay, hood scroll-topped with brass ball and spike finials (coll. Beeson).

Verge watch movement, enamel dial, bisymetrical pierced chased cock, engraved foot, cylindrical pillars, back wind, signed *Willm Drury BANBURY* 1798, ca. 1790 (coll. Beeson).

Silver cased verge watch, signed *William Drury Banbury*, 601, HM 1801.

DRURY—OXFORD (p. 95)
A Mr. Drury was made free of Oxford 14 May 1792, fine £21.

DURRAN, JAMES HOPKINS—BANBURY (p. 95)
Some clocks are signed "Durrant", e.g., 30-hour longcase, 12-inch square painted dial, gilt 19th century hands, signed Durrant Banbury; movement plated, count wheel, chain drive. Case oak with cross-banded borders, hood with scrolls and baluster turned pillars, 6ft. 8ins. high, ca. 1830 (Berry).

THE FARDONS (pp. 96 - 98)
Further information identifies more clearly clockmakers of this surname. The sequence of generations appears to be:—
- A. John (1) 1700 - 1743 = Elizabeth m. 1731, then Mary m. 1734.
- B. John (2) 1736 - 1786, son of A.
- C. Thomas (1) ? - 1838, son of B = Lydia d. 1836.
- D. John (3) 1782 - 1865, son of C = Elizabeth, 1782 - 1847.

E. Thomas (2) . . ., son of D = Anne Susanna, 1808 - 1864.
F. Thomas (3) 1841 - 1903, son of E = Amy 1842 - 1878.
G. Thomas (4) 1864 - 1941, son of F.

FARDON, JOHN (1)—DEDDINGTON (p. 96)

The Window Tax for 25 March 1749 taxed the house of widow Fardon as 14 windows.

FARDON, JOHN (2)—DEDDINGTON (p. 97)

Longcase, 8-day, arch dial 12 × 16½ inches with strike/silent, centre silvered with engraved scrolls and chinoiserie, signed on a ribbon JOHN FARDON DEDDINGTON, spandrels C. and W. 30, date lunette, chapter ring silvered. Movement rack-striking. Case pale mahogany with veneered borders on hood and case door, hood arch-topped, corinthian pillars, sides of trunk chamfered with 6 black and yellow lines, sides of panelled base chamfered, ca. 1770. Owner Peter Fardon great grandson of Thomas (3).

FARDON, JOHN (3)—DEDDINGTON, WOODSTOCK (1782 - 1865, p. 97)

Born 1782, son of Thomas (1). His wife Elizabeth, born 1782 died 25 March 1847, aged 65 years, and was buried in Woodstock Church Cemetery. A watch and a longcase signed Thomas and John Fardon indicate a working partnership but all other clocks of this period are signed by Thomas. He started a business in Woodstock and after the death of Thomas (1) in 1838 and Thomas' wife Lydia in 1836 inherited Moorey House, Adderbury.

He looked after Kidlington Church clock from 1839 but is not mentioned by Christian name in the accounts until 1851 - 1862. No clocks of his Woodstock period have been traced.

He died 6 February 1865, aged 83, and was buried in Woodstock Cemetery which suggests that he had ceased to belong to the Society of Friends.

FARDON, THOMAS (1)—DEDDINGTON (ca. 1765 - 1838, p. 97)

Presumed son of John (2) and father of John (3). The longcase with an early painted dial (p. 98) is contemporary with the deaths of John (2) and of Richard Gilkes (p. 101) and must be one of his first productions. A further link with Adderbury is shown by his purchase in March 1825 of Moorey House, part of which was then occupied by the Three Tuns Inn, and by the fact that his wife Lydia was residing there at her death in 1836. He attended the Deddington Vestry Meetings in December 1833 and March 1834 presumably in connection with the installation of the church clock (p. 37) and was also churchwarden in 1835.

In addition to maintaining Kidlington and Deddington church clocks up to the time of his death he did work on Enstone church clock in 1810 - 11.

Longcase 30-hour, 11-inch painted dial with arabic hour numerals, a date lunette, early 19th century gilt hands, signed T. Fardon Deddington. Case oak, flat topped hood, sides, doors, etc., with cross-banded borders (E. C. Fortescue).

Longcase 8-day, arch dial with painted moon work, centre silvered with central date pointer, recessed seconds dial, silvered chapter ring, unusual spandrels with bagpipes in floral scrolls, signed *Thomas Fardon DEDDINGTON Fecit*. Case figured mahogany, hood with scrolled top and 3 balls, fluted pillars, Gothic arched door, trunk with canted edges and fluted pilasters, base with canted edges, ca. 1790. Owner Charles Fardon grandson of Thomas (3). Plate 34, Fig. 65.

FARDON, THOMAS (2)—WOODSTOCK (19th Century) p. 98

Son of John (3), date of birth not traced, died after 1866, married Anne Susanna (1802 - 1864) who is buried in Woodstock Cemetery. In Woodstock possibly as early as 1823 but a house at Deddington was rated in the name of Fardon from 1839 to 1844. At a Public Vestry Meeting on 30 March 1842 the Deddington churchwardens agreed that " two pounds should be paid annually to Mr. (Thomas) Fardon for wholly attending the Church Clock and to be paid in addition for cleaning and general repairs ". This contract continued only until February 1845 when a bill from " Fardon and Gibbs " was paid; thereafter only Joseph Gibbs (p. 100) attended to the clock and Fardon interest in Deddington ceased.

In April 1842 Moorey House was mortgaged by Thomas Fardon. In November 1865 after the death of John (3) the property was sold by him.

Longcase 8-day, white arch dial, 13 × 19 inches, with painted corners and rural scene in arch, Roman hour numerals, seconds ring, semi-circular date aperture, matching moon hands, signed *Thos Fardon Woodstock;* rackstriking movement with Birmingham false plate. Case oak with mahogany veneering and banding, lion inlay on short door, hood scrolled top with metal roundels, brass-capped cylindrical pillars, bracket feet, 7' 6", early 19th century (coll. Beeson).

FARDON, THOMAS GIBBONS (3)—WOODSTOCK (1841 - 1903)

C. & W. Son of Thomas (2), born 1841, married Amy who died in 1878. Two sons, Thomas Osborne (b. 1867) and George Osborne (b. 1876). Died 17 July 1903, aged 62.

FARDON, THOMAS OSBORNE (4)—WOODSTOCK (1867 - 1941)

C. & W., son of Thomas (3), born 1867, died 22 September 1941, aged 74 years.

No clocks signed by a Woodstock Fardon (3) or (4) are known.

FINNEMORE, WILLIAM—BIRMINGHAM (1815 - 1836)

Painted dial maker, 1815 - 1828, Birmingham, 4 Edmund St., 1835 -1836, Finnemore & Son (Osborne C.A., *tit. cit.* p. 94).

FRANCIS, WILLIAM — BIRMINGHAM (1816 - 1836)

Painted dial maker. Signed on dials of clocks by Tomlinson, Bicester.

GIBBS, JOSEPH—DEDDINGTON (p. 100)

The name should be Joseph not Joshua. He moved his business from Souldern to Deddington in 1845.

GILKES, JOHN—SHIPSTON-ON-STOUR (p. 101)

Longcase, 30-hour, 10-inch square dial, 2 hands, centre with 4 rings and close zigzags, spandrels unusual rococo scrolling, signed J^{no} Gilkes Shipston. Movement 4-posted.

Longcase, 30-hour, 11 inch square dial, 2 hands, centre silvered and engraved with open leafy scrolls, date lunette, spandrels C. and W. fig. 20, signed above John Gilkes Shipston. Movement plated, count wheel.

Repaired Brailes Church Clock, Warwickshire in 1764 - 1777.

GILKES, RICHARD—ADDERBURY (p. 101)

30-hour longcase clocks with loop and spikes were sold for use as wall clocks. Sometimes these were later fitted into longcases by fixing a hook in the back board and dispensing with a seat board, e.g., Plate 33, Fig. 64, shows a 12-inch, 2-hand clock signed *Richd Gilkes Adderbury,* centre of dial with engraved scrolling and date lunette, rococo spandrels, mounted in an oak case, ca. 1765 - 70 (Miss Cherry).

He held property at Ludgershall in Buckinghamshire " by allotment under the Inclosure Act in 1777. His estate passed by the marriage of his daughter Grace to Francis Hastie, Esq., of Great Haseley, at whose death it passed in right of his daughter and heir, Mary Gilkes-Hastie, to her husband, John Ingram-Lockhart, Esq., Barrister-at-Law, D.C.L., M.P. for Oxford 1809 to 1830 " (Dix Hamilton, M.L., *Cake and Cockhorse,* Summer 1966).

GILKES, THOMAS (2)—CHARLBURY (1736 - 1779, p. 104)

Born 25 May 1736, died 14 February 1779 (Dix Hamilton, *tit. cit.*). Longcase 30-hour, 10-inch square dial, 2 hands, centre finely matt, seconds ring, date lunette, spandrels as in Plate 12, Fig. 19, signed at base of chapter ring Thos Gilkes Charlbury. Movement in brass plates connected by flat iron bars not brass pillars. Case plain painted pine (W. Thacker).

GILKES, THOMAS—CHIPPING NORTON (p. 104)

Longcase 30-hour, the brass dial signed Thos Gilks (sic) CHIPPING NORTON. Movement with the count wheel toothed on the periphery not notched (E. Ricketts).

GILKS, TOBIAS—CHIPPING NORTON (p. 105)

Repaired Wiggington Church clock in 1743 - 1744.

GURDEN, JOSEPH—OXFORD (p. 107)

Apprenticed to Edward Moor, watchmaker, Oxford, for 4 years and to a person not a freeman for upwards of 4 years. Freedom of Oxford on 20 August 1759, fine £1 1s. 0d.

HARRIS, DINAH—WITNEY (1803 - 1839)

C. & W. Widow of William Harris (p. 110), continued his business and maintained Witney Church clock from 1823 to 1839.

HARRIS, GEORGE—FRITWELL (p. 108)

Lantern clock (d). Centre of dial engraved with a rose and tulips signed above *George Harris in Fritwell Fecit,* hand with double loop head and long tail, dolphin frets; movement with verge escapement, striking the half hour on a small bell below the hour bell, converted to endless chain wind from the original Harris method of key-wound barrels. Height $15\frac{1}{2}$ inches, chapter ring $6\frac{1}{4}$ inches diameter (E. F. Bunt).

A brass memorial inscription in Deddington Church to Thomas Higgins, 1660, is signed George Harris Fecit.

HARRIS, JAMES—WITNEY (p. 109)

Maintained Witney Church clock from 1839 to 1859.

HARRIS, JOHN—WITNEY (mid-19th century)

C. & W. 30-hour longcase, 12-inch painted dial, calendar lunette; plated movement (D. Watts). The clock mentioned under James Harris (p. 109) is actually by John.

HARRIS, JOSHUA—BURFORD (late 18th century—1810)

C. & W. Longcase 30-hour, 10-inch square dial, one hand, centre matt with a basket of fruit and 3 birds, spandrels as in Plate 14, fig. 23, signed Jos^h Harris Burford; movement 4-posted, ca. 1760 (R. Warner).

Watchpaper: J. HARRIS Watch & Clock *Maker BURFORD*, with a saw-toothed border.

HARRIS, JOSHUA—WITNEY (p. 109)

Son of William Harris (p. 110). In the family business. 30-hour longcase 11-inch painted dial (D. Watts).

Maintained Eynsham Church clock 1809 to 1831.

Repaired the chimes of Witney Church clock in 1823 and supplied a new clock in 1827. Was organist of the church.

HARRIS, THOMAS—DEDDINGTON (p. 110)

Longcase 30-hour, 10-inch square Quaker dial, one hand, centre with 3 rings and zigzags, chapter ring with fleur de lys, spandrels C. and W., 23, signed *Thos Harris* DEDINGTON; movement 4-posted. Case of oak and elm, ca. 1760 (G. Wood).

HARRIS, WILLIAM—WITNEY (p. 110)

Maintained Witney Church clock from 1796 to 1823.

HARRISON, WILLIAM—CHARLBURY (p. 110)

Longcase 30-hour, 9½-inch square dial, one hand, centre with foliate scrolls and 2 birds in flight, chapter ring with fleur de lys, spandrels C. and W., 20, signed *Willm Harrison Charlbury*; movement 4-posted, ca. 1770 (Dr. R. Learner).

HASTINGS, HERCULES—BURFORD (1700 - 1730)

Longcase, month, 11-inch dial, centre with Tudor rose, engraved calendar aperture, 2-inch seconds ring, cherub spandrels, chapter ring with elaborate cross between hours, signed below *Hercules Hastings at Burford*, ca. 1700 (T. E. Ellis). Hastings had charge of the Tolsey clock, Burford, from 1709 to ca. 1730 (p. 33).

HAWTING, JOHN—OXFORD (p. 111)

Longcase 8-day, arch dial with a large seconds ring centred on the top edge of the square and overlapping half the hours ring; movement rack-striking hour and half hour with a 6-wheel going train, signed *John Hawting Oxford*. Case japanned black and gold, ca. 1770. Fully described and illustrated in *Antiquarian Horology*, IV, pp. 273 - 274. This clock was probably designed for astronomical observations. Plate 32, Fig. 62 (coll. Beeson).

Fitted a new dead beat escapement and anchor to the Blanket Hall clock, Witney in May 1776 (q.v.).

Made the church clock at Northmoor in 1785 (q.v.).

HEMINS, EDWARD (2)—BICESTER (p. 112)

Bracket clock, $\frac{1}{4}$ chiming, unusually large 11-inch square dial with an engraved laurel leaf border, centre with engraved scrolling, pendulum lunette and circular date aperture below XII, chapter ring $10\frac{1}{4}$ inches diameter, spandrels C. and W. 10, strike/silent lever at top. Movement with 7 pillars, originally verge, converted to anchor, the $\frac{1}{4}$ chime repeat altered to a Victorian set of 7 bells and an hour gong; back plate with 3 set-up ratchets, engraved scrolling, a feathered border and signed *Ed Hemins of BISETER*.

Case walnut, 20 × $14\frac{1}{2}$ × 8 inches, domed top with handle and pierced fret frieze on 4 sides, brass feet, with a Victorian carved mahogany bracket, 15 inches high (coll. Beeson). Plate 28, Fig. 54.

Lantern clock, chapter ring $6\frac{1}{2}$ inches diameter, diamond between hours, one hand, central area engraved with leafy scrolls signed above *Edw[d] Hemins BISSITER*, frets leafy scrolling; verge movement converted to anchor, ca. 1725 (coll. Beeson).

Turret clock at The Great House, Ditchley Park (q.v.).

HIGGINS, THOMAS—BURFORD (1823)

C. & W., also glass and china dealer 1823 1824.

HOLLOWAY, WILLIAM—GREAT HASELEY (? late 18th century)

C. & W., Longcase signed William Holloway Great Heasly (sic).

HUNT, JAMES—OXFORD (1851)

C. & W., In New Road in 1851.

JEFFREY, JAMES—WITNEY (p. 115)

Cleaned Eynsham Church clock in 1823.

JUKES, J.—BIRMINGHAM (mid-19th century)

Dial maker. Stamped on the back of dial of a wall clock by William Payne, Banbury (p. 133), the dial maker's name, J. Jukes Birm[n].

KEYTE, RICHARD—WITNEY (p. 117)

The Blanket Hall clock has for a setting dial the chapter ring, 2 hands and a roundel signed RICH KEYTE WITNEY taken from an arch dial bracket clock.

KINGSTON, JOSEPH—TETSWORTH

C. & W., Recorded as a watchmaker in 1786.

KNIBB, JOHN—OXFORD (p. 120)

R. A. Lee in "The Knibb Family", 1964, classifies Knibb clocks in 4 phases or time-periods corresponding to (i) before 1670, (ii) 1670 - 1680, (iii) 1680 - 1695 and (iv) after 1700. Illustrations are given in Lee's book of the following types:—

Longcases: Marquetry, (a) (ii) ca. 1675, Panelled flower with cresting, 3-train, $\frac{1}{4}$-striking, Pl. 21, 44 (hood), 152 (front plate of movement; (b) Panelled flower marquetry, (iii) ca. 1682 - 1685, Pl. 27; (c) Burr walnut with cresting, (iii) ca. 1685 - 1689, Pl. 35, 50 (hood), 110 (12-

inch dial). See also Pl. 17, Fig. 26 of this book. (d) Walnut, (iii) ca. 1685 - 1688, Pl. 36, (e) Walnut, (iv) ca. 1715, Pl. 114 (arch dial).

Hanging wall clock: Walnut with cresting, ca. 1680 - 1682, Pl. 60, 179 (movement, alarm).

Bracket clocks: Ebony cased: (a) (i) ca. 1675, Pl. 75, 129 (back plate), 157 tic tac escapement). See also Pl. 21, fig. 42 and Pl. 24, fig. 47 of this book. (b) ca. 1680 - 1690, Pl. 85 ($\frac{1}{4}$-repeat), 138 (back plate). (c) ca. 1680 - 1690, Pl. 86 ($\frac{1}{4}$-repeat timepiece). (d) ca. 1685 - 1690, Pl. 90 ($\frac{1}{4}$-repeat alarm timepiece), 140 (back plate). (e) ca. 1695 - 1700, Pl. 91, 142 (back plate). (f) ca. 1715, Pl. 186 (arch dial also signed Samuel Aldworth). Walnut cased: (a) ca. 1680 - 1690, Pl. 87 ($\frac{1}{4}$-repeat timepiece). (b) ca. 1695 - 1700, Pl. 92 ($\frac{1}{4}$-repeat timepiece).

Lantern clocks: 1669, Pl. 173, 176 (engraved doors).

The marquetry longcase signed *John Knibb Hanslap* (p. 122) is illustrated in Plate 32, Fig. 61 of this book.

Watches: Lee (*tit. cit.*, p. 22) states that " it has been suggested that the Knibb family bought and signed their watches and were not actual watchmakers ". Six cased watches by John are now known. That described on p. 121 is illustrated by Clutton, C., and Daniels, G., " Watches ", figs. 235 - 6, ca. 1700 (coll. E. Hornby).

(2). Pair case, with white enamel dial, beetle and poker hands; movement with finely pierced bisymetrical cock without a mask, extended cockfoot, signed JOHN KNIBB AT OXFORD, elaborate floral pillars, old style fusee with the great wheel arbor coming up through the winding square. Case plain silver marked II, 53 mm. diam. Outer case tortoiseshell inlaid with silver flowers, fruits, huntsman, 2 hounds and hare, a seeded rose surmounted by a crown, ca. 1690. The design is presumably without royal significance. An outer case of similar workmanship was used by T. Tompion, ca. 1675 - 80, and J. Windmills, ca. 1680 (coll. Beeson). Plate 30, Fig. 57.

(3.) Silver pair case, silver champlevé dial with gilt border, centre signed KNIBB, OXON, flanked by a crowned lion and a unicorn, beetle and poker hands; movement with a bisymetrical pierced, engraved cock showing 2 birds, a mask, side wings, rimmed cockfoot with a pointing hand beyond the setting dial, signed JOHN KNIBB OXON, plain Egyptian pillars, back wind; plain outer case, both marked I.B. in Gothic capitals, 58 mm. diam. (coll. Beeson). Plate 31, Figs. 58, 59.

(4). Silver pair case, silver champlevé dial, the centre signed *Knibb Oxon* between 2 cockle shells and a chain border, beetle and poker hands; movement with pierced, engraved, bisymetrical cock with a mask and side wings, rimmed and extended cock foot, signed *John Knibb* Oxon. Outer case plain silver; silver chain with oblong links, ca. 1700. Plate 31, Fig. 60.

(5). French style [re-]gilt metal case cast to resemble engine turning with ruby stone button. Enamel dial, front wind, beetle and poker hands; movement with pierced, engraved, bisymetrical cock with mask and side wings, cockfoot rimmed and extended, plain Egyptian pillars, signed *In⁰ Knibb in Oxford*, 55 mm. diam., early 18th century (coll. Beeson). Plate 30, Fig. 56.

KNIBB, JOSEPH—OXFORD, LONDON, HANSLOPE (p. 122)

Hanslope. The deeds of Joseph Knibb's Hanslope estate are now in Buckinghamshire Record Office (Ref. D/NO). These show that he decided to buy it more than 6 years before he sold his London business. It was bought for £1,500 from Charles Lane, gentleman, of Green End, Hanslope, whose forebears had farmed it for some generations. It consisted of a farmhouse, 1 cottage, 2 gardens, 2 orchards, 100 acres of land, 30 acres of meadow, 80 acres of pasture, 20 acres of furze and heath and common pasture in Hanslope (total area over 230 acres). The consideration was paid forebears on 28/20 May 1691.

It is worth noting that at Michaelmas 1692 Joseph and his wife, Elizabeth, together with 6 others, probably trustees, sold for £100, 2 messuages and 50 acres of land at Isham, Mears Ashby and Weekley Titchmarsh in Northamptonshire (Feet of Fines, Oxon, P.R.O.).

Joseph did not farm the Hanslope land himself. There were 9 tenants in 1691 but there are no deeds dealing with subsequent lessees or tenancies except for $2\frac{1}{2}$ acres and sward land assigned in trust in 1693 to his brother, George Knibb, Rector of Blisworth, Northants. Nor are there any records of how John Knibb managed the farm after he inherited it in 1712.

After John's death his son John, upholsterer of Oxford, with his wife Deborah, mortgaged the property on 14 January 1722/23 to William Townsend, the Oxford mason, to secure £700 and interest possibly to be able to pay the further legacies under Joseph's will. In 1726 the mortgage was transferred to Peter Hughes of the University of Oxford for £700, paid to Townsend, and £300, paid to Knibb. But in 1730 John Knibb had to secure a further sum of £320 on the property. On 19 May 1737 he conveyed the whole Hanslope estate for £2,010 to Richard Brafield of Northampton, a haberdasher of hats, and it then passed out of the control of the Knibb family.

Joseph's residence, Green End Farmhouse, which was included in the schedule of buildings of architectural or historic interest compiled under S.30 of the Town and Country Planning Act, 1947, was pulled down in 1954.

The Oxford Clocks: Joseph's clocks made during the period before 1670, all belong to Phase I as defined by R. A. Lee *op. cit.*; those illustrated are of the following types:—

Longcases: Ebony cased, (a) Pl. 9 (verge). (b) Pl 10 (verge), 52 (sliding hood), 145 (movement). (c) Pl. 11, 98 (dial), 146 (movement), 159 (pendulum cross-beat escapement of 1668 - 1669). (d) Pl. 97 (dial), 143 and 144 (verge movement). See also Pl. 17, Fig. 28 and Pl. 23, Fig. 46 of this book.

Hanging Wall Clock: Walnut cased, Pl. 58 ($\frac{1}{4}$-striking), 178 (verge movement). See also Plate 23, Fig. 45 of this book.

Lantern Clock: Pl. 171 (verge movement), 172 (front view). See also Pl. 20, figs. 39 and 40 of this book.

Turret Clock: Plate 156. See also Plate 5, Figs. 7 and 8 of this book.

The Anchor Escapement: Since publication of the first edition of this book it has been generally accepted that the first *dated* anchor escapement with a one-second pendulum was devised and completed by early 1670 for use in two Oxford turret clocks by Joseph Knibb (see pp. 62 and 65). These two examples antedate that in the clock made in 1671 for King's College, Cambridge, by William Clement, who, hitherto, has been regarded as the inventor of the escapement. R. A. Lee (*op. cit.* p. 139) considers Joseph " as a possible claimant to its first production ". He certainly was experimenting in the use of long pendulums during his period at Oxford. One longcase movement is fitted with a special escapement comprising two pallets on separate arbors linked together and linked to the crutch arbor;

they worked with a cross-beat action on a vertical escape wheel with a one-second pendulum (see adjoining text-figure). R. A. Lee and D. Parkes

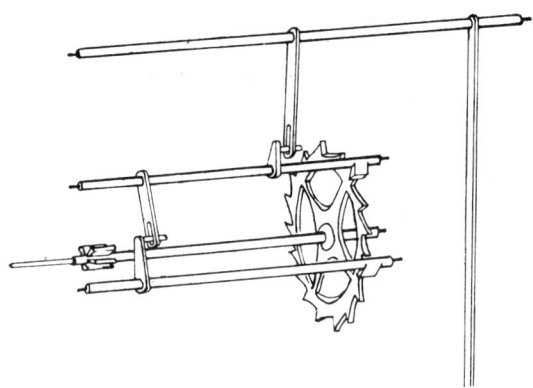

consider that this escapement antedates the recoil anchor and ask " would Joseph have bothered to make this escapement if he had seen or heard of the anchor ? " (*op. cit.* p. 145). If so it was made well before 1669.

Recently a miniature timepiece longcase by Samuel Knibb has been announced; this has a perfected anchor escapement with a $10\frac{1}{2}$ inch pendulum (*Antiquarian Horology*, p. 338, Fig. 1 and p. 323) and the date assigned to it is 1668. However, one should not regard this as the earliest anchor escapement in existence as there are certain points which, in Mr. Lee's opinion, may indicate an almost contemporary alteration, perhaps within a year or so.

The tic tac escapement does not occur in any of Joseph's clocks signed at Oxford. It was first used by him in the split second clock purchased by Professor James Gregory in 1673. Hence the tic tac should not be regarded as an early stage in the evolution of the anchor.

LAMB, JAMES—BICESTER (p. 125)

Longcases 30-hour, square 12-inch painted dials, with or without date lunette, plated movements, signed J. Lamb Bicester. One date wheel stamped WRIGHT BIRM N. Cases oak with some mahogany veneer, scrolled top to hood (D. Watts). Exhibited a 400-day skeleton clock at the Great Exhibition of 1851.

LAMPREY, B.—BURFORD (1760)

C. & W., Longcase 30-hour, 11-inch square dial, 2 hands, centre with scrolling signed B. LAMPREY BURFORD, four-seasons spandrels, date lunette; movement 4-posted. Case oak, ca. 1760.

LAWRENCE, WILLIAM—THAME (p. 126)

Longcase 30-hour, $10\frac{3}{4}$-inch square dial, centre matt, one steel hand, chapter ring with fleur-de-lys signed below *Wm Lawrance Thame*, spandrels C. and W. 21; movement 4-posted, brass plates. Plain oak case, the door inscribed " Martha Webb ", ca. 1745 (coll. Beeson).

Made the turret clock at Chilton House, Chilton, Bucks. The single hand is attached to a toothed wheel turned by a worm thread on the motion shaft. Wooden dial. The bell is dated 1756.

MACE, JOHN—DITCHLEY (ca. 1775)

Longcase 8-day, arch dial 12 × 16¼ inches, strike/silent in arch, matt centre, recessed silvered seconds dial, square date aperture, chapter ring without ¼-hour marks, irregular rococo spandrels, pierced looped hands, signed on a silvered curved ribbon *John Mace Ditchley Oxon*. Movement rack-striking. Case oak, hood a hollowed pediment with 3 urn finials, 8 ft. high, ca. 1775, at The Great House, Ditchley Park. Plate 33, Fig. 63.

This is the clock mentioned on p. 127 as owned by Viscount Dillon and appears to be London made. The case may be a copy of London style by a local cabinet maker.

MAY, EDWARD—HENLEY ON THAMES (p. 128)

Longcase 8-day, arch dial signed on silvered disc *Edwd May Henley*, dial centre matt, seconds ring and square date aperture, chapter ring without ¼ or ½ hour marks, spandrels C. and W. 26. Case figured walnut, flat-topped hood, trunk door bordered with ½-round mouldings, ca. 1755 (A. Miller).

MAY, EDWARD—WITNEY (p. 128)

The Act of Parliament clock mentioned on p. 129 is black and gilt, about 5 feet high, the dial with straight lateral sides, top convex, undersides concave; stalked fleur-de-lys between the hours, each minute marked, gilt hands; short trunk with lacquered Chinese scene on door. Movement in rectangular plates. The almost illegible signature proves to be *Edward May Witney*, ca. 1760. Previously in a Coaching House, then to the old Meeting House about 1812 and finally to the Manse, Witney.

MAY, JOHN—WITNEY (p. 128)

Longcase 8-day, 11-inch square dial, centre matt with one large bird in flight, square date aperture flanked by a basket of fruits and one bird among leafy scrolls, chapter ring brass without ¼ or ½ hour marks, spandrels C. and W. 21, signed *Jno May Witney*. Plain oak case, flat top hood (Blanket Hall).

Made clockwork roasting jacks, ca. 1755, signed JOHN MAY (Tolsey Museum, E. Bartlett Taylor).

Maintained Eynsham Church clock, 1774 to 1798.

MAY, THOMAS—WITNEY (p. 129)

Son of John May above. Signed a bill in 1772. Made clockwork roasting jacks (E. Bartlett Taylor).

MAXEY, CHARLES—WITNEY also Mazey (p. 129)

Repaired jacks for the Blanket Hall Co., Witney, in 1793-1794 and repaired the clock and dial in 1794.

Repaired Witney Church clock in 1792.

MOORE, EDWARD (1)—OXFORD (p. 129)

Joseph Gurden served as his apprentice for 4 years (?1740-1744).

MUSSELWHITE, WILLIAM—BICESTER (p. 130)

Sedan clock, engraved silvered dial, 5 inches diameter; circular oak frame; signed *Mufselwhite* BICESTER.

OSBORNE, THOMAS—BLOXHAM=Arsborn (p. 86)

Wife Jane, first child baptised at Bloxham 20 October 1830. Repaired Wiggington Church clock 1855 to 1875.

OSBORNE, THOMAS—BIRMINGHAM

Dial maker. In partnership with James Wilson, 1772 to 1777, at 2 Colmore Row; partnership dissolved 29 September 1777. Ann Osborne & Co. at Whittall Street in 1791 - 1798, A. & J. Osborne in 1800 - 1803. James Osborne at St. Mary Square in 1808. (Osborne, C. A., *Antiquarian Horology* V, p. 94.)

Painted arch dial of William Peacock, Banbury (p. 133), ca. 1790, stamped OSBORNE MANUFACTORY on the back of the moon wheel.

PAINE, JOHN—HOOK NORTON (p. 132)

Maintained Wiggington Church clock from 1835 to 1855, succeeded Thomas Webb of Hook Norton. Supplied a new turret clock to Brailes Church, Warwickshire, in 1840 for £11 10s. 0d.

PAYNE, WILLIAM—BANBURY (p. 133)

30-hour longcase, 12-inch painted dial, signed *W. Payne* Banbury; seconds ring and date lunette, gilt stamped hands; plated movement, chain drive, the date wheel stamped FINNEMORE. Case oak with mahogany veneer borders, scrolled hood, ball and eagle finial, ca. 1840.

8-day wall clock, ca. 1850, the back of the circular dial stamped J. JUKES BIRM.M. Was still in Banbury in 1855 as a member of the Baptist Church.

PEACOCK, WILLIAM—BANBURY (p. 133)

Longcase 8-day, painted arch dial, moon work, arabic hour numerals, corners with painted scenes of the four seasons, signed W. Peacock Banbury; back of moon wheel stamped OSBORNE MANUFACTORY BIRMINGHAM. Mahogany case (K. Malsbury).

Longcase 30-hour, 11-inch square painted dial, signed Peacock Banbury, 2 hands; movement 4-posted. Case entirely of figured elm, 6' 6", ca. 1790 (coll. Beeson).

PINFOLD, THOMAS—MIDDLETON CHENEY (p. 134)

Longcase 30-hour, 10-inch square dial, matt centre, fleur de lys between the hours, spandrels C. and W. , one steel hand with crossed loops and tail, signed on chapter ring *T. Pinfold Middleton*. Case oak, flat topped hood, plain pilasters, rectangular door, ca. 1745 (Miss D. Hemmings). A gravestone in Middleton Cheney Churchyard bears a brass arch dial engraved with his wife's name.

REYNOLDS, THOMAS—OXFORD (p. 136)

Longcase 8-day, arch dial 12 × 17 inches, similar to that described on p. 137, signed on boss in arch *Tho Reynolds* OXFORD. Movement rack-striking. Case oak, scrolled hood, inlaid with mahogany and light woods, ca. 1775 (R. C. Righton).

REYNOLDS and EARLE—OXFORD (1795 - 1799)

30-hour longcase, 12-inch silvered dial engraved with a basket of fruits below, and an oval cartouche above signed *Reynolds & Earle* OXFORD, applied rococo spandrels, diamond ended hands; plated chain-driven movement. Case varnished pine with a flat-topped hood, ca. 1795 (coll. Beeson). See Thomas Reynolds and Thomas Earle.

SAUNDERS, THOMAS—BANBURY (see Saunders, Charles, p. 139)

C. & W. Married Esther on 22 January 1795, child born 1810. In Parsons Lane early in the 19th century.

SINDRY, JAMES—OXFORD (1788)

C. & W., goldsmith. Longcase 8-day, engraved silvered arch dial, no applied pieces, signed *James Sindry Oxford*. Large plated movement ca. 1790 (Ricketts). Freedom of Oxford on 30 May 1788, fee 20 guineas.

SOWTER, JOHN—OXFORD (p. 141)

8-day regulator, silvered dial, centre seconds hand, separate hour and minute rings, signed John Sowter Oxford. Standard movement, jewelled pallets, Harrison maintaining power. Oak case.

Verge watch, gold hands, seconds, signed SOWTER Oxford. No. 5195 on dial and back plate, plain cock, back wind, ca. 1830 (coll. Beeson).

[STEWARD, JOHN — HENLEY IN ARDEN, Warwicks, see p. 142]

Stray papers of the Aston Cantlow church accounts record that John Steward, gunsmith of Henley in Arden, Warwickshire, was repairer-consultant of the church clock from 1717 to 1747. The name must be removed from the Oxford list.

STOCKFORD, JOSEPH—THAME (p. 142)

Longcase 8-day, arch dial, signed on boss in arch, seconds dial and date aperture. Case with arch topped hood surmounted by a dome, ca. 1770. Longcase 30-hour, 10-inch 2-hand dial, spandrels crossed univalve shells; 4-post movement, signed *Jos. Stockford* THAME. Stained pine case.

STONE, RICHARD—THAME (p. 143)

Longcase 30-hour, 10-inch square brass dial, 2 hands, chapter ring without $\frac{1}{4}$ or $\frac{1}{2}$ hour marks, spandrels rococo scrolling, signed *Richd Stone Thame*. Movement 4-posted. Case plain polished oak, 6' 6", ca. 1780.

TOMLINSON, THOMAS—BICESTER (1823)

C. & W. At Bicester in 1823 (Pigots Directory).

As many 8-day and 30-hour longcases with square painted dials are signed Tomlinson Bicester they may have been sold by Job, John or Thomas. Some dials were supplied by William Francis of Birmingham.

UNITE, MATHIAS—OXFORD (p. 146)

Bracket clock, $6\frac{1}{2}$-inch dial, chapter ring with fleur de lys, square date aperture below XII. Verge movement, timepiece repeater on 2 bells, the hammer linkage as in Knibb clocks, latched dial and plate, back with close foliate scrolling signed Mathias Unite FECIT. Case ebonised, $13\frac{1}{2}$ inches high, plain dome top, no applied mounts on door. Illustrated in Lee, R. A., 1964, plate 183.

WALFORD, HENRY—BANBURY (p. 147)

Longcase 30-hour, painted arch dial inscribed *Hy. Walford Banbury 1778*. Veneered case with broken arch and pediment hood. Mr. W. U. Hancock states that clock was a wedding gift by Henry Walford, jeweller of Banbury, to a male relative in 1778.

WALKER, GEORGE—OXFORD (p. 147)

Longcase 8-day, 12-inch square dial, ringed winding holes and date aperture, signed Geo Walker Oxon. Walnut case (T.C. Cuss.).

WALKER & HUGHES—BIRMINGHAM (1812 - 1835)

In Birmingham 1812 to 1835 as dial makers at Lower Priory in 1812, Fisher Street in 1830 (Osborne, C. A., *Antiquarian Horology*, V, p. 94).

WEBB, FRANCIS—WATLINGTON (p. 148)

Longcase 30-hour, 10-inch square dial, centre matt, chapter ring with fleur de lys and ¼-hour marks, 2 hands, spandrels C. and W. 8 signed *Fra Webb Watlington*. Movement 4-posted, brass plates, chain drive, ca. 1720 (coll. Beeson).

WEBB, THOMAS—HOOK NORTON (p. 149)

Repaired Wiggington Church clock in 1788 and onwards. On contract at £1 a year for maintenance, 1812 to 1834.

WELLS, JOHN—SIBFORD GOWER (p. 149)

Longcase, 30-hour, brass dial signed John Wells Sibford.

WESTCOTT, JOHN—OXFORD (late 18th century)

C. & W. Longcase 30-hour, 11-inch square brass dial, one double looped steel hand, centre of dial with shallow scrolling and some chinoiserie features, signed *In° Westcott Oxford*, chapter ring with ¼-hour marks, and 3 dots for ½ hours, spandrels rococo scrolls. Plated movement. Varnished pine case, ca. 1780. (coll. Beeson).

WHITE, THOMAS—WITNEY (1741 - 1743)

Carpenter, son of Thomas White and Martha of Cogges and Witney. Party to a deed of mortgage and release, 1741 - 1743 (*Oxf. Rec. Soc.*, XLV, 1966).

A one-hand wooden clock with oak octagonal dial, 18-inch diameter, black with gilt numerals, ¼-hour marks, fleur de lys and signature in large capitals, T. WHITE WITNEY; oak case 12 × 13 × 8 inches. Trains side-by-side, both main wheels with loose barrels and additional boxwood gear wheels turned by similar winding gears mounted on front plate.

Going Train: main wheel oak 7-inch diameter, ½-inch thick, arbor and barrel boxwood with 28 grooves for gut line, and boxwood ratchet. Second wheel brass, 4 straight spokes, arbor and pinion boxwood. Escape wheel brass, arbor and integral pinion iron. Anchor and crutch iron. Pendulum cock wooden. On outer face of front plate a 12-hour wheel of oak with a boxwood star wheel, arbor boxwood.

Striking train: main wheel as in going train, 13 lifting pins, barrel oak with integral ratchet. Hoop wheel oak with 2 gaps, arbor and pinion boxwood. Fly boxwood with freewheel ratchet, arbor iron with integral pinion.

Striking lever system entirely of boxwood, hammer buffer spring steel. Bell mounted below roof of case. Count wheel missing.

This clock worked with a 16-inch pendulum and stood on a wall bracket. It would seem to be about 20 years later than the wooden clocks by James Harrison. Now in Snowshill Manor, Gloucestershire (National Trust).

WILSON, JAMES—BIRMINGHAM (1772 - 1809)

Dial maker: In partnership with Thomas Osborne at 3 Colmore Row, 1772 - 1777; partnership dissolved 29 September 1777. At Great Charles St. in 1778 - 1798; at Newhall St. 1800. Died 3 April 1809 (Osborne, *tit. cit.* p. 95).

Iron false plate of a painted arch dial clock stamped WILSON.

WISE, MATTHEW—DAVENTRY (p. 151)

Admitted freeman of Daventry in 1783.

WRIGHT, CHRISTOPHER—BIRMINGHAM (1835)

Painted dial maker at 86 Summer Lane in 1835 - 1836. Name stamped on the date wheel of a 30-hour painted dial signed J. LAMB Bicester.

[YOUNG, SAMUEL — BUNBURY (p. 152)]

In the churchyard of Bunbury, 13 miles S.E. of Chester, is a sundial on a turned post, the brass dial engraved " John Davis C.W., S. Young Fecit, Roger Perry 1749 ". 30-hour longcases signed " Sam Young, Bunbury " are known. This clockmaker also produced clocks as from Baddeley, a few miles from Bunbury. The name should be deleted from the Oxford list

SUPPLEMENTARY LIST OF ILLUSTRATIONS

Plate
25. Figs. 49 and 50. DITCHLEY PARK, The Great House. Clock made by Edward Hemins of Bicester. Front and back views of pendulum suspension and sub-frame for anchor. Photos by courtesy of L. S. Northcote.

26. Fig. 51 KIDLINGTON, St. Mary. Rack-striking clock, see p. 44. Photo by courtesy of L. S. Northcote.

   Fig. 52 KIDDINGTON HALL, Stables Building. Count wheel clock. Photo by courtesy of L. S. Northcote.

27. Fig. 53 YARNTON, St. Bartholomew. Clock installed in 1641. Photo by courtesy of *The Oxford Times and Oxford Mail*.

28. Fig. 54 EDWARD HEMINS of Bicester. Bracket clock, ¼-chime repeater on 8 bells, walnut case.

29. Fig. 55 JOHN SOWTER of Oxford. Bracket clock, mahogany case, see p. 141.

30. Fig. 56 JOHN KNIBB of Oxford. Watch in gilt, cast, engine turned case, early 18th century.

   Fig. 57 JOHN KNIBB of Oxford. Watch in silver case, and tortoiseshell outer case inlaid with silver, ca. 1698.

31. Fig. 58 JOHN KNIBB of Oxford. Watch in silver pair case.

   Fig. 59 Back plate of Fig. 58.

   Fig. 60 JOHN KNIBB of Oxford. Watch in silver pair case, ca. 1700.

32. Fig. 61 JOHN KNIBB at Hanslope. 8-day clock, ca. 1712 in earlier marquetry case, see page 122.

   Fig. 62 JOHN HAWTING of Oxford. 8-day clock, ca. 1770, in black and gold case.

33. Fig. 63 JOHN MACE of Ditchley. 8-day clock, ca. 1775, in oak case. Photo by courtesy of L. S. Northcote.

   Fig. 64 RICHARD GILKES of Adderbury. 30-hour clock with loop and spikes hung on backboard of oak case. Photo by courtesy of Miss W. Cherry.

34. Fig. 65 THOMAS FARDON of Deddington. Hood of 8-day clock in mahogany case, ca. 1790. Photo by courtesy of Mr. C. Fardon.

TEXT FIGURES

Idbury church clock. Reconstruction of the Debaufre type escapement. Drawing by L. S. Northcote.   page 168

Pendulum cross-beat escapement in longcase clock by Joseph Knibb of Oxford. Drawing by M. Hurst, by courtesy of *Antiquarian Horology*.   page 186

SUPPLEMENTARY TOPOGRAPHICAL LIST

BANBURY
    Peacock, William
    Saunders, Thomas

    BICESTER
    Ball, John

BURFORD
    Harris, Joshua
    Hastings, Hercules
    Higgins, Thomas
    Lamprey, B.

CHARLGROVE
    Brown, Richard

CHIPPING NORTON
    Diston, Thomas

DITCHLEY
    Mace, John

OXFORD
    Clements, John
    Davis, Abraham
    Hunt, James
    Reynolds and Earle
    Sundry, James
    Westcott, John

TETSWORTH
    Kingston, Joseph

WITNEY
    Busby, Richard
    Harris, Dinah
    Harris, John
    White, Thomas

WOODSTOCK
    Fairbrother, James
    Fardon, John
    Fardon, Thomas

INDEX

CAPITALS indicate clockmakers & directly related craftsmen, e.g. bellfounders, blacksmiths, case makers, inventors, painters, retailers, sundial makers, etc.
Relevant professions other than clock- and watchmaker are stated
Places of residence are supplied for non-Oxfordshire craftsmen
Many dates given, along with some identifications and spellings, have been revised by the indexer from a variety of sources; discrepancies between index & text are thus deliberate corrections or clarifications
b = born; a = apprenticed (usually aged 14);   f = flourished (dates within life);   d = died
Counties cited after non-Oxfordshire place-names accord with historical definitions, before the boundary changes of 1974

Abingdon (Berks.) 19, 49, 51, 88, 91, 96, 133; St Helen, sandglass 81
Ackermann, Rudolph 80
Act of Parliament clocks 8-9, 90, 97, 98, 107, 111, 129, 130, 133, 150, 162, 187
ADAMS, Thomas (smith) (a.1614, d.1664) 61, 81, 84
Adderbury 2, 8-9, 23, 26-7, 86, 96-98, 101-103, 105, 110, 146, 150-1, 162, 179, 181; Adderbury House, clock 26; Quaker Meeting House 16, 97, 102, 110, 151; St Mary, clock 26-7, 91, 103, 143, 151, 162; St Mary, sundial 76; Moorey House 179, 180
advertisements 1, 86, 94-97, 106, 107, 112, 114, 115, 128, 133, 142-144, 148-150; see watches – lost
air-driven clock 116
Aires, Mr 61
Aked, Charles 3
alarm wall clocks 87, 120, 130, 138, 141, 148, 176, 177, 184
alarms, generally 17, 18, 86-7; as purpose of clock 6, 19; astronomical 7
ALDWORTH, Samuel (a.1673, d.c.1730) 7, 8, 22, 85, 119-20, 176; father John 85; wife Elizabeth (née Knibb) 176
Allitt, Mrs C. A., collection 102
ALMOND, William (a.1750) 85, 137; father Thomas 85
Ambrosden, St Mary, clock 27-8, 141; bell 112
American clocks 24, 140, 143
Amesbury (Wilts.), church chiming barrel 7
anchor escapement see escapements
Anne, Queen 174
ANSLEY, Richard (f.1823) 85
Antiquarian Horological Society 2-3, 17
Appleton (Berks.) 171
apprentice pieces 22, 146
apprenticeship, generally 1, 5, 13-14,20-22; boom in Oxford (1670s-90s) 21-2, 119; first specialist clockmakers' apprentices in Oxford 20; Quaker 5, 16, 22, & see Quakers; within families 5, 20, 22, & see clockmakers
ARCHER, John (smith) (f.1532-41) 42, 86; ?father John 86
architects 6, 31, 49, 53, 152, 164; see stone masons
ARESS (?Harris), --- (f.1730) 75
armour 115, 136
Arsborn see Osborne
Ashmolean Museum 3, 99, 120, 121, 125, 132, 133, 135, 138, 143, 148
Aston (near Bampton) 144
Aston (near the Heyfords) see North Aston, Steeple Aston
Aston Cantlow (Warwicks.) 189
astrolabes 1, 6, 17-18, 134; astrolabe clock 6-7
astronomers & astronomy 6, 13, 17-18, 76, 125, 152
astronomical clocks 7, 17, 125; see regulators
ATKIN(S), Joshua (f.?1780, 1823) 86
ATKINS, William (f.1764-78) 86, 92; parents William & Johanna 86; wife Sarah 86
atmospheric clock 7
ATTE LEE (or Lee), John (smith) (f.1412, d.c.1448) 41, 86, 115
AULKIN, Richard (a.1770) 86, 136; father Richard 86
automata 5, 6, 18, 106
Automaton Laboratory 106
AYER, Mr (f.1586) 71
Aylesbury (Bucks.) 4, 22, 23, 98-9, 126, 142; shop sign The Dial 142
Aynho (Northants.), St Michael, clock 25, 29, 49, 113

B., I. (watch case mark) 184
Baddeley (Ches.) 191
BAGLEY, Henry (bellfounder) (f.1680) 174
BAILEY, --- (of Salford) (19th C) 162
BAIN, Alexander (electrician, of Edinburgh) (c.1811-1877) 7
Baines, Dr H., collection 120
BAKER, John (f.1787-1800) 173
Baker, Susannah 132; father Edward 132
Balam, John (clock keeper) 41-2, 108
BALL, John (b.?1701, f.1736) 4, 176; wife Elizabeth 176
BALL, William (f.1701, ?1722-36) 4, 8, 23, 59, 86-7, 176; wife Joan 176
BALL, William (f.?1722, 1735, d.?1786) 4, 8, 23, 29, 59, 86-7, 176; wife Catherine 86
BALL, William (1738-1823) 4, 8, 46, 87,

194

176-7; wife Ann 87
balloons 7, 96, 151
Bampton 144, 149; St Mary, clock 29, 136
Banbury 8, 9, 22-24, 27, 29-30, 32, 35, 43, 48, 67, 91, 94-96, 98, 104, 107, 110, 113, 114, 116-7, 125, 126, 129, 132-134, 139, 143, 144, 146-149, 151, 178, 183, 188-190; Baptist Church 188; Borough Council, Guilds, etc. 13-14; clock, site unknown 30; earliest clockmaking in 22, 23; Historical Society 2; Quaker Meeting House 16; Reindeer Inn 2; Smiths Guild 13; St Mary, clock 29; St Mary old church, clock 48, 113; St Mary old church, sundial 76; Town Hall, clock 29-30, 126
BANKS, Henry (mason) (f.1710) 32, 76
BANNARD, Richard (f.1839) 27
Barford St Michael, St Michael, clock 30, 48, 162, 173
Barnes, John 116
Barnett, T. G. 6
barometers 96, 109, 116, 132
Barret, William (clock keeper) 69
Barrett, Maria 138
Barrett, Mary 103; father George 103
BARRITT, George (a.1800) 87, 110
BARTON, John (b.c.1659, f.1685) 22, 87; wife Mary 87
Basingstoke (Hants.) 165
BASSETT, Joseph (f.c.1780) 87-8
BATES, John (smith) (f.1613) 84
Batt, Dr B. E. A. 33
Bavaria 125
Bayley, William 122
BAYLISS, William (carpenter) (f.1844-c.1884) 38, 92, 166
BECK, I. (19th C) 88
Beckett, William 146
Beckley, St Mary, clock 162; sandglasses 81
BEDFORD, William (f.1842) 88
Beeson, Cyril Frederick Cherrington 1-9; collection 2, 4, 6, 7-9, 85, 87, 89, 90, 91, 92, 95, 97(2), 98(2), 101, 102, 105, 107, 109, 112(2), 113(2), 120(2), 122, 125, 129, 130, 131(3), 132(2), 133, 138, 140, 141(2), 143, 145, 149, 150, 151, 176, 177(3), 178(4), 180, 182, 183(2), 184(3), 186, 188, 189(2), 190(2)
Beeson, Mrs Margaret 3
Beeson, Walter Thomas & Rose Eliza 1
Beeson Collection (in Museum of the History of Science) 2, 4, 6-9
Beeson Road 3; Beeson Room 6-7
Beisley, Sidney 172
BELCHER, John (a.1723, f.1731) 88, 89; father Jonathan 88
Belcher, N. M., collection 134
bell chambers & towers see bells, campanile
Bellchambers, J. K. 1
bellfounders 13, 23, 27, 31, 33, 41, 57, 58, 69, 112, 144, 171
bellhangers 115, 142, 149, 171
bells (church & tower) 3, 6, 18-20, 26-43, 45-47, 49-53, 55-58, 60, 64, 67-71, 73, 74, 80, 107-8, 112, 115, 136, 142, 144, 153, 162, 170, 171, 173, 174, 187; bell brasses 134; bell clapper 49; bell keepers 18; Bell Lane 112; bell ropes 26, 57; relationship with history of clocks 18; ringing & tolling 17, 18, 20, 32, 36, 39, 41, 42, 108, 118
bells (clock), generally 23; see alarms, chimes
Bennett, J. G., collection 151
BENNETT, R. B. (watch repairer & jeweller) (20th C) 7
Benson, St Helen, clock 162-3
BENSON, James William (of London) (f.1857-87; company onwards) 53
Bentley, W. H., collection 138
Benwell, Nicholas (clock keeper) 57-8
Berry, --- 178
Besselsleigh (Berks.) 171; Manor, clock 171
Bettino, Mario 39, 152
Bicester 4, 8, 22, 23, 25, 28, 29, 31, 41, 44, 46, 59, 65, 69, 86-7, 91, 98, 112-3, 125, 126, 130, 145, 164, 176-7, 180, 183, 186, 188, 189, 191; bell foundry in Bell Lane 112; earliest clockmaking in 23; St Edburg, clock 31, 113; Town House & Shambles, clock 31
BIGNELL, Thomas (a.1764, f.1812) 88, 111; father John 88
billheads of London suppliers 23, 114, 142
Billings, Mary 116
BILLINGTON, Edward (?or Everard) (of Market Harborough) (18th C) 128
BIRD, Michael (a.1648, d.1689) 6, 20, 22, 88-9, 93, 109, 113, 114, 122, 126, 139, 177; ?erroneously 'Will' 88; father John 88; wife Sarah 88-9
BIRD, Michael (a.1672, f.1713) 89, ?177
BIRD, Nathaniel (a.1678, f.1693) 89
BIRD, Wright (a.1682, d.?1686) 89
BIRD, William (mason) (1624-c.1690) 77
BIRD, William (f.1683) 88 (?error for Michael)
bird scarer, clockwork 7
birds-in-flight & fruit baskets 23, 128
Birmingham (Warwicks.) 147, 178, 180, 183, 186, 188-191; Handsworth 24, 41, 138
Bix 87
blacksmiths 1, 5, 6, 13, 15, 19, 20, 23, 25, 28, 29, 33, 41, 42, 44, 48-52, 54, 56, 62-3, 69-70, 74, 84, 86, 90, 91, 93, 96, 108-110, 115, 122, 127, 134-137, 139, 141, 148, 153; and turret clocks, generally 13, 20, 23; smith of Abingdon 49; smith of Tring 69
BLANCHARD, Charles (of London) (f.1760s) 23
Blenheim Palace, clock 31-2; sundial 76
BLISS, John (painter) (a.1778, f.1813) 59
Blisworth (Northants.) 185
Bloxham 86, 110, 123, 141, 170, 173, 176, 188; St Mary, clock 30, 32, 73, 162
Bloxham, Elizabeth 134
BLOXHAM, Samuel (f.1717-1740s) 30, 173

BLUNDELL, John (a.1678, f.1700) 8, 23, 89
Board, Isaac 94
Bodicote 151; St John the Baptist, clock 32, 95, 133, 134, 143, 151; Weeping Cross, sundial 32, 76, 151
Bolton, Mr 59
Bomford, L. R., collection 124
BOMMEL, Johann Leonhard (of Nuremberg) (f.1690-1720) 7
BOSWELL, James (smith) (f.c.1757, 1761) 63; wife Mary & son Martin 63
Bourton see Great Bourton
BOWLES, Edward (f.1783) 89; wife Hannah 89
BOYS, Robert (carpenter) (f.1511) 172
bracket clocks 4, 7-10, 85, 89, 92, 112, 120-1, 124, 136, 141, 143, 145, 146, 148, 152, 174, 176-178, 183, 184, 189-191
Brackley (Northants.) 86, 95
BRADFORD, James (of Buckingham) (18th C) 166
BRADFORD, Richard (smith) (f.c.1611) 136
BRADFORD, Thomas (of Buckingham) (f.1738-75) 166
Bradford, Thomas 32, 151
BRADLEY, Langley (of London) (a.1688, f.1748) 31-2, 76
Brafield, Richard 185
Brailes (Warwicks.), church clock 178, 180, 188
brass casting 23
brass rubbings 3
BREDON, Simon (scholar) (c.1300-c.1372) 17
BREGUET, Abraham Louis (of Paris) (1747-1823) 7
Brennan, Mary 131
Brentford (Middlesex) 171
Brice, A., collection 104
BRICKLAND, Humphrey (f.c.1710, d.1750) 22, 54-5, 88-90, 119
Bridge, Hannah 126
Brill (Bucks.) 118
Bristol (Somerset) 129
British Museum 99-100, 146
British Horological Institute 17
BROADWATER, Hugh (a.1680, f.1697) 22, 90, 109; father John

Broadwater alias Griffin 90
BROGDEN, James (of London) (f.1760s) 23
Brooke, W. J., collection 128
Brooks, Mrs M. I., collection 177
BROOKS, Richard (a.1765) 90, 137; father Richard 90
Broughton, Lyonell 80
Brown, H. Miles 1
BROWN, James (& jeweller) (f.1852) 90
BROWN, Richard (f.1831) 177
BROWNE, James (a.1663, f.?1696) 90, 135; father James 90
BROWN(E), John (smith) (f.1544) 70
Bryan, C., collection 104
BRYANT (or Briant), John (& bellfounder, of Hertford) (f.1792) 24, 27, 73
Buckingham (Bucks.) 92, 166
Buckinghamshire, church bells 112, 115; Record Office 184
BUCKLAND, William (& smith) (f.1786-1821) 8, 40-1, 71-2, 90; wife Elizabeth 90
Bucknell, St Peter, clock 33, 163
BUGGIN, Richard (smith) (f.1583-1619) 48
BULL, Lionel (a.1761) 90, 111; father Jonas 90
BULL, Thomas (& smith) (f.1592) 20, 56, 90-1
BULLER, Robert (b.1705, a.1719) 91, 107; parents John & Mary 91, 107
Bunbury (Ches.) 152, 191; sundial 191
Bunt, E. F., collection 181
BURBRIDGE, William (smith & wheelwright) (f.1611) 74
BURDITT, J. W. (f.1868) 27, 91
Burford 19, 33, 96, 98, 101, 107, 109, 111, 132, 148, 163, 182, 183, 186; Kitt's Quarries 77; Quaker Meeting House 132; St John the Baptist, clock 33, 153, 163; St John the Baptist, sundial 76-7; Tolsey, clock 21, 33, 182; Tolsey Museum 187
Burton Dassett (Warwicks.) 126
BUSBY, Richard

(f.1823) 177
Byfield (Northants.) 149
CAKEBREAD, --- (sundial maker) (f.1824) 76
CAKEBREAD (or Tayler alias Cackebrede), Richard (smith) (f.1566, d.c.1592) 49, 91, 137
Calcott see Colcutt
Cambridge (Cambs.), Christ's College 124; King's College, clock 22, 185
CAMOZZI, Charles (& jeweller etc.) (f.1832-50) 91; wife Eleanor 91
campanile, detached 32
Campbell, Mrs M., collection 177
Canterbury (Kent), Cathedral, clock 17; chimes 52
Carfax Tower see Oxford churches – St Martin
carillons 18, 29, 42, 51, 61, 66, 174; see chimes
Carpenter, Alexander 18
Carpenter, Ann 126
CARPENTER, Geoffrey (carpenter) (f.1511) 172
CARPENTER, William (& jeweller etc.) (f.1834-53) 91
carpenters 14, 15, 19, 22, 23, 48, 49, 59, 61-3, 70, 80, 88, 90, 92, 93, 107, 114, 122, 137, 144, 171-2, 190
CARROW, Robert (carpenter) (f.1505) 172
CARTER, Henry (& jeweller etc.) (f.1852) 8, 91
CARTER, Richard (smith) (f.c.1605) 136
CARTER, William (a.1779) 91, 137; father William 91
CARTWRIGHT, Benjamin (of London) (f.?1669, c.1713) 92
CARTWRIGHT, William (a.1705, f.1759) 92
case makers 13, 23, 92, 114; see carpenters
cases (clock) 7, 23, 114, 120, 121, 147, 183-4, 187; Banbury 114, 147; Bayliss's 92; Knibb's 121, 183-4; London 120
Cassington 35
Castle, Ann 87
Castle, Jane 114
Chadlington 113

chainmakers 23
Chalcombe (Northants.), church clock 43
Chalgrove 40, 177; St Mary, clock 33
CHAPMAN, --- (f.c.1860-c.1885) 92
Charlbury 16, 23, 64, 96, 102-105, 110, 116, 143, 152, 181, 182; Lee Place, clock 163; St Mary, clock 33, 163; Society Museum 110
Charles I 15, 37
Charles II 122, 152
Charlton-on-Otmoor, St Mary, clock 33-4
Chaucer, Geoffrey 18
Cheney, John (architect) 34
Cheney, John (printer) 91
Cherry, Miss, collection 181
Chester (Ches.) 191
Childrey (Berks.) 85, 176
Chilton House (Bucks.), clock 186-7
chimes & chiming mechanisms 6, 7, 18-20, 26-7, 29-33, 36, 38, 41, 42, 48-53, 56-58, 61-63, 68, 86, 91, 92, 95, 113, 115, 126, 132, 136, 137, 139, 144, 148, 153, 173, 182; Canterbury 52; Westminster 29, 37, 42
Chinese fire & incense clocks 7
Chinese-style decoration 90, 98, 103, 107, 111, 125, 130, 133, 179, 182, 187, 190
Chinnor, St Andrew, sandglass 81; sundial 77
Chippendale scrolls 100, 116, 138, 140, 143
Chipping Campden (Glos.) 148
Chipping Norton 9, 16, 34, 86-88, 92, 104, 105, 108, 128, 130, 131, 133, 140, 146, 149, 173, 178, 181; Market House, clock 34; St Mary, clock 34; Town Hall, clock 34
Chipping Warden (Northants.), clock 134
Chislehampton, St Katherine, clock 164
CHRISTIE, G. (of Wallingford) (f.1864-71) 172-3
chronometers 7, 116
Church Cowley 91
Church Hanborough, Sts Peter & Paul, clock 21, 34, 38 9
Churchill, All Saints, clock 34-5, 144

Cirencester (Glos.) 106
Clanfield (Berks.) 105
Clark, Brig. W. E., collection 9, 121
CLARKE, James (surgeon) (f.1858) 7, 8, 38, 92
Claydon 85, 117, 122, 124, 126; St James the Great, clock 21, 35, 36, 96, 125, 133, 134; St James the Great, sundial 77
CLEMENT, William (of London) (a.1657, d.?1709) 66, 185
CLEMENTS, Mr (of London) (f.1810-18) 141
CLEMENTS, John (& silversmith) (f.1808-53) 8, 92, 177
Clerkenwell (London) 32, 36-38, 49, 51, 52, 64, 66, 130, 141, 144, 162, 164
'clock', early use of word 18, 41
clock bells/faces/hands/etc. see bells/faces/hands/etc.
clock club 107
clock collecting 2, 4, 6
clock houses 18, 19, 48, 49 60-62, 64, 72, 74, 165
clock keepers, generally 18-19, 25, 105, 107, & see turret clocks
clock removals (from one site to another) 19, 29, 30, 35, 36, 41, 42, 45, 48, 67, 73, 74, 162, 172, 173, 187
clock repairs & maintenance especially 7, 13, 19, 21, 23, 26, 27, 35, 40-44, 46-49, 54-57, 62-3, 65-6, 68-72, 74-5, 96, 134, 141, 143, 162, 164, 173, 189
clock retailers, generally 7, 13, 23-4; see clockmaking trade
'clockmaker', early use of word 20, 69-70, 110, 162; erroneous early use 15; first freeman of Oxford so described 110
clockmakers, amateur & academic 6, 17, 21, 38, 62, 92, 116, 152; boom in Oxford (1670s-90s) 21-2, 119; earliest specialist 22, 23, 69-70; family dynasties of 1, 4, 5, 20 (generally), 93-4 (Denton), 96-98 (Faringdon), 100-105 (Gilkes), 117-125

(Knibb), 135 (Quelch), 140 (Simms), & see Quakers; generally in Oxford 5-6, 15-16, 19-24; see blacksmiths, watchmakers
Clockmakers' Company see London
clockmaking trade, generally 5, 13, 19-24; beginnings of specialist trade in Oxford 5, 20-1, 110, 119, 138-9; generally in London 23, 24; prefabricated parts & products 23-4, 142, 147; transmission of trade & skills 5, & see apprenticeship
clocks for main types see Act of Parliament, bracket, lantern, longcase, turret, wall
clocks, air-driven 116; American, 24, 140, 143; astronomical 17, 125; monastic 7; Chinese 7; electric 7, 29, 34, 51, 52, 58, 163; Japanese 7; monumental 18; musical 26, 27, 110, 115, 141, 146; oil 7; regulators 91, 111, 114, 182, 189; rolling ball 99; sedan 188; tabernacle 9; water 17, 39, 152; weather 152; wooden 190-1
clocks, and college life 6, 19; classification 4, 183-185; earliest dated in Oxford 22, 117; earliest dated in Oxfordshire 22, 32, 46-7, 109, 117; earliest references to 17-19, 41, 68; invention of 6, 17; old names for parts of 26, 48, 54-56, 67, 69-71
clockwork, other uses of 5, 7
Clutton, Cecil 3
cock, as symbol of Michael Bird 22, 89; for sundial 80; weather 40
Cogges 190
Coggins, C. J., collection 87
COLCUTT (or Calcott), John (f.1658-81) 48
COLE, Benjamin (sundial maker) (1667-1729) 65, 80
Colegrave, Penelope 65, 99
COLES, Richard (a.1771, d.1819) 86, 92
Collier, R., collection 127
Collingtree (Northants.) 176

Cologne 125
Combe, St Lawrence, clock 6, 9, 21, 35-6
Constantinople 110
Cooper, Henry 28
COOPER, John (painter) (f.1818) 41, 92
Cooper, Mary 143
Cope, Sir Anthony 39
Copleston, Edward 53
CORBET, Hugh (smith) (f.1589-1616) 20, 54, 56, 91-93
Corbet, Richard 60
Cornish, William 117
Cornwall, clockmaking 1
COSIER, John (smith) (f.c.1754) 44
COSTER, Charles (& jeweller etc.) (f.1853) 93
COSTER, James (& jeweller) (f.1798-1830) 93, 150
Cothele House (Cornwall) 109
COTTISWOLD, Thomas (smith) (f.1458-64) 19, 69, 93
COWLYS, --- (locksmith) (f.1571) 70-1
COWPER, John (f.c.1470) 69
Cox, H., collection 140
Cox, Mary 96
COX, William (f.1680s) 57
craft organisation see apprenticeship, clockmakers, guilds, privileged tradesmen, Quakers
crafts connected with clockmaking, generally 6, 13, 20, 23-4; see architects, bellfounders, blacksmiths, carpenters, case makers, chainmakers, engravers, glaziers, goldsmiths, gunsmiths, ironmongers, jewellers, locksmiths, painters, plumbers, scientific instrument makers, silversmiths, stone masons, wheelwright, whitesmiths
CREED, Thomas (a.1657, f.?1668-99) 93, 135; father Henry 93
Crewe, Sir Thomas 80
CROKER, Emanuel (f.1621) 162
Cromwell, Oliver 106
Cropredy 122; St Mary, clock 20, 35, 36, 42, 130, 148
Crosier, --- (clock keeper) 59
Crosier, Thomas 79

crossing-out (decoration characteristic of Hemins) 28, 31, 65, 175
crown wheel escapements see escapements
Croydon (Surrey) 32, 53, 172
Crusader castles 3
Cuddesdon 9, 23, 126, 128; All Saints, clock 36
Cullimore, H., collection 104
Cumnor (Berks.), clock 175
curfew 20, 36, 66, 70
CURSIAN, Nicholas (?of London) (f.1540) 125
CURTICE (or Curtis), Greenaway (a.1688, d.1702) 22, 89, 93, 126; father Thomas Curtis 93
Cuss, T. C., collection 190
cutlery 148, 174; Bicester cutler 86

D., G. (clock face mark) 111
DALE, John (sundial maker) (17th C) 79
DALLAM, Thomas (smith & organ maker, of London) (f.c.1600) 110-1
Daniell, J. A. 128
Daventry (Northants.) 151, 191
DAVIDSON, Daniel (?of Edinburgh) (f.c.1770) 115
DAVIS, Abraham (& jeweller etc.) (f.1850-58) 8, 177
DAVIS, John (of Windsor) (f.?1678-89) 24, 66
DAVIS, John (of Windsor) (b.1739, f.1779, ?1784) 66-7
Davis, John 191
DAVIS, John (?silversmith) (f.1792) 127
Day, Mary 87
Day, Nathaniel (clock keeper) 62
Deane, Elizabeth 142
Debaufre escapement 167-169
decorative features of clocks & cases, generally 23; see birds-in-flight, Chinese-style, Chippendale scrolls, crossing-out, faces, Gibbons garlands, magical symbols, painters, Sheraton cases
Deddington 8, 23, 27, 37, 44, 96-98, 100, 102, 105, 110, 143, 165, 179-80, 182; Sts Peter & Paul, clock 37, 98, 144, 179, 180; Sts

Peter & Paul, memorial brass 181
degree plate for pendulum 7, 9
Dehra Dun 1
DENT, Edward John (of London) (1790-1853) 165; E. Dent & Co. (f.1870 onwards) 44, 63, 66
DENT, Frederick (of London) (f.1857-81) 165
DENT, William (of London) (f.1674-1704) 131
Denton family 93-4
DENTON, Robert (f.1730, d.1769) 22, 55, 65, 93, 94, 176, 177
DENTON, Samuel (f.1756, d.1795) 8, 94, 177-8; wife Mary 94
DENTON, Samuel (& silversmith) (b.1776, f.1827) 16, 94, 141; wife Elizabeth 94
DENTON, William (f.1756, d.?before 1774) 94; William & Samuel Denton (f.1756) 94
Derby (Derbys.) 27, 29, 30, 33, 34, 36, 37, 42, 51, 67, 68, 72, 74, 163, 165
DERRELL, Thomas (f.1535) 42
DETOUCHE, Constantin Louis (of Paris) (f.1820-50) 7
'detting' 54
DEVELAY, Abram (of Lausanne) (f.1789) 9
DEVERELL, John (of Winslow) (18th C) 128
Devizes (Wilts.) 104
Devon, clockmaking 1
Dewe, John 62
'dial', the word 5
dial as shop sign 22, 113, 142
dials see faces, sundials
Dillon, Viscount 127, 187
DISTON, Thomas (d.1724) 178
Ditchley 127, 187; Ditchley Park, Great House, clock 113, 164, 183, 187; Model Farm, clock 164
Dobinson, Mary 148
Dolton (Devon) 126
Dorchester 70; Sts Peter & Paul, clock 37, 130
DOWNING, Humphrey (of London) (f.1648, d.c.1662) 133
D'Oyley, Bray 16, 102
DRURY, --- (f.1792) 95, 178
DRURY, Charles William (& jeweller) (f.1823, ?1847) 94-5, 143, 178

DRURY, James (f.c.1780-1800) 8, 95
DRURY, William (& jeweller) (f.1773, d.?1810) 8, 95, 178; wives Mary & Elizabeth 95
Dry Sandford (Berks.) 9, 131
DURRAN, Eustace (& jeweller etc.) (f.1877) 95; Durran & Smith 95
DURRAN(T), James Hopkins (& jeweller etc.) (f.1832-54) 34, 95-6, 139, 178
Ducklington, St Bartholomew, clock 37, 164-5
Duns Tew, clock 143
DUTTON, T. (f.1841-2) 96
Dyer, John (clock keeper) 61

Eales, Margaret 126
EARLE, Thomas (f.1796-1812) 55, 59, 66, 96, 137, 150, 189
East Hendred (Berks.), church clock 20
EDES, John (f.1582) 48
Edward III 15
Edward VI 42, 69
Edwards, H. N. 165
EDWARDS, Thomas (f.1773) 96, 97
EDWARDS, William (f.1776) 27
Egglestone, Phoebe 133
electric clocks 7, 29, 34, 51, 52, 58, 163
Elizabeth I 78, 80
ELLIS, Griffith (of London) (f.1760s) 23
Ellis, T. E., collection 182
engravers & engraving (except of prints) 23, 92, 100-1, 106, 123, 127, 128, 148, 174, 181
ENOCK, Ezra (1799-1860) 96; parents John & Elizabeth 96; wives Eliza & Ann 96
ENOCK, John (1834-1883) 96
Enstone, St Kenelm, clock 165, 179
entomology 1-4
equal hours 18, 76, 80
equatorium 17-18
escapements, anchor, conversions to 22, 40, 50, 57, 61-2, 75, 152, 153, 173, 175; anchor, invention & earliest examples of 2, 21, 22, 57, 61-2, 66, 152, 185-6; crown wheel & pendulum 62; crown wheel & verge 21, 22, 38, 39, 45, 73, 173; Debaufre 167-170; experimental by Joseph Knibb 185-6; Graham 164; Huygens verge pendulum 21, 48; Lepaute 63, 162; tic-tac 120, 184, 186; types invented by James Clarke 38, 92
Espinasse, Prof. P., collection 149
ESTE (or Easte), William (mason) (f.1505, d.c.1526) 19, 51, 79, 96
Eustace, Martha 142
EVANS, William Frederick (of Handsworth) (f.1854, d.1899) 24, 41, 138
Evens, G. S., collection 102
Ewelme, St Mary, clock 37, 142, 165
Ewen, Ursula 139; father Stephen 139
Ewes, John 73
Exeter (Devon) 139
Eynsham, Abbey, clock 165; St Leonard, clock 165, 182, 183, 187
Eyre, Charles 165

faces (clock), generally 5, 6, 23, 24; ?as shop signs 22, 113, 142; as symbol of Joseph Knibb 22, 123; balloon decoration 96, 151; birds in flight decoration 23, 128; lack of them on towers 18, 34, 35, 43-45, 47, 73, 166, 175; makers & suppliers 23, 24, 147, 180, 183, 188-191; on gravestone 188; once handed 19, 22, 26, 28, 32, 33, 38, 51, 53, 56, 62, 63, 74, 163, 174, 186-7, 190; painted, especially 24, 95, 96, 98, 103, 129, 147, 151, 179, 180, 188, 189, 191, & see painters; Quaker 7, 23, 43, 100-102, 107, 131, 182; skeleton 28, 49, 63, 68, 144; slate 64, 67, 172
factory clock 7
FAIRBROTHER, James (f.1823) 96
Fairfax, Mary 95
Falkland, Lady (wife of Henry, Viscount Falkland) 15, 133
Fardon family 44, 96-98, 100, 110, 178-9
Fardon, Charles, collection 180

FARDON, John (c.1700-1743) 23, 96-7, 101, 104-5, 110, 178, 179; ?parents Thomas & Hannah 16, 96; wives Elizabeth & Mary 96-7, 178, 179
FARDON, John (1736-1786) 8, 96, 97, 110, 178, 179
FARDON, John (1782-1865) 97, ?100, 178, 179, 180; wife Elizabeth 178
Fardon, Mary 110; parents Richard & Mary 104, 110
Fardon, Peter, collection 179
Fardon, Sarah 104
FARDON, Thomas (f.1791, d.1838) 8, 27, 37, 44, 97-8, 101, 102, 165, 178-180; Thomas & John Fardon ?44, 97-8, ?100, 179; wife Lydia (1768-1836) 97, 178
FARDON, Thomas (f.1823-66) 8, 98, 179, 180; Fardon & Gibbs 100, 180; wife Anne Susannah (1802-1864) 179, 180
FARDON, Thomas Gibbons (1841-1903) 179, 180; son George Osborne 180; wife Amy 179, 180
FARDON, Thomas Osborne (1867-1941) 179, 180
Faringdon (Berks.) 148
Farnborough (Warwicks.) 118, 123, 124
Farthinghoe (Northants.), church clock 21, 175
FAULKNER, John (a.1790) 98, 148
Fenny Compton (Warwicks.) 146
Fergie, J. A. K., 26; collection 97, 102, 105, 108
FERRIMAN, William (f.1680s) 62
Field, C., collection 143
FIELD, Thomas White (a.1787, d.1832) 98, 130, 142
Fielding, Lady Diana 85
Finmere 7, 8, 37-8, 92, 166; Finmere House 38, 92; St Michael, clock 37-8, 99, 166; Water Stratford House 38, 92 (erroneously Waterstock House)
FINNEMORE, William (clock face maker, of Birmingham) (f.1815-

199

36) 180, 188; Finnemore & Son 180
Finstock 96
fire engines 27, 90
fishing tackle 148
Fitzclarence, Lord Augustus 45
'flee' 70
'fliers' 71
FLOWERS, John (f.1797-1814) 98; wives Ann & Esther 98
fly catcher, clockwork 7
flying machines 6
foliot 21, 36, 48, 59, 62
FOOSE, Louis (painter, of Abingdon) (f.1505) 19, 51, 96
FORD, John (a.1682, f.1725) 8, 22, 38, 98-9, 119; father William 98; Mr Ford's man 38
Forest Research Institute 2
forestry & forest entomology 1-2, 3
Fortescue, A. J., collection 110
Fortescue, Charles 170; Fortescue-Turville family 170
Fortescue, E. C., collection 178, 179
fortifications 136
Foskett (Bucks.) 123
FOSTER, Joseph (of London) (a.1684, f.1704) 93
Foulkes, Henry 51
foundries 23, 112; see bell-founders
FOUNTEYNE, Geoffrey (f.1537) 69
Fowler, Mary 107
FOWLER, William (a.1667) 99, 135; father William 99
Fox, George 16, 146 (?another)
Fox, Richard 78
FRANCIS, William (clock face maker, of Birmingham) (f.1816-36) 180, 189
Francklin see Ranklyn
FREE, John (a.1696, d.1726) 22, 65, 99, 119; father John 99; son John 99; wife Penelope (f.1704-34) 65, 99
Fremantle, Lady, collection 98
French, Mary 149
FRIDAY, Gabriel (f.c.1850) 166
Fritwell 22, 25, 39, 46-48, 74, 84, 108-9, 115-6, 181; church clock 21, 145
FROMANTEEL, Ahasuerus (of London) (1607-1693) 7
Fuller, Hugh 70
Fyfield (Berks.) 131, 171

'gabel rope' 26
GAINSBOROUGH, Humphrey (minister) (1718-1776) 99-100; father John 99; brother Sir Thomas 100
Gardiner, Elizabeth 94
Gardner, Mrs A. K., collection 149
GARDNER, William (a.1755) 100, 137; father William 100
GARLAND, John (of London) (f.1760s) 23
'Garnors' 48
Garsington, St Mary, clock 24, 38, 133, 144
GENT & CO. (of Leicester) (1878 onwards) 58
geometrical solids 80
Gibbons garlands 32
Gibbs, James (architect) 164
GIBBS, Joseph (f.1805-55) 100 (erroneously Joshua), 180; Fardon & Gibbs 100, 180
GIBBS, Richard (f.1852) 100
Gilkes family 23, 100-105 181
Gilkes, Grace 102; parents William & Anne 102
GILKES, John (f.1740-77) 8, 101, 180; ?wife Mary 101
GILKES, John (f.c.1800) 101
GILKES, Richard (of London) (a.1678, f.1703) 105
GILKES, Richard (1715-1787) 7, 8, 23, 27, 98, 101-105, 146, 179, 181; daughter Grace 181; wife Grace (née Gilkes) & children 102
GILKES, Richard (of Devizes) (1745-1822) 104
GILKES, Richard (f.c.1800) 103
GILKES, Thomas (b.?1665, f.1743) 22, 96, 101, 103-105; parents Thomas (b.1645, f.1681) & Mary 16, 104-5; wife Anne 101, 103
GILKES, Thomas (1704-1757) 23, 102-105; wife Mary 103, 104
GILKES, Thomas (1736-1779) 104, 181; wife Sarah (née Fardon) 104
GILKES, Thomas (f.1758-c.1770) 104, 181; wives Mary & ?Alice 104
GILKES, Thomas (of Shipston) (f.c.1760) 104
GILKES, Tobias (f.1743-c.1770) 105, 128, 173, 181
Gilkes-Hastie, Mary 181
Gill, Mrs, collection 177
Gillett, M. M., collection 143
GILLETT, Thomas (a.1698, f.1702) 105, 119; father George 105
Gillett, W. R., collection 146
GILLETT & BLAND (of Croydon) (f.1862-80) 32, 53, 172, 174; Gillett & Johnston (c.1880 onwards) 29
GILPIN, Edmund (or Edward, of London) (f.1632-77) 88
glass & china dealers 125, 183
glaziers 49, 59, 151
globe, celestial 145; globe clock 6; globe dials 77, 79, 80; Globe Room 2
Gloucester (Glos.) 106, 170
GLOVER, Boyer (of London) (f.1756) 110
Glympton 108
GODFREY, Henry (a.1676, f.1707) 8, 105, 135; father George 105
Goggs, C. L., collection 102
goldsmiths 13, 24, 126, 127, 142, 145, 148, 150, 151, 189; see jewellers, silversmiths
Gomes, William (clock keeper) 55, 105
Goring (Berks.) 132
Goshawk, Mrs E. 170
GOWETH, John (b.1669, f.1694, d.?1734) 22, 105-6, 119; father John (gunsmith) 105-6
GRAHAM, George (of London) (c.1673-1751) 7, 164
gramophones 7
Grant, Ann 98
Graunden, William 15
Gravesend (Kent) 142
GRAYSON, John (& jeweller etc.) (f.1823-c.1850) 42, 106
GRAYSON, William (& jeweller etc.) (?a.1780, f.1839-60) 106; wife Jane 106
Great Bourton 36; All

200

Saints, clock 32; Chapel School, clock 32
Great Dassett (Warwicks.) 126
Great Haseley 23, 38, 40-1, 92, 114, 142, 181, 183; St Peter, clock 40-1, 90, 114, 116, 138; St Peter, sundial 77
Great Marlow (Bucks.) 93
Great Milton 38, 41, 141-2; St Mary, bell 142; St Mary, clock 46-7, 109
Great Rollright, St Andrew, clock 172
GREEN, --- (painter) (f.1741) 62
GREEN (or Smith), George Smith (f.1750, d.1762) 106
Green (or Grene), John (clock keeper) 69
Green, John 107
GREEN, John (f.1794-1823) 106; wife Mary 106
GREEN, Samuel (a.1754, f.1823) 106; father Henry Greene 106
GREEN, William (b.?1722, f.c.1770-?1780) 8, 107; ?parents Isaac & Joan 107
Gregory, James 186
Gregory, Richard 73
GREGSON, --- (either Jean or Pierre, of Paris) (f.c.1785) 24, 138
GRIFFIN, Richard (& jeweller) (f.1853) 107
Grimthorpe, Lord 50
Guildhall Museum 121
guilds 1, 5, 13-15; see London, Oxford City, Witney
'guge' 70
GUNN, --- (painter) (f.1783-1808) 166
GUNN, William (of Wallingford) (f.1714-c.1740) 91, 107; wife Mary 107
gunsmiths 15, 41, 54, 92, 105, 112, 145, 151, 189
Gunter's quadrant 134
Gunther, R. T. 1, 76
GURDEN, Joseph (f.1755, d.1772) 107, 181, 187; wife Martha 107
GUTKNECHT, N. C. (engraver, of London) (f.1760s) 23

H., I. (watch case mark) 94
Haddenham (Bucks.) 90
HADDOCKE, George (smith) (f.1685) 147
Hagbourne (Berks.) 29, 136
HAINES, Robert (f.1777-99) 16, 107

Hambridge, Dinah 110, 173, 181
HAMILTON, --- (painter) (f.1681) 74
Hampton Court Palace (Middlesex), clock 49, 125
Hanborough see Church Hanborough
Hancock, W. U. 190
HANDLEY & MOORE (of Clerkenwell) (f.1870s-80s) 24
hands (clock) especially 27, 43, 46, 74, 84, 100; see faces
Handsworth (Staffs.) 24, 41, 138
Hanslope (Bucks.) 8, 118, 120, 122-125, 184-186; Green End Farm 185; spire 118
Hanwell, Castle, waterclock 39, 152; St Peter, clock 22, 39, 109
hardware merchants 24; see ironmongers
Harington (or Clerke alias Harington), Thomas (clock keeper) 19, 41-2, 107-8, 115
Harper, William 38
Harris see Aress
Harris, Eliza 96
HARRIS, George (& smith) (1614-1694) 22, 25, 39, 48, 74-5, 84, 108-9, 181; father Jeffrey 108; wife Betteris 108, 109
HARRIS, James (f.?1795, 1839-59) 109, 173, 181, 182; ?wife Mary 109
HARRIS, John (f.1631) 109
HARRIS, John (a.1668, f.1696) 8, 22, 89, 90, 109, 146; father John 109
HARRIS, John (19th C) 109, 182
HARRIS, Joshua (f.c.1760-1810) 182
HARRIS, Joshua (f.1809-31) 109, 165, 173, 182
HARRIS, Nicholas (& smith) (1657-1738) 47, 48, 108, 109; wife Elizabeth 109
HARRIS, Thomas (& locksmith) (f.1642) 71
Harris, Thomas 28
HARRIS, Thomas (1732-1797) 110, 182; parents Joseph & Mary 110; wife Mary (née Fardon) 110
Harris, William 122
HARRIS, William (a.1756) 110; father John 110

HARRIS, William (f.1794, d.c.1823) 87, 110, 173, 181, 182; wives Elizabeth & Dinah 110, 173, 181
HARRISON, James (maker of wooden clocks) (18th C) 191
Harrison, Sir Robert 118
HARRISON, William (f.c. 1770-1792) 64, 110, 182
Harrison maintaining power 189
Hart, R. C., collection 152
Hart, Richard (clock keeper) 61
HARVEY, Robert (f.1588-c.1600) 20, 110-1
Haseley see Great Haseley
Hastie, Francis 181
HASTINGS, Hercules (f.c.1700-c.1730) 33, 182
HAWKINS, Richard (painter) (f.1637-81) 111, 127, 137
Hawksmoor, Nicholas (architect) 53
HAWTING, John (a.1745, d.1791) 5, 8, 9, 25, 64, 66, 88, 90, 111, 137, 171, 174, 182; father John 111; wives Eliza & Mary 111
Hawting, Sarah 5, 113
HAWTING, William (a.1770) 111
Haynes, Thomas (clock keeper) 174
HEAD, James (f.1757) 111
Headington, St Andrew, clock 41, 45
Hearne, Thomas 99, 118
HEATH, John (a.1756) 111; mother Elizabeth 111
Hedges, T. 72
Hegge, Robert 78
heliostats 7
HEMINS, Benjamin (?gunsmith) (f.1739-45) 112
HEMINS, Edward (f.1699) 4, 44, 112
HEMINS, Edward (& smith & bellfounder) (f.c.1720, d.1744) 4, 8, 23, 25, 28, 29, 31, 33-4, 44, 49, 65, 112-3, 164, 183; mother 112; wife Elizabeth & children 112
HEMINS, Edward (f.1744-c.1770) 4, 8, 112, 183
HEMINS, Joseph (f.1739-45) 29, 48, 112, 113
Hemmings, Miss D., collection 188
Henley-in-Arden (Warwicks.) 189

201

Henley-on-Thames 13, 22, 41-2, 86-88, 93, 99-100, 106-108, 115, 128, 142, 148, 150, 187, 189; Borough Council, Guilds, etc. 13; Congregational Meeting House 99; earliest clockmaking in 22; Merchant Guild 13; St Mary, clock 13, 18-20, 41-2, 106
Henry I 15
Henry VII 60
Henry VIII 36, 76, 125
HERBERT, Edward (f.1613) 71
Herbert, G. 139
HERBERT, John (f.1742-94) 5, 22, 62-3, 113, 144; son John 113; wife Sarah (née Hawting) 5, 113
Hertford (Herts.) 24, 27, 73
HETH, Robert (f.1573) 70, 113
HEWITT, Owen (f.1823) 113
Heyford 98; see Lower Heyford, Upper Heyford
HEYLEY, Thomas (f.1586) 71
HEYTEN, Robert (a.1656) 89, 113; father Robert 113
Heythrop House, clock 74
HICKMAN, Edward (f.?1818, 1823-6) 113
HICKMAN, Richard (& siversmith) (f.1776) 145
HIDE, Charles (f.1757, d.1773) 113
Higgins, Thomas 181
HIGGINS, Thomas (& glass & china dealer) (f.1823-4) 183
HIGGS, James (of Wallingford) (f.1795) 162
Hillend (Berks.) 105
HINE, John (d.?1777) 114
Hine, Thomas 73
Hinksey (Berks.) 138
Hinton family (William snr & jnr, Thomas, Samuel) 40
Hinton, Elizabeth 110
historical research 1-5
Hitchcock, Mr 43
HITCHCOCK, William (a.1675) 114, 119; father Thomas 114
Hitchcox, John 67
Hitchins, Miss E. M., collection 145
HOBDELL, Henry Ba(n)shard (& silversmith) (f.1832, d.?1844) 114; wife Emily 114
HODGES, Anthony (a.1664, f.1672) 22, 89, 114; father Anthony Hodges alias Hedges 114
Hollis, G. 58
HOLLOWAY, Edward (case maker) (f.1833-42) 114, 147
HOLLOWAY, Thomas (f.1736, d.?1764) 23, 40, 114, 142
HOLLOWAY, William (?late 18th C) 183
HOLLOWELL (or Holloway), Charles (cabinet maker) (19th C) 114
HOLMES, John (of London) (f.c.1780) 150
Hook Norton 88, 132, 149, 173, 188, 190; St Peter, clock 42
HOOKE, Robert (scholar) (1635-1703) 152; Hooke's joint 48
'horecudium' 19, 41, 107-8
Horley 8, 23, 89; St Etheldreda, clock 36, 42-3, 47, 148, 172
Hornby, E., collection 184
Hornton 43, 148; St John the Baptist, clock 25, 43, 47, 172
'horologium' 17, 19, 60, 64, 69
Horspath, St Giles, clock 43
Howard, Lady 28
HOWSE, Charles (of London) (a.1750, f.1804) 23, 114, 142; father William 114
Howse, James 165
Hudson, Mr 37
Hudson, Mrs G., collection 121
Hughes, Peter 185
HUMPHRI(E)S (or Humphreys), ?William (f.1850) 114
HUNT, James (f.1851) 183
HUNT, John (f.?1795, 1813-23) 114; wife Jane 114
HUNT, Thomas (f.1777-c.1800) 16, 115
Husbands Bosworth (Leics.) 170
HUYGENS, Christiaan (scholar, of The Hague) (1629-1695) 21, 48, 50, 62
HYDE, Thomas (smith) (f.1464-87) 41, 115
hydraulic machines 99, 150; see waterclocks

I., V. R., (clock face mark) 68
Idbury, Manor House 170; St Nicholas, clock 7, 9, 44, 166-170

Iffley 90
Imperial (Indian) Forest Service 1
India 1-2, 3
Ingram-Lockhart, John 181
innkeepers 2
Irish, Dr 59
ironmongers 15, 24, 91, 97, 144
Irons, Thomas 59
ironwork 20, 54, 141; for fortifications 136; for sandglasses 81, 84; see blacksmiths, turret clocks
Isham (Northants.) 185
Islip, St Nicholas, clock 44, 112, 174

jacks (clock) 18-20, 56-58, 91, 134, 137, 144, 187; see roasting jacks
JACKSON, John (mason) (f.1643) 79
JACKSON, John (f.1783-86) 115; ?wife Mary 115
JACKSON, Martin (of London) (f.1697-1721) 124
Jagger, Catherine 86
James II 88, 118
JAMES, William (?early 19th C) 73
JANVIER, Antide (of Paris) (1751-1835) 7
Japanese clocks 7
JEEVES, Anthony (f.1744-c.1770) 115
JEFFREY, James (f.1823-53) 115, 183
JEFFREYS, John (of London) (f.1735) 116
JEFFS, John (bellhanger) (f.1640-50) 61, 115
Jenkyns, Richard 49
Jennings, John (clock keeper) 56, 105
Jennings, Matthew (clock keeper) 56
JENNINGS, Thomas (& smith) (1722-1773) 115; father Robert 115; wife Philippa 115
JENNINGS, William (1716-1780) 115-6
JENNINGS, William (f.c.1803) 46
Jerome's American clocks 24, 143
jewellers 24, 90-95, 106, 107, 116, 126, 133, 134, 138, 140-143, 147, 148, 151, 177, 190
Johnson, Manuel John 49
Johnson, R. A., collection 102, 105
JOHNSON, Robert (f.1553) 70
Jones, Alice 139
Jones, Inigo (architect) 80

202

JONES, John (scholar) (1645-1709) 116; father Matthew 116
Jones, Mrs L. V., collection 149
JONES, William (& silversmith) (f.1772) 145
Jordan family 40, 116
JORDAN, James (b.c.1751, f.1776) 116; wife Mary 116
JORDAN, Thomas (f.1770-90) 8, 40, 116
JOYCE, John Barnett (of Whitchurch, Salop.) (b.1826, f.1888, company onwards) 50, 172
JUKES, J. (clock face maker, of Birmingham) (f.c.1840) 183, 188
JULLION, Francis (of Brentford) (f.1760-c.1780) 171

KALABERGO, John (& jeweller) (f.1832, d.1852) 116
KEENE, Richard (bellfounder) (f.1676) 57 (erroneously John), 58
Kelley, M., collection 102
KEMPSTER, Christopher (mason) (1627-1715) 76-7
KENDALL, Larkum (1721-1795) 116
KENNING, William (b.1648, f.1687), 22, 116-7; ?brother John 116; parents Martin & Alice 116
KENNINGTON, John (f.1772) 117, 138; mother 117
KEYTE, Richard (f.1770) 117, 174, 183
Kibble, J. 44
Kiddington Hall, clock 44, 170
Kidlington 75; St Mary, clock 44, 98, 179
King, Sarah 86
Kingham, St Andrew, clock 44
Kings Sutton (Northants.) 26; church clock 21, 26, 134
KINGSTON, Joseph (f.1786) 183
Knibb family 2, 117-125, 184-5
Knibb, Elizabeth 117, 176
Knibb, George (of Farnborough) 123, 124, 185; son Isaiah 123; wife Jane 124
KNIBB, John (1650-1722) 6, 8-10, 22, 40, 54, 65, 75, 85, 89-90, 93, 98, 99, 105, 114, 117-124, 126, 146, 150, 176, 183-185, 189; parents Thomas & Elizabeth (née Wise) 117, 122; wife Elizabeth & children 117-119, 123-4, 176, 185
Knibb, John (d.1754) 117-119, 123-4, 185; wife Deborah 119, 185
KNIBB, Joseph (1640-1711) 2, 6, 7, 15, 22, 25, 61-2, 65, 66, 85, 117-120, 122-126, 141, 152, 184-186; sale of his stock 85, 120; son Thomas (d.1703) 123; wife Elizabeth 119, 123-4, 126, 185
KNIBB, Joseph (b.1695, f.1722) 124
Knibb, Mary 126
KNIBB, Peter (b.1651, f.1679) 123, 124; wife Katherine 124
KNIBB, Samuel (& instrument maker) (1625-c.1670) 122-125, 186; namesakes 124; parents John & Warborough 124
Knightcote (Warwicks.) 126
KRATZER, Nicholas (scholar) (1487-?1547) 60, 62, 76, 78-80, 96, 125, 127, 134

LAMB, Benjamin (of London) (f.1760s) 23
Lamb, Dinah 132
LAMB, James (& glass & china dealer) (f.1818-53) 125, 186, 191; wife Ann 125
LAMBERT, Nicholas (of London) (18th C) 128
LAMPREY, B. (f.c.1760) 186
LAMPREY, Benjamin (f.c.1696, d.1721) 22, 23, 125; wives Jane, Elizabeth, & another 125
LAMPREY, John (1704-1759) 9, 23, 35, 125, 126; wife Mary 125
LAMPREY, John (b.1734, f.1771) 29, 126; wife Ann 126
Landes, David 3
Lane, Charles 185
Lane, John 165
LANE, Thomas (?locksmith) (f.1604) 92
LANE, Wright (a.1676, f.1689) 89, 93, 126; father Nathaniel 126
Langford, St Matthew, clock 21, 45
LANGLEY, Thomas (a.1673, f.1687) 126, 135; father Thomas 126
lantern clocks 6, 8-10, 20, 22, 85, 93, 108-9, 112, 117, 119, 120, 124, 125, 130, 132, 142, 146, 147, 181, 183-185; earliest dated 22, 109, 117
lantern pinion drilling machine 7
lathes, watchmaker's 7
Latcham, Dr P., collection 105, 129, 152
Laud, Archbishop William 61
Launchbury, C. S., collection 97
Launton 176
Lausanne 9
Law, Mary 111
LAWRENCE, Thomas (a.1759) 126
Lawrence, Thomas Edward (Lawrence of Arabia) 1, 3
LAWRENCE, William (f.1744-64) 9, 23, 126, 128, 186-7; wife Margaret 126
Lawrence Hinksey (South Hinksey, Berks.) 138
Lea, W., collection 129
Learner, Dr R., collection 182
Leatherbarrow, Esther 98
Lee see Atte Lee
Lee, R. A. 183-4, 185-6; collection 120
LEE, Roger (of Leicester) (f.1691-c.1700) 128
Leeds (Yorks.) 28
Leeds, E. T. 88
Leicester (Leics.) 58, 106, 128; Museum 106
Lepaute escapement 63, 162
Lester, Mrs F., collection 102
LEVI, Benjamin (f.1770) 126
LEVI, Israel Morris (& jeweller etc.) (f.1853) 126
Lewes (or Lew), Thomas (clock keeper) 42, 108
LEWIS, C. A. L. (f.c.1950) 46
Lewis Evans Collection 6
Lewknor, St Margaret, clock 45
LIDBROOK, Thomas (a.1679) 119, 126; parents Robert Lidbrooke or Ladbrooke & Mary (née Knibb) 126
Line, Susannah 142

Llandaff (Wales) 116
LOCK, Edward (& silversmith) (1729-1813) 9, 24, 126-7; wife Hannah 126
LOCK, Henry (a.1795, f.1802) 127, 133
LOCK, Joseph (?silversmith) (f.1782) 127
locks 6, 70, 71, 80, 112, 151
locksmiths 6, 15, 20, 70, 71, 80, 84, 92, 112, 135-6, 139, 141, 147, 151, 153
Logan (or Loggin) family 170
Logan, Elizabeth 170
Loggan, David 76-80, 122
Lombardy 116
London 1, 2, 5-8, 15, 17, 20-24, 28, 30-32, 44, 49, 51-53, 55, 64-66, 70, 76, 77, 85, 86, 89, 90, 92-94, 96-98, 105, 106, 110, 114, 116-7, 120-126, 128-131, 133, 138, 141, 144, 146-7, 150, 151, 165, 172, 174, 176, 184-187; British Museum 99-100, 146; City Companies 5; Clerkenwell 32, 36-38, 49, 51, 52, 64, 66, 130, 141, 144, 162, 164; Clockmakers' Company 32, 85, 86, 88-93, 105, 106, 109, 110, 114, 116, 117, 122-126, 131, 136, 142, 144, 146, 151; Clockmakers' Company Museum 129; clockmaking in, generally 23, 24; College of Physicians 116; earliest clocks in 17; Great Exhibition (1851) 186; Gresham College 152; Guildhall Library & Museum 121, 142; London Museum 93; Merchant Tailors' Company 146; old St Paul's Cathedral, clock 17; Quakers 146; Royal Society 152; St Paul's Cathedral 77; Science Museum 2; Westminster 118; Westminster Palace, clock 17; Whitechapel 50, 96
Long Compton (Warwicks.) 9, 23, 131
longcase clocks 2, 4, 7-10, 22, 23, 38, 84-87, 89-98, 100-107, 109-116, 119-122, 124-134, 137-140, 142, 143, 145-147, 149-152, 176-191; electric 7; with lantern clock movements 128, 131; with turret clock movements 7, 38, 92; see regulators
longitude, finding the 99
Loomes, Brian 4
Lord, Alice 104
Loughborough (Leics.) 34, 37, 58; bellfoundry 144
Lovegent, Isabella 64
LOVELL, Mr (painter) (f.1792) 27
Lower Heyford, St Mary, clock 42, 45
Ludgershall (Bucks.) 181
Lupton, Roger 36
Luther, Martin 80
Lydall, Richard 171
LYNCHE, James (f.1505) 172

McDonald, J. W., collection 87
MACE, John (f.c1775) 127, 187
MACE, John (smith) (19th C) 127
MACE, Robert Harris (smith) (f.1853) 127
machine for tagging laces 139
Maclean, Dr B., collection 9, 138
Maddison, Francis 3
MAELZEL, Johann Nepomuk (mechanician, of Vienna) (1772-1838) 7
magical symbols 110, 128, 184
'making' & 'new making', the words 5, 55, 65
Malsbury, K., collection 188
Mapledurham, St Margaret, clock 45
Margerisson, M., collection 107
MARGETTS, George (1748-1804) 4, 6, 7, 9
Market Harborough (Leics.) 128
Marlborough, John, Duke of 31, 76
MARRINER, William (a.1774, f.1782) 127, 137; father John 127
MARSH, Anthony (of London) (f.1750) 114
Marsh Gibbon (Bucks.) 28
MARSHALL, Thomas R. (& jeweller etc.) (f.1852) 127
Marston see Old Marston
Mary I 13
MASEY, Thomas (smith) (f.1550) 60, 127
Mathews, Eliza 111
Mathews, L. W., collection 132

MAT(T)HEWS, Christopher (painter) (f.1667, d.c.1700) 75, 80, 111, 127
MAXEY (or Mazey), Charles (f.1771-95) 129, 173, 174, 187
May family 5, 23, 128-9
MAY, Edward (f.1725-c.1760) 128, 144, 187
MAY, Edward (f.1755-?1795) 22 (erroneously c.1680), 128, 187
MAY, John (of London) (?f.1693, ?d.1738) 5, 128, 129
MAY, John (f.?1725, c.1755-1798) 9, 128-9, 165, 174, 187
MAY, Thomas (f.1772-90) 129, 174, 187
MEAKINS, George (f.1767) 129
MEARS & STAINBANK (bellfounders, of Whitechapel) (1865 onwards) 50
Mears Ashby (Northants.) 185
mechanical inventions 6, 39, 99, 116, 152
'mechanical microcosm' 133
medal 151
MEDCALF, --- (f.c.1800) 129
Mellet, Mary 115
memorial brasses (made by clockmakers) 181, 188
Mercer, Dr R. V., collection 138
MERCHANT, Thomas (f.1847-51) 129
MERN, William (of Bristol) (f.1774) 129; father Daniel 129
Merton, St Swithin, clock 46, 87
Mesopotamian campaign 1
metal worker 151
meteorology 152; see barometers, weathercock
metronome 7
Metropolitan Museum 146
Middleton Cheney (Northants.) 134, 151, 188
Middleton Stoney, All Saints, clock 46, 170-1
MIDWINTER, John (f.1764-72) 129; wife Hannah & son John 129
Miles, R., collection 138, 140
Miles, W. F. 72
Miller, A., collection 187
Miller, David B., collection 87
milling machine 5
Milton (near Abingdon, Berks.) 107

Milton (near Adderbury) 96, 110
Milton see Great Milton
Milton, L., collection 104
Milton-under-Wychwood 8, 16, 107, 132; Quaker Meeting House 132, 146
Minchin, Susannah 132; father Anthony 132
ministers 33, 146; Congregational 13, 99; Quaker 23, 102, 104
MINN, Henry James Hall (1870-1961) 35
mirrors 96
MITCHEL, John (bellfounder, of Wokingham) (f.1493) 41
MITCHELL, John (of London) (f.1753) 136
Mixbury, All Saints, clock 47, 171; sandglass 81
Mollington, All Saints, clock 43, 47, 53, 172
Monconys, Balthazar de 152
monumental clocks 18
moon dial (sundial) 76
moon model 152
Moore, Anne 143
MOORE, Edward (f.1714, d.1774) 9, 10, 22, 49, 50, 57, 129-30, 181, 187; father Solomon 129
MOORE, Edward (f.1751-c.1785) 10, 22, 130
MOORE & SON(S), John (of Clerkenwell) (f.1831-81) 24, 36, 37, 42, 51, 130, 138, 164
Morgan, Octavius 49
Morrell, T. S. 67
MORRIS, John (f.c.1770) 130
Morten, 172
Mose, Dinah 150
MOSS, Thomas (f.1800-30) 173
MOUNTFORD (or Montfort, Mumford), Zachariah (?a.1677, f.c.1730-38) 4, 58
movements, Black Forest 148, Quaker 23, 100-103; use of prefabricated & Londonmade 23-4
Mumford see Mountford
Museum of the History of Science, generally 6-10, 78; Barnett Collection 6; Beeson Collection 2, 4, 6-9, & see Beeson; clocks & related instruments 6-10, 130, 148; Combe clock 6, 9, 21, 35-6; Hawting regulator 6, 9, 111; Kratzer sundial 125; Lewis Evans Collection 6; locks 6; Wadham College clock 6, 9, 64-66, & see Oxford colleges – Wadham
musical boxes 5, 7
musical clocks 26, 27, 110, 115, 141, 146; watches 141, 145; see chimes, tunes
musical instruments 38, 91, 96, 149; see organs
MUSSELWHITE, William (f.1787-91) 98, 130, 188

National Trust 191
'natural magic' 6
NAU, George (of London) (f.1675, d.1698) 89
Neale, Ann 125
NEALE, Henry (bellfounder) (f.1635) 33
Neithrop 91
Nether Worton 47; St James, clock 74
NETHERCOT(T), --- (f.1707) 23, 131
NETHERCOTT, --- (mid 18th C) 131
NETHERCOTT, George (of Wantage & Fyfield) (f.c.1770) 130
NETHERCOTT, John (f.c1750) 4, 9, 23, 101, 131
NETHERCOTT, John (?of Dry Sandford) (18th C) 4, 9, 131
NETHERCOTT, William (of Long Compton) (f.c.1750) 131
Nettell, D. F. 3
Nettlebed 70; St Bartholomew, clock 47
New York 146
Newbury (Berks.) 107, 128
Newington see North Newington, South Newington
Newport Pagnell (Bucks.) 99, 124-5
Newton Purcell, St Michael, clock 49, 166, 171
Nicholson, John 49
nocturnals 7
Norris, Muriel Maud 68
North, Charles 75
North, Mary 151
North Aston, St Mary, clock 28
North Newington 16, 96, 104, 110; St Giles, clock 47
Northampton (Northants.) 185
Northamptonshire, church bells 112
Northcote, L. S. 162-164, 166-170, 173, 176

Northmoor, St Denys, clock 171, 182
Norwich (Norfolk) 86

Oakley, Esther 165
OAKLEY, James (f.1731, d.1749) 9, 22, 65, 131; wife Mary 131
OAKLEY, John (a.1685, f. 1704) 9, 22, 57, 62, 131-2
oil clock 7
Old Marston, St Nicholas, clock 41, 45-6, 143
Oldham, Bishop Hugh 78
Onibury (Salop.), church clock 7
opticians 24, 127
organ maker 110-1; organist 182
organs 38, 60, 69, 110-1
orreries 5, 7, 133, 152
ORTELLI, A. (f.c.1790-1825, ?1846) 132; A. & D. Ortelli (f.c.1825) 9, 132; Ortelli & Co. 132
OSBORNE & CO., Ann(e) (clock face makers, Birmingham) (f.1791-98) 188; A. & J. Osborne (f.1800-03) 188; ?son James (f.1808) 188
OSBORNE, Thomas (clock face maker, of Birmingham) (f.1772-77) 24, 188, 191; Osborne & Wilson 188, 191
OSBORNE (or Arsborn), Thomas (f.1830-75) 86, 173, 176, 188; wife Jane 188
OSMOND, G. H. (f.c.1840) 4, 9
Osney, Abbey, clock 19, 49
Over Worton 66; Holy Trinity, clock 74
Oxford for general references see Oxford City, Oxford University
Oxford churches, All Saints 93, 94, 105, 107, 126-7, 135, 138-9, 145, 146; Christ Church Cathedral, clock 19, 49-50, 91, 130, 137, 139, 150; Osney Abbey, clock 19, 49; St Clement, clock 53; St Cross 117, 119, 124; St Cross, sundial 79; St Frideswide's Priory (Christ Church), clock 19, 49; St Giles, clock 53, St Martin (Carfax Tower), clock 4, 19, 20, 23, 55-58, 63, 91, 92, 105, 115, 118, 130, 132, 136-139, 144, 147, 153; St

205

Mary Magdalen, clock 20, 23, 57, 58-60, 87, 92, 136-138; St Mary Magdalen, sundial 79; St Mary the Virgin 33, 60-63, 104, 107, 111, 129, 131, 145; St Mary the Virgin, clock 19, 20, 22, 23, 25, 54, 57, 60-63, 65, 84, 113, 115, 123, 124, 127, 131, 136, 144, 152, 153; St Mary the Virgin, sandglass 81; St Mary the Virgin, sundial 76, 79-80, 96, 125, 127, 134; St Michael, clock 19, 64; St Peter in the East 99, 113, 133, 138; St Peter in the East, sandglass 81; St Peter in the East, sundial 80, 134

Oxford City 1-10, 14-25, 31, 33, 34, 37, 38, 40, 41, 45, 46, 49-66, 68, 74, 75, 84, 85, 88-96, 98-100, 105-107, 109-127, ?128, 129-141, 144-148, 150-153, 164, 171-2, 174, 176-178, 181-190; alternative capital 6; Automaton Laboratory 106; Bocardo Gate 15; Broad Street 55, 66, 150; Castle Mills 118; Catte Street 33; City Church 55-58, & see Oxford churches – St Martin; City Ditch 118, 150; City Wall 15, 52, 122, 150; Clockmakers and Watchmakers Guild 15, 122; clockmaking generally 5-6, 15-16, 19-24; Cordwainers Guild 15; Cornmarket Street 65, 88; Council, Guilds, etc. 14-16, 20, 56-58, 84, 88, 90, 106, 117-8, 122-3, 127, 135, 136, 147, 150, 153; Council, dispute with University 15-16; Council, sandglass 81; Council, sundial 78, 152; earliest clockmaking in 17-20; earliest dated clock 22, 117; first freeman described as 'clockmaker' 110; fortifications (Civil War) 136; Gloucester Green 4; Guildhall 139; High School 1; High School, clock 7; Holywell Street 22,
50, 55, 59, 65, 77, 111, 117-8, 122; Locksmiths, Gunsmiths & Farriers Guild 15; Merchant Guild 14; New Parks 118; Northgate 64, 118; Oxford Historical Society 14; Oxford Times 143; Radcliffe Infirmary, clock 111; Radcliffe Observatory 49; Radcliffe Observatory, clock 6, 111; St Clement's (street) 117, 122, 138; St Edward's School, clock 92; shop sign The Dial 22, 113; Smith Gate 118; Smiths Guild 15; Smiths & Watchmakers Guild 15, 122; South Bridge sundial 78, 152; Tailors Guild 127; time-measuring systems 18; Town Hall 58; Turl Street 90; waterworks 150

Oxford colleges, All Souls 59, 106, 152; All Souls, sundial 77; Balliol 18, 96; Balliol, clock 49, 138, 144; Balliol, sundial 77; Brasenose, sundial 77; Christ Church 60, 144, & see Oxford churches – Christ Church Cathedral; Christ Church, sundial 77; Christ Church, Tom Tower, clock 7, 9, 50, 130, 137, 138, 153; Christ Church, Tom Tower, bell 50, 60; Corpus Christi 76, 79, 80, 96, 125, 145; Corpus Christi, sundial 76, 78, 96, 125, 145; Exeter, clock 50-1, 138; Exeter, sundial 78; Jesus 91, 116; Jesus, clock 51, 130, 138; Lincoln, clock 130; Magdalen 96; Magdalen, clock 19, 51, 96, 138, 139, 171-2; Magdalen, St John's Chapel, clock 171; Magdalen, sundial 78; Merton 17-18, 117, 122, 137, 144; Merton, clock 17-19, 51-2, 144; Merton, sundial 78; New, clock 7, 9, 52-3, 153, 172; New, sundial 79; Oriel, clock 53, 141, 146; Oriel, sundial 79;

Queen's 152; Queen's, clock 53; Queen's, sundial 79; St Edmund Hall 79; St Edmund Hall, sundial 79; St John's 1, 152; St John's, clock 19, 25, 53-55, 90, 92, 93, 122, 136, 138, 139; St John's, sundial 79; Trinity 15, 117, 122-3, 152; Trinity, clock 64, 111; Trinity, sundial 80; University, clock 24, 64, 110; Wadham 6, 152; Wadham, clock 2, 6, 9, 22, 25, 57, 62, 64-66, 68, 93, 99, 111, 122-124, 131, 137, 138, 152, 185; Wadham, sundial 80; Worcester, clock 66, 138, 144

Oxford University 1, 5, 6, 15-17, 55, 60-63, 78, 80, 84, 88, 93, 94, 107, 111, 117, 122-3, 127, 136, 145, 153, 185; Ashmolean Museum 3, 99, 120, 121, 125, 132, 133, 135, 138, 143, 148; Bodleian Library 79, 84, 153; Botanic Garden 77, 80, 117, 151, 153; Botanic Garden, sundial 77, 80; Convocation House 84; dispute with City 15-16; Divinity School 111, 153; Imperial Forestry Bureau 2; Museum of the History of Science 2, 4, 6-10, 21, 35, 66, 78, 111, 125, 130, 148; Old Ashmolean 111, 153; privileged (tradesmen) members 15-16, 22, 55, 93, 94, 107, 115, 117, 122-3; Schools Quadrangle 84, 111; Sheldonian Theatre 6, 139, 152; University Chest 60; University Church & clock, see Oxford churches – St Mary the Virgin

Oxfordshire, Archaeological Society 30, 87; church bells 52; earliest dated lantern clocks 22, 109, 117; earliest dated turret clocks 32, 46-7; earliest mention of sandglass 81; Lord Lieutenant of 15, 133; Quakers of north, generally 5, 16, 22, 100-1, 105,

206

146, & see Quakers; Record Society 41; Victoria County History 46, 112

PACKER, John (of Tingewick) (f.?1810-1825) 166, 171
PACKER, William (of Buckingham) (f.?1776-1810) 166
PADBURY, Matthias (b.1751, f.1785) 132; parents Matthias & Hannah 132; wives Susannah & Susannah 132
PAGE, Edward (a.1684) 132, 135; father John 132
Page, Mary 149
PAGET, William (f.?1771) 132
Paine, James 86
PAINE, John (f.1824-55) 132, 173, 188; parents Robert & Ann 132; wife Dinah 132
painters 19, 51, 74-5, 92, 111, 127, 137; painting & gilding (clock faces, sundials) 27, 40-1, 43, 53, 59, 62, 75, 77, 79, 80, 92, 111, 127, 134, 137, 148, 166, 172
Palmer, L., collection 132
Palmer, W. D. 143, 152
Pangbo(u)rn, Oliver 28
PANTIN (or Panton), Nicholas (a.1651, f.1663-?1682) 15, 133
Paris 24, 138
PARIS, Nicholas (of Warwick) (f.c.1669, d.1716) 54
PARIS, Thomas (of Warwick) (1688-1753) 54-5, 62, 63
PARKER, William (f.1657-75) 40, 71
Parkes, D. 186
PARSONS, Joseph (f.1853) 133
PAVIER, Thomas (gunsmith) (f.1737-46) 59
PAYNE, George Septimus (f.1880) 133; Payne & Son (1880 onwards) 58, 133
PAYNE, Thomas (of Abingdon) (f.?1813, 1818) 59, 133
PAYNE, William Petty (& jeweller etc.) (1818-1897) 133, 185, 188
'payses' (?pieces) 70
PEACOCK, William (f.1788-1823) 9, 32, 35, 133, 188; wife Elizabeth 133

Pearson, Hannah 89
PEARSON, Hawtin (f.1825-53) 133
PEARSON, Henry (f.1827-62) 133; wife Phoebe 133
PEARSON, Richard (a.?1778, f.1823) 38, 63, 127, 133, 144; sons Richard & John 133
PEISLEY, Bartholomew (mason) (f.1689, d.1715) 80
pendulum, adjustments to 50; application to verge escapement 21, 48; degree plate 7, 9; earliest examples of & conversions to 2, 22, 48, 57, 59, 61-2, 66, 75, 152, 185-6; pendulum weather clock 152; see escapements
Pentyrch (Wales) 116
Peres (or Perry), Thomas (clock keeper) 69-70
PERIGAL & SON, Francis (of London) (f.1763-1805) 30
perpetual motion device 39
Perpignan, Royal Castle, clock 3
Perry, Roger 191
Peter 70
Pettifer, Mrs, collection 104
PHILIPPS, John (carpenter) (f.1505) 172
PHILIPPS, Joshua (f.1897) 48
PHILLIPS, Charles (& jeweller etc.) (f.1832) 134
PHILLIPS, Michael (f.1853) 134
Phillipson, W., collection 85
Piercy, Mary 94
Pinfold, Ann 151
Pinfold, Elizabeth 133
PINFOLD, Thomas (smith, of Middleton Cheney) (f.1751, 1762) 134
PINFOLD, Thomas (f.?1751, 1762, d.1789) 9, 32, 35, 134, 151, 188; ?daughters Ann & Elizabeth 151, 133; wives Mary & Elizabeth 134, 188
Pinwill, W. J. 42-3
PITTAWAY, --- (smith) (f.1705) 62; wife (f.1713-18) 62
PITTAWAY, Edward (?smith) (f.1718-41) 62
Plot, Robert 39, 116, 152
plumbers 70, 80, 151
'plummets' 56
plush makers 95
Poles, Mary 109

POTTER, Francis (sundial maker) (1595-1678) 80
POTTINGER, John (smith) (f.1731) 96; daughter Elizabeth & son George 96-7
POTTS, William H. (of Leeds) (f.1872, company onwards) 28, 29
Poundon (Bucks.) 28, 141
POWELL, James (f.1852) 44, 134
POWLEN, Thomas (smith) (f.1488-1500) 69, 134
Prestbury (Glos.), clock 172
Price, D. J. 17
Price, F. D. 173
Pricket, Hannah 129
Princeton University, sundial 78
privileged tradesmen (of the University) 15-16, 22, 55, 93, 94, 107, 115, 117, 122-3
Prophett, Ann 96
PRUJEAN (or Prijon, etc.), John (instrument maker) (a.1646, d.1706) 80, 134
Purcell, Henry 42
Purefoy, Henry 86-7
PYKE, George (of London) (f.1754) 106

quadrant 134
Quakers 5, 13, 16, 22, 23, 86, 91, 96-98, 100-105, 107, 110, 132, 140, 146, 149-151, 179; beginnings of clockmaking tradition among 22, 96, 101-2, 104-5, 110; see Adderbury, Sibford Gower
QUARE, Daniel (of London) (c.1648-1724) 7
Quelch family 5, 135
QUELCH, John (a.1652, d.c.1695) 20, 22, 105, 122, 126, 132, 135, 138
QUELCH, Joseph (f.1684) 135
QUELCH, Martin (a.1650, f.1653) 135
QUELCH, Richard (a.1608, d.1652) 20, 135, 139; father Richard 135
QUELCH, Richard (f.1652, d.?1667) 20, 22, 90, 93, 99, 122, 135
Quill, Humphrey 3

R., I. (watch case mark) 149, 177
Radcliffe, John 111
Radcliffe Infirmary, clock 111

Radcliffe Observatory 49; clock 6, 111
Ramsdell (Hants.), Rawlins Farm 165
RANDALL, Henry Pearce (f.1852) 135
RANKLYN (or Francklin, Ranckle), Thomas (smith) (f.1604, d.1658) 20, 56-7, 74, 99, 135-6, 139, 165; Widow Ranklyn 57
Ratley (Warwicks.) 114
Rawlinson, Richard 28, 53, 73, 79
RAYE (or Rayer, Wray), John (smith) (f.1617-48) 20, 54, 56-58, 61, 62, 84, 115, 136
RAYE, Richard (smith) (f.1631) 136; ?father 136
Reading (Berks.) 71, 148; St Lawrence, clock 18, 153
Reading, Rev. 98
regulators, astronomical 6, 7, 9, 91, 111, 114, 182, 189
REYNER, Thomas (f.c.1740) 136
REYNOLDS, Henry (a.1753) 136; father Robert 136
REYNOLDS, John (smith) (f.1702-33) 5, 29, 136
REYNOLDS, Thomas (& smith) (c.1724-1799) 5, 7, 10, 22, 25, 49, 50, 55, 58, 59, 65-6, 68, 85, 86, 90, 91, 96, 100, 111, 127, 136-7, 188, 189; Reynolds & Earle (f.1795-99) 9, 96, 137, 189; wife Hannah 137
Richard II 18
RICHARD of Wallingford (scholar) (c.1292-1336) 17
Richmond (U.S.A.) 93
Ricketts, E., collection 181, 189
Ridge, Martha 144
RIGGINS, Edward (carpenter) (f.1725-?1730) 59, 137
Righton, R. C., collection 7, 10, 188
RIXON, John (painter) (f.1660-76) 57, 111, 137
roasting jacks 5, 174, 187
Robins, S. M., collection 102
Robinson, L., collection 94
ROBINSON, Richard (smith) (f.1714) 59
Robinson, T. O. 3
Robinson, Thomas (clock keeper) 174

ROLEWRIGHT, Thomas (smith) (f.1584-88) 49, 91, 137
rolling ball clock 99
Rollright see Great Rollright
ROMLEY, Robert (clock-bell maker, of London) (f.1760s) 23
ropes (clock) especially 17, 26, 70-1
ROSE, T. (d.1772) 117, 138
Ross, Elizabeth 90
Ross-on-Wye (Herefs.) 106
Rotherfield Peppard, All Saints, clock 172
Rousham House, clock 24, 66-7
ROWELL, George (1770-1834) 9, 24, 51, 55, 59, 66, 138; son George Augustus (1804-1892) & his wife Maria 138; wife Mary 138
ROWELL, R. S. (& jeweller) (f.1898-1915) 9
ROWELL, Richard Rouse (& jeweller etc.) (b.c.1813, f.1834-74) 7, 41, 49-50, 55, 59-60, 66, 138; Rowell & Son (c.1835 onwards) 7, 10, 55, 60, 138
ROWLEY, John (instrument maker, of London) (f.1698, d.1728) 32, 76
Rowntree, R., collection 117
ROY, Julien le (of Paris) (1686-1759) 169
Royal Army Medical Corps 1
Royal Society 152
RUDHALL, Abel (bellfounder, of Gloucester) (1714-1760) 170
RUSTIN, Joseph (a.1686) 135, 138; father Henry 138
RYSSON, Thomas (smith) (f.1587) 91

S., W. (sundial mark) 79
Sadler, James 151
SAGE (or Sayer), Matthew (f.c.1760) 139
SAINSBURY, Henry (of Walthamstow) (f.1875-85) 163
St Albans (Herts.) 4; clock 17
St Neots (Hunts.) 144
SAINT PAUL, John de (or John the Frenchman) (f.1576, d.1596) 5, 19, 20, 49, 53-4, 138-9; wives Elizabeth (née Smyth) & Alice 5, 138, 139

SAINT PAUL, Richard de (a.1639) 139
SAINT PAUL, Triumph de (f.1601-51) 20, 51, 54, 135, 139; wife Ursula & son Lewes 139
Salford (Lancs.) 162
Salisbury see Sarum
Salter, H. E. 1
SAMSON, George (f.1547-54) 55
Sandford St Martin, church clock 74
sandglasses 7, 81, 84, 162; on pulpits 81, 84
Sarum (Wilts.) 150; St Edmund, clock 18
Saunders, Catherine 141
SAUNDERS, Charles (f.1794-1823) 95, 139, 189; wife Kitty 139
SAUNDERS, Richard (a.1674) 89, 139; father Ralph & grandfather Thomas 139
SAUNDERS, Thomas (f.1795-1810) 139, 189; wife Esther 189
Savage, Elizabeth 95
Savage, N. S., collection 97
SAVAGE, T. (f.?1750) 139
SAYER, Matthew (of Exeter) (f.1763) 139
Sayer see Sage
'sayll' 69
Science Museum 2
scientific instrument makers 76-80, 124-5, 134; see sundials
scientific instruments, generally 6-7, 17-18, 152; see astrolabes, barometers, equatorium, orreries, quadrant, sextant, sundials, thermometers, weathercock
Scott, Sir George Gilbert (architect) 49
scribings & doodlings (except signatures) 23, 132; see crossing-out
scroll finials especially 25
sedan clock 188
Sedgley (Staffs.), clock 37
Sen-Sarma, P. K. 3
sextant 100
Seymour, A., collection 114
SEYMOUR, Henry (of Wantage) (f.1525) 20
Seymour, N. W. & T. 72
Shalstone (Bucks.) 86
Shelswell Park, clock 67
Shepard, A. E., collection 104
Sheraton cases 109, 116; Sheraton inlay 178
SHEWELL (or Showells), John (smith) (1632-?1689) 57, 136, 139

208

Shipston-on-Stour (Warwicks) 8, 101, 104, 149, 180
Shipton-under-Wychwood, St Mary, clock 67, 172
shop signs 22, 113, 142
SHORT, Mr (f.1727) 57
Shrewsbury, Katherine 124
Shutford 16, 107
Sibford Ferris 110
Sibford Gower (& Sibford) 16, 22, 67, 96, 100-105, 132, 149, 190; Holy Trinity, clock 67, 143; origins of Quaker clockmaking in 16, 22, 104-5; Quaker Meeting House 16, 102, 105, 149
silversmiths 13, 24, 91-95, 100, 106, 114, 126, 133, 134, 138, 141-143, 145, 147, 148, 177
Simms family 140
SIMMS, Benjamin (19th C) 140
SIMMS, Charles P. (1820-1910) 140
SIMMS, Daniel Rutter (1864-1954) 140
SIMMS, Frederick (1816-1894) 140
SIMMS, John (f.1772-79) 140
SIMMS, John (1757-1823) 140
SIMMS, Samuel (& jeweller) (1790-1869) 9, 34, 140
SIMMS, William (1785-1844) 9, 140
SINDRY, James (& goldsmith) (f.1788) 189
SINFIELD, T. (f.1777) 140
Skelton, Joseph 64
SLATFORD, John (smith) (f.1592) 91
SLATFORD, Thomas (f.1629-33) 56
SLY, Robert (f.1823-53) 140
SMITH, Abraham (a.1771) 141, 144; father Abraham 141
Smith (or Smyth), Elizabeth 5, 138
Smith, J. E. H. 163
SMITH (or Smyth), John (smith) (f.1474) 69
Smith, Mary 106
SMITH, Richard (smith) (f.1641) 71
SMITH, Richard (of Brill) (b.?1765, f.1810-12) 46
SMITH, Robert (smith) (f.1442-55) 19, 69, 141
SMITH (or Smyth), Thomas (smith) (f.1455-c.1470) 69, 141

SMITH, Thomas (a.1669) 123, 141; father John 141
SMITH, Vincent (smith) (f.1711-18) 28, 141
Smith, W. J., collection 100
SMITH, William (smith) (f.1502) 69
SMITH, William (smith) (f.1625-30) 71
SMITH & SONS, J. (of Clerkenwell) (f.1835-42) 32
SMITH & SONS, John (of Derby) (f.1925 onwards) 27, 29, 30, 33, 34, 36, 37, 42, 51, 67, 68, 72, 74, 163, 165
Smyth, William Henry 49, 61
Snowshill Manor (Glos.), clock 191
'solarium' 60
Souldern 100, 180; Manor House, clock 113
South Hinksey (Berks.) 138
South Leigh, St James the Great, clock 172
South Newington, St Peter ad Vincula, clock 20, 25, 29, 30, 47-49, 67, 109, 113, 173
SOWTER, John (& jeweller etc.) (f.1810-53) 9, 141, 189
Sparkes, Mary 104
Spelsbury 127; All Saints, clock 20, 67; gravestone 127
Spencer, Sir Thomas 74
Springold (or Springall), Thomas 70-1
Stadhampton 8, 40, 116, 177; church clock 142
Standlake 9, 131
Stanford (Berks.) 126
STANTON, William (of Bloxham & Buckingham) (b.1804, f.1847) 166
Stanton St John, St John the Baptist, clock 67, 142
steam engine 99
Steane Park (Northants.), sundial 80
STEELE, Joseph (& jeweller (f.1831-46) 94, 141; wife Frances 141
Steeple Aston, St Peter, clock 7, 9, 28-9, 34, 67, 113
Stevens, Hannah 137
STEVENS, John (f.1784-1819) 41, 141-2; wife Catherine 141
STEVENS, Robert (f.?1781-1795) 142
STEWARD, John (& gunsmith, of Henley-in-Arden) (f.1717-47) 142, 189
Stockford, Edward 59
STOCKFORD, Joseph (smith & bellhanger) (f.1770-74) 37, 142, 165, 189
STOCKFORD, Thomas (f.1764-85) 40, 114, 142
Stockwell (Surrey) 146
Stoke Lyne, St Peter, clock 67
STONE, John (& jeweller etc.) (b.c.1740, f.1764-?1802) 4, 23, 142; wives Elizabeth & Martha 142
STONE, John (of Aylesbury) (b.1768, f.1789) 4, 142; wife Susannah 142
STONE, John (f.1795) 142
STONE, Nicholas (mason) (1586-1647) 77, 80
STONE, Richard (a.1761, f.1783) 23, 41, 46, 114, 142-3, 189; father Edward 142
STONE, Thomas (f.1801-09) 143
stone masons 14, 15, 19, 22, 31, 51, 62, 76-80, 96, 99, 116, 152, 185; see architects
Stonesfield, St James the Less, clock 20, 67
Stonor (Berks.) 72
Stonor, Thomas 72
Stott, Lady, collection 120
STRANGE, Thomas (f.1823, d.?1866) 27, 32, 67, 94, 143, 151
Strause, Henry P., collection 93, 147
STREET, Richard (f.1795) 143; wife Mary 143
STRIPLINGE, --- (f.1640) 61
Stubbs, John (clock keeper) 59
Sudbury (Suffolk) 99
SULLY, Henry (of London & Versailles) (1680-1728) 167, 169
Summers, W., collection 120
sundials 5, 6, 7, 17, 32, 33, 35, 60, 62, 65, 76-80, 96, 100, 111, 125, 127, 134, 145, 151, 152, 162, 191; ?as shop signs 22, 113, 142; globe 77, 79, 80; pillar 76, 79-80, 96, 125, 127, 134, 145; polyhedral 76, 78-80, 125; used to check clocks 62, 76; with moon dial 76; wooden 79
Surman, J. M., collection 108

209

SUTTON, Henry (instrument maker, of London) (f.1654-64) 125
'swaipe' (foliot) 48
Swalcliffe, Sts Peter & Paul, clock 25, 65, 68, 137
Symonds, John (clock keeper) 51

tabernacle clock 9
Tadmarton 107
Talbot, Charles 74
talking statues 6
Taploe (Bucks.) 100
TAPPIN, --- (painter) (f.1808) 166
TASKER, William (& jeweller etc.) (f.1813-53) 9, 143; wife Anne 143
Taston 127
TAWNEY, Robert (a.1755, f.1776) 113, 133, 141, 144; father Robert 144
Taylor family 34, 144
Taylor, E. Bartlett, collection 187
TAYLOR, John (painter) (f.1688-95) 79
TAYLOR, John (carpenter) (f.1761) 63
TAYLOR, John (bellfounder) (1797-1858) 144; John Taylor & Co. (of Loughborough) (1839 onwards) 34, 37, 58, 144
Taylor, Philippa 115
TAYLOR, Robert (bellfounder) (1759-1830) 144; Taylor & Sons (f.1826) 34, 144, 164
TAYLOR, Thomas (of London) (a.1638, d.1684) 88
TAYLOR, William (& bellfounder) (1795-1854) 24, 34, 37, 63, 144; W. & J. Taylor (f.1830s-54) 37, 46, 58, 144
telescope driving mechanisms 7
TERRY, John (f.1700, d.1736) 144; wife Patience 144
Tetsworth 183
Thacker, W., collection 181
Thame 4, 8, 9, 19, 20, 22, 23, 37, 40, 41, 46, 68-72, 89, 90, 93, 110, 113, 114, 126, 134, 136, 141-143, 145, 165, 186-7, 189; earliest clockmaking in 23; Moot Hall, clock 20, 70, 72; St Mary, clock 19, 20, 68-72, 90, 113, 134, 141, 145; St Mary, sandglass 81
thermometers 96

THOMPSON, John (engraver, of London) (f.1760s) 23
THOMSON, Adam (of London) (f.1839-60) 172; Thomson & Profaze (f.1860-81) 172
THORNDELL, Richard (& ironmonger) (f.1762-68) 144; wife Martha 144
THORP, Benjamin (a.1726) 128, 144; father Thomas 144
THWAITES, Aynsworth (of Clerkenwell) (a.1735, f.1792) 98; Aynsworth & John Thwaites (f.1792) 24, 64, 110, 144
THWAITES, John (of Clerkenwell) (a.1772, d.1826) 38, 51-2, 64, 110, 133, 144, 162; Thwaites & Reed (c.1827 onwards) 24, 30, 49, 55, 59, 64, 66, 73, 144, 162
tic-tac escapement see escapements
tide mill 99
time measuring, early methods & devices 6, 7, 17-18; hour systems & equal hours 18, 76, 80; in academic life 6, 19; solar time & use of sundials 62, 76
Tims, Mrs T., collection 102
Tingewick (Bucks.) 166, 171
TIPPING, Richard (?smith) (f.1685) 80
Titchmarsh (Northants.) 185
tokens 22, 89, 123
Tolsey Museum 187
Tom Tower see Oxford colleges – Christ Church
TOMLINS, --- (f.1640) 61
Tomlinson family 41, 145, 189
TOMLINSON, Job (& gunsmith) (f.1819, d.c.1865) 41, 71-2, 145, 180, 189; wife Mary Anne 145
TOMLINSON, John (& smith) (f.1772-1819) 71, 145
TOMLINSON, John (f.1843) 145, 189
TOMLINSON, Thomas (f.1823) 189
TOMPION, Thomas (of London) (1639-1713) 7, 184
TONGE, George (& silversmith) (1730-1803) 9, 24, 145; father Henry 145

tools, clock- & watchmaker's 7, 108, 109, 112
Torbay (Devon) 118
towers see bells, campanile, turret clocks
Townesend, John (mason) 31, 76, 78
Townesend, John (mason) 62
Townesend, William (mason) 185
trade card 91
TREADWELL, Richard (f.1832-63) 165
Tring (Herts.) 19, 69, 143
'trochiliack horloges' 106
TUCKER, --- (of London) (f.1867) 28
TULL, Jethro (of Newbury) (18th C) 128
tunes 29, 61, 153, 174; Purcell's 'Britons strike home' 42; see carillons, chimes, musical
Turkey, Sultan of 110
TURNBULL, Charles (scholar) (b.1556, f.1605) 78, 145
Turner, Mary 134
TURNER, Thomas (& silversmith) (f.1784) 145
turret clocks 1-4, 6-10, 19-76, 84, 86, 87, 90-93, 95, 96, 98, 99, 101, 103, 106, 108-117, 122-127, 130-139, 141-146, 148-153, 162-175, 179-183, 185-188, 190; 'bedstead frame' 127; comparison of variations in design 4, 25; earliest generally 3, 18; earliest signed & dated 32, 46-7; flatbed 7; turret movements in longcase clocks 7, 38, 92; unusual design of 45; wood-framed 7, 44, 166-170
TYLER, Richard (1733-1800) 146
'tympanum' (of waterclock) 39

umbrellas 96, 151
UNITE, Matthias (a.1681) 119, 146, 189; father Matthias 146
University see Oxford colleges, Oxford University
Upper Heyford, St Mary, sandglass 81

Vanbrugh, Sir John (architect) 31
VANS, Patrick (of London) (a.1672) 123

210

VAUGHAN, Richard (a.1732) 146, 148; father William 146
Veldin (Essex) 99
VERNON & SON, Joseph (f.1852) 146
VESEY (or Veasey), Robert (a.1682, f.1694) 109, 146; father Robert 146
'vice whele' 69
Victoria, Queen 48, 73
Victoria & Albert Museum 85, 105, 120
Vincent, Richard 33
Virginia Museum 93
Vivers, Edward 16
VULLIAMY, Benjamin Lewis (of London) (1780-1854) 24, 32, 44, 45, 49-50, 53, 146, 170

Wade, R. H., collection 109
WAGSTAFF, Thomas (1724-1802) 146-7; parents Thomas & Sarah 146
Wainfleet (Lancs.) 146
WALFORD, Henry (& jeweller) (f.1778) 147, 190
WALFORD, Henry (& jeweller etc.) (f.1853-61) 147; Henry Walford & Son 147
WALFORD, John George (& jeweller etc.) (f.?1790, 1814-53) 9, 27, 114, 147; Walford & Son (1814 onwards) 147
Walis, Ralph 73
WALKER, George (f.1689) 9, 147, 190; ?father George 147
Walker, John 112
WALKER, Thomas (smith) (f.1665-1715) 57, 147
WALKER & HUGHES (clock face makers, of Birmingham) (f.1815-35) 24, 132, 147, 178
walking sticks 96
wall clocks 8-10, 87, 110, 111, 119, 120, 124, 130, 134, 138, 140, 141, 143, 147, 148, 176, 177, 183-185, 188
WALLEN, J. (f.c.1790) 148
WALLEN, William (f.1725-?1756) 148
Wallingford (Berks.) 91, 107, 133, 135, 162, 173; Lamb Inn, clock 162; see Richard
Walls, Richard 112
WALSHE, Thomas (carpenter) (f.1496) 171-2
WALTER, Edmund (& jeweller etc.) (f.1853) 148, 173; E. Walter & Son (f.1870s) 173
Walthamstow (Essex) 163
WANGLER, Luke (& silversmith) (f.1852) 148; L. Wangler & Co. (f.1881) 10
WANGLER, M. (f.1842) 148
Wantage (Berks.) 130; clock 142
Warborough, St Lawrence, clock 172-3
Ward, Kitty 139
WARD, Robert (of London) (f.1760s) 23
Wardington, St Mary Magdalene, clock 72
War(e)d, John (carpenter) 70
WARFIELD, Alexander (of London) (a.1683, f.1719) 92
Warner family (of Chipping Campden) 148
Warner, John 165
Warner, R., collection 182
WARNER, Richard (f.1790) 98, 148
WARNER, William (f.1673-76) 62
Warwick (Warwicks.) 54, 62
watch cocks 151
watch repairer 7
watches 6, 8-9, 20, 24, 89-91, 94, 98, 99, 101, 105-107, 109, 113-115, 118, 120-122, 124-126, 131-133, 135, 138, 140, 142, 143, 146, 148, 149, 174, 177-179, 184, 189; lost 87, 91, 93, 98, 111, 113, 115, 116, 126, 129, 130, 132, 140, 144; prices of (19th C) 24
watchmakers especially 4, 5, 13-15, 20, 22-24, 41, 43, 87-90, 93, 97, 99, 105, 109, 113, 114, 122, 126, 127, 130, 132, 133, 135, 139, 144, 146-148, 150, 151, 181, 183; see clockmakers, jewellers
watchman's clock 7
watchpapers 86, 88, 90-95, 97, 98, 100, 106, 107, 111, 114-116, 133, 134, 137, 138, 141, 142, 145, 147-149, 151, 182
Water Stratford House (Bucks., near Finmere) 38, 92 (erroneously Waterstock House)
waterclocks 6, 7, 17, 39, 152
Waterhouse, Mrs N., collection 141
waterworks, municipal 41-2, 107-8, 150
Watlington 9, 10, 23, 72-3, 113, 114, 128, 146, 148-9, 190; Town Hall (Market House), clock 72-3
Watson, Dr N., collection 9, 147
WATTS, Charles (& silversmith) (f.1823-29) 148; wife Elizabeth 148
Watts, David 176; collection 182, 186
Watts, Jane 106
weather clock 152
weathercock 40
WEBB, --- (f.1776) 149
WEBB, Charles (f.1850-53) 43, 148
WEBB, Francis (f.c.1710-32) 9, 10, 23, 128, 146, 148-9, 190; wife Mary 148
Webb, Martha 186
WEBB, Thomas (f.1788-1834) 149, 173, 188, 190
Weekley (Northants.) 185
Weideman, R., collection 120
weighing machine 99
WELLER, Richard (f.1603) 54
WELLS, John (f.1774) 149, 190; wife Mary 149
WELLS, John (of Shipston) (f.1785, d.1809) 149; parents Thomas & Elizabeth 149; wife Mary 149
WELLS, John (1787-1847) 149; John Wells & Co. (Thomas & John) 149
Wells, Thomas 174
WELLS, Thomas (of Shipston) (1786-1855) 149
WELLS, Thomas (& bellhanger) (f.1832-53) 149
WENTWORTH, George (& goldsmith) (1692-c.1747) 9, 22, 49, 119, 150, 151; wife Dinah 150
WENTWORTH, Thomas (of Salisbury) (f.1675-?1727) 150
WESBOURNE, John (carpenter) (f.1545) 19, 49
West, Diana 123
West Bromwich (Staffs.), clock 37
WESTCOTT, John (f.c.1780) 9, 190
Westminster (London) 118; chimes 29, 37, 42;

Palace, clock 17
wheel & teeth cutting 23; engines 7
wheelwright 74
'wherle' 67
Whitchurch (Salop.) 50, 172
White family (bellfounders, of Besselsleigh & Appleton) 171
WHITE, Bartholomew (?smith) (f.1625) 81
WHITE, James (a.?1804, f.1812) 96, 150
White, Sir Thomas 88
WHITE, Thomas (& carpenter) (f.1741-43) 190-1; parents Thomas & Martha 190
WHITE, W. (& bellhanger, of Fyfield) (f.1863) 171
Whitechapel (London) 50, 96
Whitehead, C. L., collection 120
WHITERN, William (a.1806) 93, 150
whitesmiths 13, 85, 86, 88, 90, 91, 109, 111, 115, 127, 136-7, 142, 145, 151; see blacksmiths
Whitfield (Worcs.) 139
Whorwood, Brome 117, 122-3
Wickham, William 68
Wigginton, St Giles, clock 21, 30, 32, 53, 73, 149, 162, 173, 181, 188, 190
Wild, Edward 61
WILD, Thomas (of London) (f.1760s) 23
Wilkes, H. I., collection 85
WILKES, John (of Brackley) (f.c.1715) 101
Wilkes, Richard 67
WILKINS, George (f.1768-98) 150
WILKINS, John (scholar) (1614-1672) 6, 80
WILKINS, John (goldsmith) (f.1750s) 145
William III 46, 118
William IV 45
WILLIAMS, Joseph (1762-1835) 7, 9, 27, 96, 98, 102, 150-1; parents William & Hannah 150; wife Hannah 150
Williams, T., collection 101, 107
WILLIAMS, Thomas (of Kings Sutton) (f.1723-44) 26-7
Williams, W. 76, 78-80
WILLIAMS, William (& jeweller etc.) (1793-1862) 27, 151
Williamscote Hill 116
Williamson, Martin 19, 51, 96
WILLSHIRE, James (of London) (f.1760s) 23
Wilson, Dr 61
WILSON, James (clock face maker, of Birmingham) (f.1772, d.1809) 178, 188, 191
WINCKLE, John (smith) (f.c.1597) 135
Winckle see Wynkyll
windlass 5, 56
WINDMILLS, Joseph (of London) (f.1671-1720) 184
Windsor (Berks.) 24, 66-7; Castle, Curfew Tower, clock 66
Winslow (Bucks.) 128
wire drawing 95
Wise family 32, 151
WISE, Doiley (& plumber & glazier) (a.1720, d.1788) 150, 151; father William 151; wife Mary 151
Wise, Elizabeth 117
WISE, John (of London) (b.1658, a.1676, f.1723) 32, 151
WISE, Matthew (b.1759, f.1786) 134, 151, 191; parents Matthew & Rachael 151; wife Ann (née Pinfold) 151
Wise, Richard 32, 151
Wise (or Wyse), Richard 123
WISE, Richard (of London) (f.1730) 32, 76, 151
WISE, Thomas (19th C) 152; son Samuel 152
Witney 5, 9, 22, 74, 75, 85, 87, 96, 106, 109, 110, 115, 117, 126, 128-9, 134, 135, 140, 148, 165, 173-4, 177, 181-183, 187, 190-1; Blanket Hall, clock 174, 182, 183, 187; Congregational School (Old Meeting House), clock 129, 187; Manse, clock 187; St Mary, clock 173-4, 181-2, 187; Weavers Guild (Blanket Hall Company) 174, 187
Wittenburg 125
Wokingham (Berks.) 41, 69
Wolsey, Cardinal Thomas 15, 125
WONTNER, John (of London) (f.1790) 94
Wood, Anthony 53, 60, 84, 88, 118, 122
Wood, G., collection 182
WOOD, Richard (mason) (c.1640-1700) 78, 152
Woodbridge, Martha 107
Woodeaton, Holy Rood church, clock 174
wooden clocks 190-1
wood-framed turret clocks 7, 44, 166-170
Woodstock 4, 6-9, 22, 31, 58, 67, 73, 76, 97, 98, 129, 179, 180; St Mary Magdalene, clock 24, 73; see Blenheim Palace
Woodward, Thomas 117
Wootton, Judd's Garage, clock 67; St Mary, clock, 74
Worcester (Worcs.), Cathedral, clock 50
workshops, especially 7, 23, 102, 108, 109, 112
Worton see Nether Worton, Over Worton
WOTEN, --- (carpenter) (f.1676) 40
WREN, Christopher (scholar) (1632-1723) 6, 62, 65, 66, 77, 152
WRIGHT, Christopher (clock face maker, of Birmingham) (f.1835-42) 186, 191
Wroxton, All Saints, clock 74
WRYGHT, Roger (carpenter) (f.1518) 172
WYLDGOOSE, Henry (painter) (f.1689-98) 80
WYNKYLL (or Winckle), Thomas (smith) (f.c.1545) 49
Wytham (Berks.) 114

Yarnton, St Bartholomew, clock 20, 21, 74-5, 109, 122, 127, 136, 175
Yerke, Jane 67
YOUNG, Samuel (of Bunbury, Ches.) (f.1749) 152, 191
YOUNG, William (& locksmith) (f.1656, d.c.1695) 6, 25, 33, 40, 50, 52, 57, 61, 80, 122, 153; wife Jane & son Thomas 153

zodiac 85

212